The Atomic Archipelago

Intersections: Histories of Environment, Science, and Technology in the Anthropocene

SARAH ELKIND AND FINN ARNE JØRGENSEN, EDITORS

The Atomic Archipelago

US NUCLEAR SUBMARINES AND TECHNOPOLITICS OF RISK IN COLD WAR ITALY

Davide Orsini

UNIVERSITY OF PITTSBURGH PRESS

Published by the University of Pittsburgh Press, Pittsburgh, Pa., 15260

Manufactured in the United States of America
Printed on acid-free paper
10 9 8 7 6 5 4 3 2 1

Cataloging-in-Publication data is available from the Library of Congress

ISBN 13: 978-0-8229-4718-9
ISBN 10: 0-8229-4718-8

Cover art: Tatiano Maiore
Cover design: Melissa Dias-Mandoly

A mia madre, che mi ha insegnato a non arrendermi mai

Contents

Acknowledgments

This book is the result of a long journey. Many people inspired, accompanied, and helped me throughout my research preparation. At the University of Michigan, I found an incredible group of scholars, who became my friends and colleagues. I want to thank Kimberly Powers, Edgar (Jack) Taylor, Ismail (Aji) Alatas, Joseph Viscomi, Ali Sipahi, and Jeremy Johnson for being there with me and for me. Paul C. Johnson has been a great director of the Anthrohistory Program, providing exceptional insights and encouraging us to think above and beyond disciplinary boundaries. David Cohen kindly agreed to shepherd me through the meanders of history making after his retirement. I am grateful. Geoff Eley and Ronald Suny have provided examples of leadership, both in their institutional roles and through their intellectual force. The staff of the History Department has offered invaluable assistance. Without Diana Denney, Lorna Alstetter, Kathleen King, Terre Fisher, and many others my life would have been terribly complicated. When I first arrived in Ann Arbor as a visiting student in 2005, Andy Markovits gave me the comforts of his generous friendship and later supported my decision to apply to doctoral programs in the United States. I will always be grateful for his encouragement and infinite generosity.

Dario Gaggio has seen this project take off, witnessing my uncertainties, doubts, and discoveries. I landed on La Maddalena in the summer of 2009 thanks to his excitement and knowledge. He read early versions of my chapters, providing comments and suggestions that greatly improved my thinking and my writing. Without his empathy, I could not have achieved so much. Paul N. Edwards's advice and stimulating comments have given me the opportunity to reshape important aspects of my theoretical approach. Among other things, he has shown me how to make presentations without reading a paper. I am indebted to him. Stuart Kirsch has given me confidence. He has listened to me for hours in his office, taking notes, questioning my assumptions, and pushing me to find my own voice. He gave me intellectual freedom and continues to be one of the best examples of engaged anthropologists I can think of. Gabrielle Hecht taught me how to teach and write. She believed in me even when I felt uncertain about the feasibility of this project. Her intellectual sharpness and inspiring work are benchmarks for scholarship at its best. I had the privilege of learning from her how to make sense of my intuitions, how to persevere, and how to transform my uncertainties into research questions. Her generosity and unconditional support gave me strength and made a difference in my life. I am looking forward to continuing my conversations and collaborations with such exceptional mentors, and—if they agree—colleagues.

The Science, Technology, Medicine, and Society program has been my other intellectual home at the University of Michigan. Dan Hirschman, Emily Merchant, Kevin Donovan, Adam Johnson, Daniel Williford, and Emma Park have contributed exciting discussions and comments in seminars, workshops, and conferences. Shobita Parthasarathy reminded me that if academic scholars want to make a difference, they need to communicate their research outside restricted intellectual circles. It was a humbling lesson, and I thank her for taking the time to teach me that.

The list of people who made my journey in Ann Arbor special is long. Here I want to thank Robyn D'Avignon, Bruno Renero-Hannan, Shana Melnysyn, Jen Bowles, Fede Pous, Patty Mullally, Cyrus O'Brien, Emma Thompson, Aaron Seaman, Brady G'Sell, Luciana Aenasoaie, Adriana Chira, Ananda Burra, Ashley Rockenbach, Stephen Sparks, Nafisa Essop Sheik, and Roberto Vezzani for their friendship.

Sonja Schmid gave me the opportunity to work on her project on nuclear emergency responses as a postdoc in the Science and Technology Studies program at Virginia Tech, National Capital Region. Her important work, exceptional generosity, and collaborative spirit helped me throughout my stay in DC. Her continuous friendship is precious. Hugh Gusterson read parts of my research and provided astute comments during private conversations and conference panels.

Finally, I would like to express my gratitude to all the students and colleagues at Mississippi State University, where I was an assistant professor of history from 2016 to 2021. During the last two difficult years, I made the decision to move back to Europe. I was fortunate enough to be granted a Marie Sklodowska Curie Fellowship to conduct a research project on nuclear decommissioning in collaboration with the Rachel Carson Center for Environment and Society in Munich. My thanks go to Helmuth Trischler and Christopher Mauch for their encouragement and warm welcome. I look forward to opening another exciting chapter of my professional career.

Without the help and guidance of many Maddalenini, writing this book would have been simply impossible. Antonio Conti, Salvatore Sanna, Gian Carlo Tusceri, Giovanna Sotgiu, Adriano Tovo, Mario Birardi, Pasqualino Serra, Michele Serra, Fabio Presutti, Marco Leoni, Frascesco Nardini, Salvatore Abate, Andrea Nieddu, Alberto Sega, Luca Ronchi, Stefano Giorgi, Franco Ugazzi, Massimiliano Guccini, Erio Fiori, and Giancarlo Fastame, among others, have shared their archives and their thoughts with me since the beginning of my research. Their advice and sincerity have been invaluable. Special thanks go to Tatiano Maiore for making his exceptional photographic archive available to me. I am particularly indebted to Antonello Tovo for his help and friendship throughout the past twelve years and for sharing with me his master's thesis "L'America a La Maddalena," a brilliant political and cultural history of the US Navy presence in the Archipelago of La Maddalena. Without any doubt, Antonello's manuscript has been the single most important source of bibliographical and archival references for this book.

I also want to thank the Municipal Administration of La Maddalena, especially former mayor Angelo Comiti for giving me access to local archives and for allowing me to work in the city hall during my research trips. La Maddalena has given me joy and comfort. Many other people, including retired US Navy personnel who agreed to collaborate with me through interviews and informal conversations, would deserve to be mentioned. I want to thank the families Demontis and Rossi for making me feel at home in their house. Nena Rossi's generosity went beyond any imagination. I will be eternally grateful to her for her hospitality and for her graciousness. Michele Demontis knew how to make me happy. He took me on board his boat when he went out fishing and showed me many hidden corners of the archipelago. The memories of those places stay always with me.

At the ENEA Center for the Study of Marine Environments of Lerici, Roberta Delfanti and Carlo Papucci have been my mentors and research partners. They have shared their knowledge and memories of La Maddalena, opened archives, listened to my ideas, and answered my questions. They know how grateful I am for their help and friendship. All I learned about radioecology comes from them and Arrigo Cigna. Of course, they are not responsible for the flaws that remain in this book. I want to thank Carlo Papucci further for the incredible support he gave me throughout all the phases of my research. By now he must know how much I admire him, both as a scientist and as a citizen.

My work has been possible thanks to generous funding. I was the recipient of a National Science Foundation Grant (SES-1155907) and several research grants available at the University of Michigan.

Finally, to my friends Valio, Mario, Stefano, Emanuele, Enrico, and Bibo: You have always been with me as if my departure had never set us apart. Thanks for understanding. To my family: thanks for accepting my decisions, even when it was hard to understand them. I hope that one day I will be able to give back at least in part what I have taken away from your lives.

Elisa came into my life unexpectedly, in the sense that I did not think she would accept my proposal to take a flight and visit me in Mississippi. The courage, patience, and imagination for thinking the duet are all hers.

List of Abbreviations

APAT	Agenzia per la Protezione dell'Ambiente e del Territorio
CAMEN	Centro Applicazioni Militari Energia Nucleare
CGIL	Confederazione Generale Italiana dei Lavoratori
CISAM	Centro Interforze Studi per le Applicazioni Militari
CNEN	Comitato Nazionale Energia Nucleare
CNR	Consiglio Nazionale delle Ricerche
CNRN	Consiglio Nazionale delle Ricerche Nucleari
CoCiS	Comitato Civico Spontaneo
COMIPA	Comitato Misto Paritetico della Regione Sardegna sulle Servitù Militari
CRIIRAD	Commission de Recherche et d'Information Indépendantes sur la Radioactivité
DC	Democrazia Cristiana
DISP	Dipartimento Sicurezza e Protezione
ENEA	Ente Nazionale Energia e Ambiente
ISPRA	Istituto Superiore per la Protezione e la Ricerca Ambientale
ISS	Istituto Superiore di Sanità
PCI	Partito Comunista Italiano
PSDI	Partito Social Democratico Italiano
PSI	Partito Socialista Italiano

The Atomic Archipelago

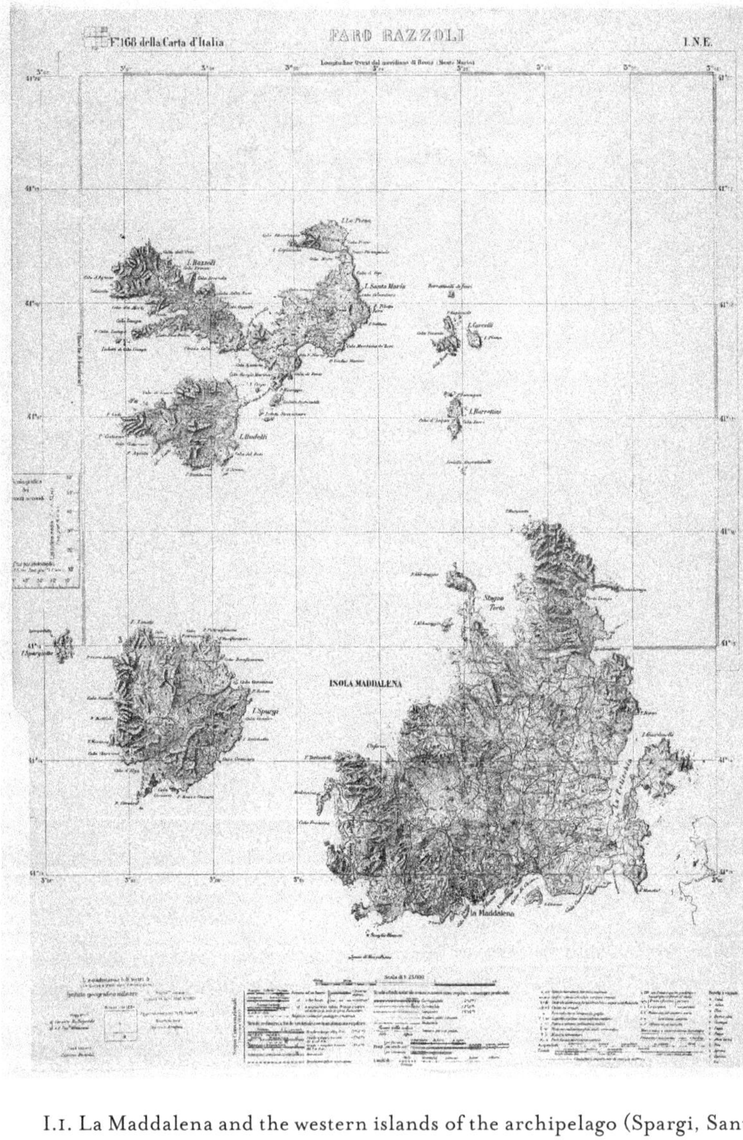

I.1. La Maddalena and the western islands of the archipelago (Spargi, Santa Maria, Razzoli, and Budelli). *Source:* Istituto Geografico Militare, 168 I-NE (Faro Razzoli), Year 1931, Series: 25V. Authorization number: 7091.

INTRODUCTION

The Atomic Archipelago

The Archipelago of La Maddalena is a group of seven islands and islets located offshore the northeastern corner of Sardinia, on the Italian side of the Strait of Bonifacio. In the early nineteenth century, the archipelago became an important naval station. During the Napoleonic Wars the small Sabaudian fleet established its hub in La Maddalena, and a few decades later the royal navy transformed it into a maritime fortress, with an annexed arsenal. After World War II, the permanence of the Italian navy bases secured the economic future of the islands and the continuity of their military traditions.[1]

The Cold War added another layer to the history of this strategic naval outpost, which is the subject of this book. In 1972, under an executive secret agreement with the US government, Italy allowed the US Navy to install a Naval Support Activity base in the bay of Santo Stefano, where a ship's tender performed refit and repair operations on "fast attack" nuclear submarines associated with the Sixth Fleet.[2] The mission of the Sturgeon and Los Angeles class submarines—the "hunter killers," as they were named, for their ability to chase down the enemy—was to surveil Soviet underwater activities in the Mediterranean.

US and Italian defense officials determined that La Maddalena offered favorable conditions for the installation of a new base due to its

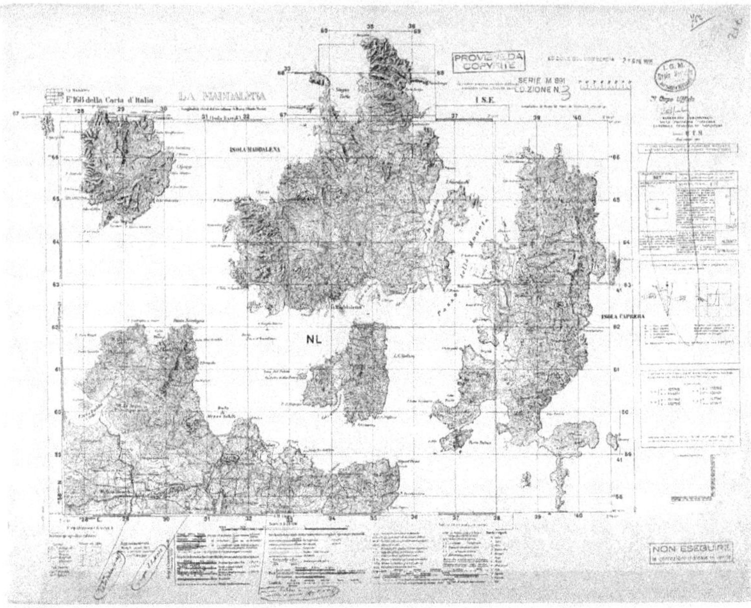

I.2. La Maddalena, Santo Stefano, and Caprera. *Source:* Istituto Geografico Militare, 168 I-SE (La Maddalena), Year 1931, Series: 25V. Authorization number: 7091.

strategic position and the local community's habituation to living in a militarized place, which seemed to guarantee a peaceful coexistence with the new "guests." However, when the agreement between the two governments was revealed, the Italian left, environmental activists, and sectors of the scientific community raised questions about the radiological safety of northern Sardinia. Accused of having made the decision to host a foreign nuclear base without any parliamentary debate, the Italian government defended its choice by reasserting the country's obligations to the North Atlantic Treaty Organization (NATO) alliance and provided risk assessment reports (with the collaboration of conniving experts) that swiftly denied the existence of hazards for the local population and the environment. The effort to demonstrate the innocuousness of the US nuclear submarines proved unsuccessful as several radioprotection experts expressed doubts about the neutrality and the scientific validity of the government's sources. Most antibase parties and environmental activists followed suit and instead of using the leitmotif of US imperialism, decided to bolster their arguments by focusing on radiological safety and the lack of transparency of the US Navy operations in the archipelago. Between 1972 and 1974 La Maddalena became one of the most controversial US Navy posts of the Cold War.

In the following years, expert and public debates revolved around the design and the implementation of an environmental monitoring system that the Comitato Nazionale per l'Energia Nucleare (CNEN, Italian Committee for Nuclear Energy) and the Istituto Superiore di Sanità (ISS, Higher Institute of Health) established with many difficulties only in 1979. The initial lack of environmental data pushed Italian experts to conduct a lengthy survey of the archipelago and later to adopt unusual research strategies to cope with organizational shortcomings, scarce human and technical resources, and classified information regarding the reactors and operations of the US nuclear submarines. Military secrecy and the haphazardness of the Italian nuclear program delayed and ultimately limited the efficacy of the monitoring system, which worked under severe technical and political constraints. In this context, rumors of malfunctions and allegations of connivance between expert agencies and military authorities fed distrust toward the radiosurveillance program among La Maddalena's residents.

Relying on the strength of popular antinuclear sentiments—culminating in the post-Chernobyl referendum that halted the operations of Italian nuclear plants—in 1988 the Region of Sardinia promoted a consultative referendum over the permanence of the US Navy base. Worried about the political implications of the initiative, the Italian government appealed to the constitutional court, which ruled that a regional constituency could not express their views about an international agreement between the Italian state and another sovereign country. Even after suspending its nuclear program, Italy denied Sardinians the possibility to vote on the permanence of a foreign base for nuclear submarines in their region.

In the 1990s the strategic objectives of the US Navy changed. Established as an antisubmarine warfare and surveillance node at the height of Cold War tensions over the control of the Mediterranean and the Middle East, the base of La Maddalena was repurposed to assist military operations in the Persian Gulf and in Africa.[3] Local opposition, usually minoritarian among long-term inhabitants, became particularly active in 2003 when the nuclear submarine USS *Hartford* ran aground in the archipelago's waters. The accident—which US and Italian military authorities kept secret for two weeks—did not have environmental consequences but provoked unprecedented and widespread concern due to the contradictory results of the first radioecological reports that state and independent researchers produced one month after the event. The mismanagement of the postaccident phase, besides discrediting military and scientific institutions, offered local antibase protesters the opportunity to discuss openly the risks involved in the operations of the US submarines and the limits of the environmental monitoring program established in the mid-1970s. Following a major reassessment of its global basing strategy in 2005, the Bush administration—almost unexpectedly—decided to close the base, which was finally decommissioned in 2008. Like other communities strongly dependent on militarized economies, after the decommissioning of the US base and the progressive downsizing of the Italian navy presence, La Maddalena entered a painful and seemingly endless transition: loss of jobs, depopulation, and a drastic reduction of vital services make it hard to reimagine the future of the archipelago.

Several expert agencies, both military and civilian, have monitored environmental radioactivity levels in La Maddalena for over thirty years and always found that the quantity of radioactive isotopes in the area was negligible from a sanitary point of view. In sum, no single study has ever shown that the presence of nuclear submarines caused a relevant increase of radioactivity. In May 2017, five years after I completed my fieldwork in the archipelago, a group of local historians invited me to La Maddalena to present my study in public. At the end of the conference—the *municipio* was filled with people—a former mayor of La Maddalena, who was also chairing the panel, tried to wrap up my presentation: "So, we heard what Dr. Orsini found and everybody, I think, should feel reassured that the base did not have any negative effects on our community, as many of us kept repeating for decades, despite the alarms created for political and ideological reasons." I did not have the time to reply then, but many in the audience responded with a buzzing noise of disapproval and the shaking of their heads.

So, why study the case of a base for nuclear submarines that did not have negative environmental and public health consequences rather than leave it to the past, where it belongs? One obvious answer is that the risk implications imposed on the local community in the name of Italy's Atlantic position cannot be silenced or justified after the fact just because the base did not cause serious environmental and health effects. The elementary and fundamental rights of the Maddalenini to know what risks they were running and to feel safe were not always guaranteed. The story I tell in this book is, at least in part, a way to acknowledge that and to analyze aspects of Italy's Cold War history that have so far remained unexplored.

In *The Atomic Archipelago* I examine the environmental monitoring system set up by Italian expert agencies as a prism to explore the ecological, political, and social implications of the US military presence in Italy during the Cold War and beyond. I document the multiple—and sometimes conflicting—meanings of nuclear power and technology that emerged in everyday encounters with radiological risk, from the perspectives of long-term residents, local administrators, US Navy personnel, and Italian experts who lived and worked in La Maddalena from the early 1970s until 2008. Using radiological risk as a historical tracer, I endeavored to reassemble the sociotechnical systems that sustained the deployment of US nuclear weapons and military outposts abroad, to explain the politics of knowledge production in Cold War Italy, and to shed light on practical aspects of the Italian nuclear program, such as the work of radioecology experts in the field, that so far have not received the attention they deserve.

The submarine base offered Italian radioecologists and radioprotectionists a unique training opportunity, as they mounted a laboratory to monitor potential radiation releases around the site. Experts worked under severe constraints of military secrecy—jointly if unevenly maintained by the US and Italian militaries—and the radiosurveillance system suffered from the inadequate organization of the Italian nuclear program, including lack of technical infrastructures and expertise in peripheral places like Sardinia. These conditions shaped which and how knowledge was produced and how it did or didn't travel. The radiosurveillance system of La Maddalena, I argue, was the result of profound compromises between public safety and military security; it was not just a technical solution to the problem of radiological safety but the material instantiation of a technopolitical compromise between military security and public safety, in the sense that it was neither the result of a strategic political design to conceal military secrets nor a strictly technical solution, but combined elements of both.[4] For one, Italian radioecologists working in La Maddalena adapted routine radiosurveillance protocols to an unusual situation in which information was limited because of external constraints. Furthermore, the design of the radiosurveillance system resulted from the institutional arrangements of the Italian nuclear program and the epistemological approaches of Italian radioecologists during the 1970s.

The case of La Maddalena was not strictly unique. During the Cold War, the US Navy installed other bases overseas through the legal sanction of bilateral agreements with allied governments.[5] There, similarly to La Maddalena, the imperatives of military security often obstructed the application of standard radiosurveillance protocols implemented around civilian nuclear sites.[6] My analysis, however, runs against the assumption that the global outreach of the US basing strategies had uniform effects everywhere.[7] Transnational and global approaches to Cold War studies have finally started to recognize the importance and active role of local communities, organizations, and environments in coproducing knowledge and systems of power, including the experiences of actors who often disappear in the flat, bipolar world of top-down diplomatic histories.

Exploring the "lumpy world" of knowledge production and movement (or lack thereof) during the Cold War—and in other periods—involves attending to the material entanglements of technoscientific practices; explaining their meanings, causes, and effects in different contexts; and including case studies outside the domain of the anglophone world (and its archives).[8]

This study moves a step in that direction by offering a close examination of expert and nonexpert understandings of risk in the making. The story I tell in the following pages reveals the obstacles that nuclear experts found in applying their methods in the archipelago's political and environmental conditions. By following the work of radioecologists and radioprotectionists in labs and in the field and comparing the situated and practical applications of technoscientific protocols with standard national and international regulations, this research reveals the contradictions, the organizational features, and the epistemic cultures that characterized the trajectory of the Italian nuclear program in distinctive ways. This allows me to take into account the mutual effects and interactions of international and national nuclear regulatory regimes and scientific practices as well as the contributions and uses of technology to embodying and enacting political goals and to shaping power relations between Italy and the United States, and (within Italy) between center and periphery.[9] Finally, I advance a novel phenomenological approach to the study of risk, aiming to overcome established assumptions about the role of essential epistemic divides in sociotechnical controversies between experts and nonexperts. The semiotic approach to risk that I propose here focuses on practical processes of signification in which both experts and nonexperts use material signs to objectify and represent risk in tangible ways.

The narrative and the organizational structure of the book reflect my effort to include local, regional, and transnational levels of analysis. The sources and the stories mobilized for this study emerge from peripheral (to the canonical institutional and diplomatic histories of the Italian nuclear program) places and archives to reveal the daily work and struggles of unknown experts, anonymous soldiers, Italian citizens, and bureaucrats whose role has been as important as that of known scientists and preeminent political figures, that appear more frequently in studies of Cold War Italy. In the rest of this introduction, I describe the central themes of the volume, its major arguments, and its organization.

NUCLEAR ITALY UNSEEN

In *The Atomic Archipelago* I bring Italy back into a global and more theory-oriented discussion of the meanings and effects of nuclear power and its applications during and after the Cold War. Existing histories of the Italian nuclear program have focused almost exclusively on the institutional life of expert agencies and on diplomatic and domestic struggles among scientific and political elites over the development of nuclear technology. There are two main historiographical interpretations of the parable of the short-lived Italian nuclear program: The first attributes its failure to conflicting political and economic interests (both domestic and international) behind Italy's energy production. The second, usually advanced by convinced nuclearists, maintains that irrational fears fomented by antinuclearist groups—especially after the Chernobyl accident—guided the majority of Italian citizens to vote against the continuation of the nuclear program in the national referendum of 1987. Behind these explanations are research traditions that, with a few exceptions, look alternatively at elite struggles and democratic protest in isolation and ignore the tensions that—even inside nuclear agencies—existed between technocratic and democratic views of nuclear power since its inception, and even more so during the 1970s.[10]

Looking at radioprotection and radioecological practices in the environment of La Maddalena allowed me to overcome the most important limitation of current Italian nuclear studies: the treatment of nuclear technology as a black box, a mere political instrument, whose technical characteristics are relegated to the rank of historical curiosity. In this study, I instead look inside nuclear technology as the embodiment of complex sociotechnical systems that allowed its development, applications, and deployment in one place, including the disputed claims about its effects on the environment and public health. Moving beyond the mere documentation of formal institutional arrangements and elite storytelling involves adopting a more localized approach that allows us to adopt different scales of analysis and to have a better grasp of the everyday experiences of communities of experts and nonexperts simultaneously.

While I agree that bottom-up approaches are not inherently superior to diplomatic histories, I am convinced that looking at the deployment of nuclear technology and the implementations of regulatory regimes on the ground offers a perspective that so far has been sorely missing in Italy. I deem it crucial for scholars of science and technology to move beyond mere discursive analyses of the history of nuclear politics and to engage more directly with the material entanglements, experiences, and affordances that contributed to shaping expert and nonexpert conceptions of nuclear technology.

LA MADDALENA AND THE POLITICS OF SCIENCE IN COLD WAR ITALY

The widespread identification of the archipelago's inhabitants with military institutions did not prevent a minority of Maddalenini from opposing the US nuclear base immediately after its installation. With the help of expert activists who denounced the inadequate safety protocols set up to monitor the archipelago's environment, they were able to open a controversy over risk assessment. The articulation of these critical positions by both expert activists and local opposition underscored immediate objectives, such as the removal of the US base from the archipelago, but also reflected widespread social concerns. Between the end of the 1960s and the beginning of the 1970s, analogous to other public interest movements in the United States, inside Italian research centers and agencies, experts started to interrogate their role in society and pushed for a democratization of technoscience.

The politics of science emerged as a central theme in Italian public debates especially during the years of contestation, although the problem was not new.[11] Normative and formal understandings of objectivity in technoscientific knowledge production underwent a scrutiny in the late 1960s as the political ferment of student revolts, the movement against the Vietnam War, and the birth of political ecology shaped more critical perceptions of the presumed neutrality of technical knowledge and of the relationship between science and progress.[12] This critical rethinking of the social and environmental responsibility of technoscience became politically influential with the formation of what Kelly Moore, in discussing the United States, has called "public interest science organizations," like the Scientists' Institute for Public Information (led by biologist and political ecologist Barry Commoner), the Union of Concerned Scientists, and Science for the People.[13]

In Italy, where the articulation of social conflicts and the life of political institutions was much more structured around parties and labor unions, groups who advocated a democratization of science were often active members of national political organizations within professional societies, public companies, and state agencies. In research laboratories, scientists, technicians, and workers created union chapters and party cells, sometimes in coexistence with autonomous and more radical collectives, who contested the hierarchical structures of research institutions and universities, and their subservience to American capitalism and its military-industrial complex. Political occupations of university buildings, laboratories, and research centers were quite common. In 1969, for example, a group of occupants started a five-month protest at Casaccia, CNEN's biggest research center, against the exploitation of workers and technicians and the subordination of the agency to the "capitalistic mode of knowledge production."[14] Also inside ISS, the instances of change promoted by the 1968 movements inspired political initiatives and vibrant debates about the future of the institute and its role within Italian society.[15]

The activity of contestation and the proposals of expert activists were more frequently—but not exclusively—associated with parties, unions, and collectives on the left of the political spectrum. Among more organized and institutionalized parties, the Partito Comunista Italiano (PCI, Italian Communist Party) discussed systematically the necessity to reform the school system and the health system, and to reorganize technoscientific research around strategic national goals.[16] The conference "La ricerca scientifica e la società in Italia" organized in Ariccia (outside of Rome) on March 20–22, 1970, was the first organic attempt by the PCI to catch up with the wave of youth protest and to establish a structural alliance between students, researchers in the public sector, and workers.[17] In presenting the objectives of the event in the pages of *L'Unità*, Giovanni Berlinguer explained that Italy's science policy needed to move past the legacy of fascist autarchy and the current subordination to American technoscientific power. Berlinguer affirmed that technoscientific research could find its fundamental raison d'être in "the human needs of the masses to imagine an alternative use of science. This is why an alliance of workers, students, and all citizens with the movement of researchers and scientists is an urgent necessity."[18] Almost two years later, another important conference titled "Uomo, Natura, Ambiente: Ecologia e Rapporti Sociali," organized at the PCI's headquarters of Frattocchie, signaled the party's shift of attention to the emerging world of environmentalism.[19] But a strategic and organic alliance between the labor movement (traditionally concerned with occupational safety) and political ecology had to wait until the second half of the 1970s, after the Seveso disaster of 1976 and the formation of Lega per l'Ambiente in 1979, of which Berlinguer was one important promoter.[20] The mid-1970s were also fundamental years for the formation of the ecological movement in Italy, which built its political platform and consensus around the antinuclear protests that local communities launched against the Piano Energetico Nazionale (PEN, National Energy Plan) of the Ministry of Industry and the government's plan to install twelve to twenty new nuclear power plants in ten years.[21]

The divisions and heated debates within Italian research institutes and agencies can be understood only by considering this general context of political mobilization around central questions concerning the role of science and technology in society. Thus, throughout the chapters that follow, the readers will find many instantiations of ideological cleavages within ISS and CNEN and other expert communities. Conflicting ideological orientations and different political affiliations did not prevent individual experts and groups from collaborating with one another, necessarily, but alternative views of the role of science in society undoubtedly catalyzed alliances between scientists and political activists in distinctive ways. Parties, unions, collectives, and the church organized discussions, aggregations, and cultural elaborations, which defined strategies, platforms, and geographies of knowledge production tied to their worldviews. In the chapters that follow I show how the controversy around the nuclear status of La Maddalena brought to the fore these conflicts, which found expression in two alternative views of nuclear science and technology: the technocratic outlook of high cadres, who conceived of technoscience as an exclusive expert business, and the democratic approach, typically advanced by leftist researchers working in the public sector.

ITALY AND US MILITARY BASES

Scholars of the Cold War largely agree that the United States exercised its hegemonic role (with various outcomes) through a variety of tools: military, diplomatic, cultural, economic, which do not necessarily fit within formal definitions of colonial empires. For example, historians of science and technology have documented how the United States' active participation in rebuilding Europe's technoscientific capabilities after World War II was not just inspired by the universal principle of sharing knowledge for the betterment of human life, or by an idealistic appeal to universal progress, but was primarily guided by more mundane objectives like shaping Europe's development according to US interests.[22] More traditional Cold War historiography has offered representations of US hegemony in Europe mostly in terms of "Americanization," cultural colonization, and economic and military dependence.[23] These studies present only cursory descriptions of the environmental problems that local communities often experienced in relation to the deployment and operation of US military personnel and infrastructures.[24] To move beyond formalistic comparisons of the United States' global military and economic outreach after World War II with other empires, in this study I examine empirically the deployment of US military bases overseas to, first, document the localized "imperial effects" of US global military, geopolitical, and economic strategies; and, second, introduce themes and actors usually excluded from prevalently top-down diplomatic histories of the Cold War.[25]

Joining the now abundant literature on US overseas military bases, my analysis foregrounds the importance of islands as strategic nodes for the formation of the "US networked empire."[26] US overseas basing strategies during the Cold War (and earlier periods) responded to two related needs. On the one hand, islands allowed for the deployment of armaments, troops, communication networks, and surveillance systems, and provided necessary refueling spots for global intervention. On the other hand, islands were strategic because they enabled the formation of an invisible infrastructure, a deterritorialized network of stations that appeared far removed from conventional colonial practices of territorial annexation.[27] Military islands are crucial loci for understanding the flexible modes of operation and the legal and extralegal arrangements that allowed the United States—like other empires before World War II—to build its global military outreach and protect its economic and commercial interests, while claiming for itself the status of an anticolonial power.[28]

The Italian government ceded the use of La Maddalena to the US Navy without prior parliamentary debate or approval. This was in part was because in 1954 Italy and the United States stipulated a bilateral infrastructure agreement in implementation of the North Atlantic Treaty. Based on these executive provisions, whose terms remain classified, Italy allowed—and sometimes invited—the deployment of US military personnel and the use of bases on its territory for defensive purposes within NATO's coordinated strategies. After the signature of the bilateral infrastructure agreement, several US and NATO military bases opened on the national territory.[29]

During the Cold War, pro-NATO Italian political elites perceived the deployment of American nuclear weapons on the national territory as a guarantee of defensive capabilities and used it as a diplomatic shortcut for reaching higher international status.[30] For this reason, when the Eisenhower administration proposed "nuclear sharing" to its European allies, Italy actively pursued a special partnership with the United States by offering logistic support and collaboration—such as the installation of nuclear submarine bases on Italian soil and seas. This is in marked contrast to other European countries, such as France, especially after the advent of the Fifth Republic, which were reluctant to enter such agreements.[31]

As early as 1955, Italy and the United States started an exchange of secret executive notes concerning the storage of nuclear weapons and in 1962 the two governments formalized a more detailed Atomic Stockpile Agreement.[32] The deployment of the Jupiter missiles in Gioia del Colle, Puglia, in 1961–1962 and of the Pershing II missiles in Comiso, Sicily, between 1983 and 1987 are well-known examples of the strategic importance that Italian political elites assigned to military collaboration with the United States.[33] Given their visibility and the internal and international opposition they triggered, governmental decisions to welcome the installation of US missile ramps could not bypass public scrutiny. But less visible infrastructures—many of them still operational and repurposed after the end of the Cold War for the "War on Terror"—made possible the implementation of NATO strategies in Italy. Like the small airbase of Ghedi, in Lombardy, used since 1963 as a storage facility for nuclear warheads, a myriad of "technical facilities" were disseminated across the Italian peninsula for radar interception and communication, submarine surveillance, radio communications, and satellite tracking.[34]

The history of the US Navy presence in La Maddalena is part of this larger Cold War military infrastructure.[35] The US personnel operating at the base could pursue their assigned antisubmarine warfare and surveillance objectives thanks to their coordination with other installations. In northern Sardinia the US Air Force ultra high frequency (UHF) antennas of Mount Limbara, and the US Navy very low frequency (VLF) communication station mounted on the Island of Tavolara (just in front of the Gulf of Olbia) in 1962 (guarded by small US military contingents) guaranteed aerial and underwater communication transmission to and from the base of La Maddalena and other facilities.[36] As was the case with other US military bases during the Cold War, Italian pro-NATO elites delegated technical decision-making about the installation of the US Navy station in La Maddalena to restricted circles of military and diplomatic personnel.[37]

THE POLITICS OF KNOWLEDGE PRODUCTION

The installation of the US Navy base in 1972 did not automatically bestow on La Maddalena the status of a nuclear site. For reasons of military security, the Italian government and the US Navy initially claimed that there was no need for Italian expert agencies to conduct radiation monitoring activities normally adopted for inland nuclear plants. Establishing the nuclear status of the archipelago entailed two years of intense debates, during which only a minority of La Maddalena's residents seemed to be actively involved—namely, members of the local chapter of the PCI, sectors of the Partito Socialista Italiano (PSI, Italian Socialist Party), and a few intellectuals critical of the traditional subservience of the local community to military authorities.

Interested or distracted observers of La Maddalena have often interpreted the apparent passivity of the local community as due to its traditional acquiescence to military authorities. In the words of Massimo De Carolis, a conservative member of the Democrazia Cristiana (DC, Christian Democratic Party) in the 1970s: "The tranquility of the Maddalenini is easy to explain. Over the past two centuries the island has always been a naval base and now the presence of the submarines leaves them indifferent. The Americans spend five million dollars per year and have always been fair and polite."[38] This interpretation, like others of the same tenor, is based on the simplistic assumption that silence or lack of protest means acceptance.

Such swift explanations neglect the fact that assessing risk is a difficult and often controversial task, which has to do with probabilistic calculations of the consequences of adverse events that may or may not take place. In addition, some sources of risk are not immediately perceptible, as in the case of radiation, and their appraisal is significantly influenced by environmental variables and the socioeconomic conditions of the communities living near them. For these reasons, thinking about risk and defining it in concrete terms requires work, expertise, and infrastructures.[39] The observation that the Maddalenini seemed at ease with the presence of US nuclear submarines due to their traditional identification with the Italian navy glosses over the complex individual and public epistemologies (formal and informal as they might be) through which risk is rendered visible or remains invisible. In sum, cost-benefit explanations alone do not consider the complexities of knowledge production about risk and the "politics of (in)visibility."[40] This book digs into this complexity and documents how in La Maddalena the uncertain status of US nuclear submarines as nuclear things, military secrecy, and the epistemic traditions and the (dis)organization of the Italian nuclear agencies contributed to shape a particular "regime of perceptibility" that made certain aspects of nuclear risk visible and others invisible.[41]

Nuclear Submarines as Nuclear Objects

CNEN and ISS experts found it difficult to convey to local administrators (and nonexperts in general) the gravity of the US Navy base's safety implications in La Maddalena because nuclear submarines pose specific challenges to risk assessment. Unlike big, visible nuclear plants with their reactor domes dominating the landscape and conditioning in many concrete ways the everyday life of entire communities around them, nuclear submarines can be quickly and quietly removed from the local context: mobility and invisibility play a crucial role in making them ambiguous nuclear things. As subsequent chapters will detail, in the archipelago nuclear submarines gave local administrators the illusion that risk was controllable precisely because its sources could be removed at any time: "If they start creating problems, we can push them away," said Giuseppe Deligia, the mayor of La Maddalena, during a conference on the US base organized in February 1975. Therefore, at least initially, the ontological status of the submarines as "nuclear things" needed to be demonstrated and established through comparisons and analogies with more stereotypically nuclear artifacts like nuclear power plants or, in more alarming tones, nuclear bombs.[42] Given that until 1979 Italy did not have formal and consistent regulations of the transit and mooring of nuclear military vessels in national waters, ISS and CNEN experts treated the US base in La Maddalena as if it were a nuclear plant.

Gabrielle Hecht calls the contested technopolitical category of being nuclear *nuclearity*, to point out to the geographically contingent and historically shifting attribution of radiological risks and their potential consequences to objects and practices.[43] Nuclear submarines are not only technological artifacts for conducting war. In La Maddalena, different groups with direct or indirect knowledge about their characteristics developed different understandings of nuclear submarines' risk and safety implications.[44] For example, mayors and other members of the local elites were occasionally invited on board for short cruises underwater. As I was told during interviews, it was on those occasions that the US Navy used nuclear submarines as diplomatic tools to demonstrate the innocuousness of the most advanced war machines in the world. Additionally, the belief that all US Navy personnel working on the base were aware of their occupational risks circulated among the Maddalenini since the base was installed. This was reassuring to civilian workers and long-term residents, who repeatedly told me: "If US Navy servicemen were at ease with their job, why should we worry? They brought their families with them. Do you think they would do that if there were a real danger?" However, interviews with retired US Navy servicemen reveal a different story: only select personnel knew the technical details of submarines' equipment and were authorized to access specific sectors of the base. Only partially visible to the rest of the population, safety measures against radiation exposure defined sociotechnical and professional hierarchies through thresholds of accessibility and areas of exclusion inside the base.[45] As dosimeters, gloves, clothes, shielding procedures, and time of exposure defined the daily encounters of specialized US Navy personnel with nuclear risk, long-term residents coped with their lack of experience and information by constructing a reassuring image of the competent and self-aware American soldier.

Nuclear things acquire different meanings (dangerousness, security, development) according to who uses them.[46] In La Maddalena, like in the rest of Europe, images of Hiroshima and Nagasaki, as well as other public notions of the potentially disruptive power of nuclear technologies, were in circulation well before the installation of the US Navy base. The Maddalenini often interpreted information and evidence about the risks generated by the presence of nuclear submarines through their political lenses: Christian Democrats and other pro-NATO party members usually defended the American presence in the archipelago, while PCI and PSI members retained an inherent distrust toward America's foreign policy and intentions. Thus, political, and ideological affiliations inflected controversies about safety in La Maddalena and made technical assessments of nuclear risk hardly separable from other considerations about the trustworthiness of expert agencies, of the Italian navy, and US intentions.

Consider again the common interpretation circulating among the Maddalenini of antibase campaigns and concerns from leftist parties as alarmist propaganda: "If the Americans were here with all their families it means that they knew they were safe, otherwise they would not expose themselves to the risk of contamination. So, this means that we were safe too!"[47] Consolidating this sense of safety and protection was the lionization of American technological prowess and competence—the idea that nuclear submarines represented the apex of US technical mastery associated with rigorous individual and collective training. As I will explain further, these narratives about American technological competence were not simply the result of false consciousness or of materialistic calculations of the costs and benefits of the US Navy presence but emerged from a wider field of interactions between actors (experts, US personnel, local residents, political authorities), things (submarines, the environment, radiosurveillance instruments), and ideologies (scientific knowledge, radiological training, local traditions) that carried and produced more or less coherent objectifications and representations of nuclear risk.

The political relevance and power of nuclearity—the quality of being nuclear—as a technopolitical category is directly tied to *nuclear exceptionalism*, the idea that nuclear things are essentially different from ordinary ones.[48] After World War II nuclear exceptionalism has been a recurrent theme in public discourses. The ability to exploit the power of fission and radioactivity became the material demarcation of a historical "rupture" with the world as it existed before the nuclear era. Utopias of limitless progress, made possible by infinite sources of energy, coexisted with dystopias of apocalyptic scenarios, nightmares of annihilation, and thermonuclear wars. Nuclear exceptionalism thus "transcended political divisions" and was filled with contradictions. It could be manipulated for opposite political purposes: either to create a sense of collective empowerment and futuristic enthusiasm or to warn the public that nuclear threats required special safety measures and more nukes to dissuade the enemy from launching an attack.[49]

The apparent contradictions of nuclear exceptionalism coexisted also in La Maddalena, where the Italian government and US Navy officers justified the maintenance of secrecy as a necessary limitation to scientific inquiry in the name of absolute security imperatives. At the same time, they had to represent nuclear technology as safe, benign to the local population, and tightly controlled by the technical mastery of the US Navy. The attempts of the Italian government and the US Navy to assuage public anxieties by naturalizing nuclear technology and making it appear innocuous, harmless, and safe is an example of what Stephen Hilgartner and others have called "nukespeak," the set of rhetorical devices and edulcorated language that the nuclear establishment (military, industrial, and scientific) uses to make the potential negative effects of nuclear technology invisible.[50] Thus, in La Maddalena different registers of nuclearity shaped the initial controversy over the presence of the US submarine base. According to the Italian government, nuclear submarines were to protect La Maddalena from the Soviet enemy, yet the archipelago did not need radiosurveillance measures, as nuclear submarines were apparently less dangerous than nuclear plants. On the other hand, Italian radioecologists and radioprotectionists, supported by a variegated antibase front (mostly from outside La Maddalena), did see nuclear submarines as nuclear objects—and demanded acknowledgment of the risks associated to their operations. Consequently, they argued, the archipelago should have been monitored as any other nuclear site to guarantee the safety of the local population.

Secrecy, Bureaucracy, and the Production of Ignorance

Several studies have documented how radiation and its effects are often rendered invisible by hiding, challenging, and classifying information.[51] Stated otherwise, the "invisibility" of radiation is not simply a natural phenomenon related to its physical characteristics but is also produced by the very material acts of confining knowledge and data only to experts and military authorities.[52] Works on agnotology—the production of ignorance—have demonstrated how strategic acts of knowledge removal impeded access to information by the public in various fields: from tobacco industry's concealment of data about the health effects of smoking to the production of uncertainty concerning global warming data.[53]

In La Maddalena, military secrecy—imposed jointly by the US State Department and sectors of the Italian government—conditioned and limited the application of scientific protocols that Italian experts routinely implemented at civilian nuclear sites. CNEN and ISS personnel voiced these contradictions and explained how restricted data about US nuclear submarines' reactors and lack of information about environmental characteristics of the archipelago forced them to adopt alternative, more complex, and time-consuming research strategies. The radiosurveillance system of La Maddalena, I argue, was a technopolitical compromise between military security and public safety, which epitomized the constraints and contradictions of Italy's sovereignty during the Cold War. As mentioned above, Italy was not an exception or a unique case.

During the Cold War, secrecy became an integral component of the national "security state" in the United States, transforming the very nature of American democracy into a "compartmentalized knowledge society."[54] The US Atomic Energy Act (1946) and the National Security Act (1947) separated national security from state security and officially "introduced a new kind of information—nuclear weapons data—that did not need to be formally classified" because it was born secret.[55] The exclusion of activities and information of strategic interest from public scrutiny created a typical Cold War phenomenon: Removing knowledge entailed the proliferation of extensive bureaucratic apparatuses exclusively devoted to maintaining and producing secrets.[56] "How did secrecy and bureaucracy become so entwined—a vast secrecy system almost wholly hidden from view?" asks Daniel Moynihan in his book devoted to this subject.[57] For Joseph Masco, the proliferation and ramifications of secret apparatuses in the social fabric of American democracy have provoked the dislocation of secrecy itself from a clearly definable center of power to a pervasive and yet fragmented and uncoordinated security system in which "knowledge is rendered suspect."[58] As Masco has ably demonstrated, secrecy not only consists of acts of knowledge removal, concealment, and sequestration but is also a ruling technique shaping "a pathological administrative form."[59] The national "security affect" as a form of social technology allowed the American state to achieve collective "perception management and control."[60] Masco's analysis foregrounds two complementary aspects. The first is the organization of secrecy as a bureaucratic infrastructure that became incorporated into democratic institutions.[61] The second is the role of secrecy as a social device that influenced individual and collective interactions, expectations, and rights.

In his book *The Worst-Kept Secret*, Avner Cohen explores Israel's nuclear policy—identified with the Hebrew term *amimut*, which means "opacity"—by showing how the state successfully maintained an ambiguous posture with regards to its nuclear status.[62] *Amimut* required special diplomatic arrangements (the US active contribution to the protection of Israel's nuclear secrets and Israel's decision not to sign the Non-Proliferation Treaty). Internally, the politics of opacity involved the creation of an invisible bureaucratic infrastructure: special constitutional provisions; direct and exclusive control of the executive and the military commands over the nuclear program; and a vast apparatus of surveillance, censorship, and classification.

These and other works have revealed that the priorities of the national security state made secrecy an essential component of the everyday life of entire nations. As I document in the case of La Maddalena, the Italian executive left the area of military nuclear applications under its control to achieve a separation between military activities and civilian radiosurveillance programs and to preserve the autonomy of the first from public scrutiny. The result of this informal institutional arrangement was the creation of a dual system of radiosurveillance that contributed to the compartmentalization of nuclear knowledge.

Let's consider now the second aspect of secrecy as a pervasive phenomenon that shapes interpersonal relationships and the very content of citizenship—that is, the relations between individuals and groups with the state. Amimut, as Cohen argues, could not have worked out if, in addition to ad hoc institutional arrangements, the existence of the bomb had not been transformed into a taboo, a form of tacit knowledge that has been individually interiorized and collectively expunged from public discourses. This is an example of what Michael Taussig calls a "public secret," which he defines as "that which is generally known, but cannot be articulated."[63] For Taussig, secrecy also involves knowing what not to know, which he calls the "labor of the negative," the deployment of ignorance as a strategy for surviving within a context in which knowing is risky. An inherently social device, secrecy creates bonds and solidarity; conversely, it excludes those who are not given privileged access to the secret.[64] Wondering what is concealed is part of the secret's allure, which also invites its transgression.[65]

Secrecy, confidentiality, and opacity can be a reflex of technocratic approaches to public participation in technical matters when citizens are deemed unprepared, immature, not educated enough, and too emotional to be able to deal with the culture of risk in a rational way. Multiple strategies for excluding the public from decision-making processes include formalized rituals such as public audits and hearings whose goal is to neutralize critiques through the rationalizing (and patronizing) techniques of administrative practices.[66] Secrecy, as a form of social regulation, morphs into a variety of knowledge removal techniques. But what is the reception of secrecy within society? How does the public in democratic systems respond to the fact that secrecy lies at the very core of state power?[67]

Secrecy is not an exclusive tool of knowledge control in the hands of power elites. In Italy, for example, state secrecy (*segreti di stato*) has become a popular interpretative category that citizens use to make sense of opaque, unintelligible, or unexplainable actions and inactions of public institutions.[68] Italian jurist and philosopher Norberto Bobbio argued that state secrecy was a pervasive threat to the life of democracy.[69] He used two concepts to describe Italian state secrecy: *sotto governo* (subterranean government) and *criptogoverno* (cryptogovernment). The first term referred to the growing intervention of the state in the economy through which political elites exercised an inscrutable control over centers of power—such as banks and nationalized and state-subsidized industries. By *cryp-*

togovernment he meant "the totality of actions carried out by paramilitary political forces which operate behind the scenes in collaboration with secret services."[70]

In La Maddalena, military secrecy and diffused technocratic views of nuclear technology as an exclusive expert field created a substantial exclusion of the local community from important decision-making processes concerning the radiosurveillance system. This systematic divide between decision-makers and citizens contributed to establish an atmosphere of distrust and diffuse sense of resignation. Some members of the local community embraced conspiracy theories to explain otherwise unexplainable bureaucratic delays, malfunctions, and silences by expert and military authorities.

Secrecy, however, does not fully explain the limits of the radiosurveillance system installed around the US Navy base of La Maddalena. Framed within a larger national historical context, this case study offers ample evidence of the shortcomings and contradictions that characterized the Italian nuclear program since its inception. I argue that both secrecy and knowledge gaps resulting from epistemic traditions and regulatory practices produced ignorance about the environmental status of La Maddalena.

Sociologist Scott Frickel observes that scholars working on agnotology have generally explored secrecy as an active removal of existing knowledge, or "knowledge sequestration," while leaving other aspects of the production of ignorance aside.[71] Inspired by David Hess's work on "undone science," Frickel proposes to focus instead on the multilayered and cumulative selective effects of epistemological assumptions and regulatory practices that on the one hand stabilize and define objects and scope of scientific research and on the other exclude other areas of investigation and underrepresented/nonprivileged sectors of the population from expert intervention.[72] In doing so, the "new political sociology of science" aims to move beyond the individualistic paradigm of agnotology, which examines the "nonproduction of science" only as a result of state agencies, corporations, and military interests in concealing the "truth" or as the result of bad science.[73] Form this point of view, I see Frickel's proposal as an institutional variant of Michelle Murphy's "regimes of perceptibility," which openly relies on the analytical tradition of "historical ontology."[74] But Frickel's "institutional approach" does not exclude taking into account larger historical and political trends that—for example, during the Cold War—inevitably shaped institutional processes and collective trajectories of knowledge (and ignorance) production.

Instead of considering the "knowledge sequestration" and "selective attention" approaches as alternative research strategies, I adopt them simultaneously to grapple with the political economy of knowledge production shaped by interested acts of knowledge removal and institutional arrangements that created and reproduced knowledge gaps in La Maddalena. I show that the Italian government made specific institutional choices concerning the radiation surveillance program (what I described as the Cold War technopolitical compromise between public safety and military security) and that within the limits of institutional arrangements and epistemic traditions of Italian regulatory agencies some experts voiced their disagreement and proposed solutions to achieve more inclusive—and consequently more accurate—radioprotection practices.

SEMIOTICS OF RISK

Especially during the six years following the installation of the US base (1972–1978), incomplete data prevented expert agencies from providing a concrete definition of the risks that long-term residents were facing. Lack of previous experiences with radiological hazards left many Maddalenini wondering about what could go wrong (accidental and routine discharges of radioactive material), the likelihood of an odd event taking place, and the possible consequences of accidental releases of radioactive substances around the US base. Expert assessments, citizens' conjectures about the potential consequences of radiocontamination, and practical knowledge of radiological hazards among specialized US Navy personnel shaped different understandings of risk in the archipelago. In the third part of the book, I address two questions: First, how do nonexperts make sense of radiological risk in the absence of previous experiences with it? Second, how do objectifications of risk change and get stabilized within the same community over time?

As a field, science and technology studies (STS) has successfully overcome some severe limitations of cognitive and psychometric approaches to risk perception in vogue until the end of the 1980s.[75] While one "traditional" branch of "public understanding of science" (PUS) has substantially maintained the public deficit model,[76] since the beginning of the 1990s "interpretationist" approaches have shifted the STS community's attention to questions of trust in technoscientific institutions, epistemic recognition of lay knowledge, and public participation in technoscience.[77] Here the influence of Ulrich Beck's theses on "risk society"—especially the proposition on "relations of definitions"—is hardly disputable.[78]

Initially conceived of as a breakaway from traditional PUS, the analytical leverage of community-based approaches has become an almost teleological frame to study power relations in sociotechnical controversies. Within STS many have assumed the existence of an essential incompatibility between scientific knowledge production and other forms of experiential knowledge.[79] Works on street science, citizen science, and popular epidemiology, to cite a few critical PUS approaches, tend to reify "a dichotomy between 'scientific' and 'public' forms of knowledge and understanding . . . and a level of homogeneity and coherence within both 'scientific' and 'public' understandings."[80] "The possibility of disagreements and contradiction within each form of understanding is played down [while] the use of 'knowledges' in this context can suggest a static account, rather than directing attention to the kinds of sense-making activities engaged in by public and other groups."[81] In addition, both traditional and critical PUS tend to use the concept of knowledge as a catchall category to describe a variety of cognitive, affective, and cultural processes in which local communities and communities of experts are involved when trying to define and objectify risk.[82]

Brian Wynne's point that "the most germane risks are (social) *relational*"—that is, perceptions of risk depend in large part on evaluations of the trustworthiness, competence, and independence of expert institutions—is now taken for granted but leaves the question of materiality and the role that it plays in processes of signification largely unexplored and undertheorized. This is even more problematic, I would argue, because it is through the analysis of the material entanglements of communities' experiences with unprecedented and concerning issues that we can fully grasp the ethical implications of Wynne's relational approach.

Object-oriented analyses of public involvement in technoscience have helped STS move beyond schematic analyses of lay/expert controversies.[83] For example, Noortje Marres explicitly shows how domestic devices and other objects allow different modes of participation, whether the role of material objects is consciously articulated in the enactment of participatory practices or remains unaccounted for and "under-articulated."[84] Affected publics formulate hypotheses about what is going on and what's next in moments of disorientation, surprise, ambiguity, and extraordinary attention solicited by unexpected events. If publics, as in John Dewey's conceptualization, could be defined as "communities of affected interest" and if, as Bruno Latour suggests, we need to explore how "matters of concern" catalyze assemblies of heterogenous actors, then the problem is: what happens after "issues spark a public into being"?[85] To move beyond the recognition that objects actively enable, enhance, and constrain forms of participation through their affordances, we need to attend to the material processes of signification in which publics are involved when trying to make sense of the circumstances that concern them. In sum, how do publics move from being concerned/affected (by something they do not know about) to making sense of their experiences?[86]

Phenomenological studies of risk perception and assessment suggest that to make invisible risks (such as radiation) visible, nonexperts and experts alike rely on a variety of evidentiary practices through which they can objectify and represent risk in tangible ways.[87] Without "opportunities for articulation," risk signs remain unnoticed.[88] In other words, expertise, monitoring devices, and technical infrastructures are necessary to make sense of bodily symptoms, environmental changes, and sensorial alterations that would otherwise remain in the background, unexplained.

Building on these insights, I propose that instead of assuming lay/expert epistemic divides as the basic explanation for the existence of divergent definitions of risk, we should first attend to the processes of signification through which local communities and communities of experts and practitioners pay attention to signs and interpret them as indicators of risk or lack thereof. This allows us to explore the interdependence of experts and nonexpert understandings of risk in and across specific contexts, but to do so we need to renounce the assumption—as STS scholars have done for expert knowledge—that lay or local knowledge is always already *in place*, as if it were a constitutive feature of any community by virtue of an osmotic (almost deterministic) relationship with the surrounding environment.[89] Instead, we should ask how and when objectifications and representations of risk arise from the material entanglement of experts and nonexperts with the environment, broadly defined.

To explain how residents of La Maddalena, local administrators, experts, and US Navy personnel interpreted or ignored signs of risk I advance a semiotic approach that attends to material processes of signification through which communities of experts and nonexperts establish and challenge definitions of risk over time. The semiotic approach to risk is based on the following theoretical and methodological propositions:

1) Both experts and nonexperts use material signs to make invisible risks like radiation visible.

2) Understandings of risks change over time and are highly contingent on local power dynamics. Objectifications and representations of risks change and get stabilized over time as signs of risk become available for interpretation (i.e., environmental changes, unprecedented events, health effects).

3) Experts and nonexperts are not bounded and homogeneous groups. Especially in situations of uncertainty, when knowledge about risk is not established yet, epistemic and communicative strategies of experts and nonexperts emerge out of contradictions and conflicts, and influence one another, at times through forms of active collaboration.

4) When experts and nonexperts elaborate arguments in public controversies, members of both groups engage in activities that aim at reducing internal conflicts and contradictions, which can undermine their credibility. I call this set of activities "the politics of coherence."

The word *semiotics* derives from the ancient Greek *σημειόν* (semeión; sign, miracle) and *σημειόω* (to observe). One common use of the word *σημειωτική* (semeiotiké) denoted the act of observing signs that post-Hippocratic medical practitioners used to make diagnoses about their patients' illnesses.[90] This original phenomenological approach to the use of signs as vehicles of knowledge (through the definition of their contextual meanings) is retraceable in the semiotic theory of Charles Sanders Peirce, one of the founders of the science of semiotics: "It appears, then, that the rule for attaining the third grade of clearness of apprehension is as follows: Consider what effects, that might conceivably have practical bearings, we conceive the object of our conception to have. Then our conception of these effects is the whole of our conception of the object."[91]

In this formulation Peirce sketches out the logical implications of the use of signs as vehicles for knowing about the world, which is an inductive process. Knowledge is always partial and inductive because it is based on our experiences of the world through signs as vehicles: we know the causes of phenomena because we learn to pay attention to and to interpret their effects. The interpretation of signs (from natural phenomena to social habits, from technological artifacts to events) requires that the interpreters acquire a set of strategies to delimit their potential meanings, which may vary according to material circumstances, to the context (historical, cultural) of interpretation, and to the communities of interpreters, who may share different interests in and definitions of a phenomenon. This is the role of what anthropologist Webb Keane calls *semiotic ideologies*—that is, "basic assumptions about what signs are and how they function in the world."[92]

The concept of semiotic ideology, among other things, helps us explain just this: how we move from the realm of affectedness to that of representation, or in Peircean terms, how we move from the level of secondness (being struck by and reacting to something/someone) to that of thirdness (thinking about, conceiving of, and representing something, such as radiological risk, trust, or others' intentions, for example). The strike of thunder may be taken to signal the imminence of a storm, but it may be interpreted as a sign of divine intervention if we believe that gods and spirits are real agents. As Keane underlines, the word *ideology* in this context should not be intended as false consciousness but rather as indicating "a fundamental reflexive dimension of the general human capacity to use signs" subject to history and struggle.[93] According to Keane, semiotic ideologies provide instructions for how to use abductive reasoning in processes of signification but involve much more than this, as "semiotic ideology mediates between abductive inference or interpretation, which are general cognitive processes, and the more specific material and conceptual circumstances that prompt them, the forms of judgment to which they give rise, along with hopes and anxieties that attend them."[94]

An example of the reflexive dimension of semiotic ideologies and their consequences is the possibility for different interpretations of the same signs as iconic or indexical. For example, in Wynne's classic analysis of the interactions between experts and Cumberland farmers the problem of incommunicability resulted from underlying assumptions held by the latter about the untrustworthiness of the former, given the history of secrecy and reticence surrounding the activities of Sellafield. In this light the ambiguous and contradictory behavior of experts on the ground was taken by the farmers to be iconic, a reflection of the true essence (untrustworthy) of that group, rather than an index of the difficult applicability of standard scientific protocols under the specific environmental circumstances of their region.[95]

The role of semiotic ideology is to furnish us with instructions for how to connect signs with objects through guessing practices. This is due to the very assumptions upon which Peircean semiotic theory is built: in particular, the ontological presupposition that we have only partial encounters with objects in our experiences through their sign vehicles, and this happens in different contexts and under specific material circumstances. To use a simplification, we can say that our knowledge of the external world involves inferring the cause of an event by interpreting its effects as we perceive them. So, the potential causes of a phenomenon can be numerous, and our guessing efforts could be potentially infinite if shared cultural, historical, and cognitive codes wouldn't delimit the set of possible interpretations. This is the role of semiotic ideologies.

But what happens when a sign cannot be interpreted based on our past experiences? What if we found ourselves in front of something never experienced before and for which we lack a code for interpretation? In this case we enter the terrain of the conjectural, as we would probably start formulating hypotheses about what could have caused the unprecedented phenomenon before our eyes.

As I document in the final part of this book, nonexperts generated hypotheses about the presence or absence of radiocontamination in La Maddalena. These were not just based on fears, or ideological preconceptions. Residents made inferences, like scientists do, but with different material at their disposal: They relied on common images of nuclear things and their effects, repeated observations of the environment, and unprecedented events, like episodes of birth defects, which they interpreted as possible consequences of radiocontamination due to the proximity of the US Navy installation. Unprecedented events prompted new potential understandings and conceptions of risk that became the object of public controversies. In an effort to establish their credibility vis-à-vis Italian governmental authorities, expert agencies, and the US Navy, local antibase elites adopted a set of strategies to delimit the proliferation of hypotheses about the radiological causes of birth defects, with the help of allied experts. The semiotic regimentation of potential risk signs shows that conceptions of risk are subject to "history and struggle," and that static representations of expert and lay knowledges gloss over the complex mechanisms that shape controversial understandings of risk in moments of uncertainty.

ORGANIZATION OF THE BOOK

This book is organized both chronologically (loosely) and thematically; therefore, it can be read in various ways, not necessarily from cover to cover. The introduction provides readers with a review of key research problems and approaches within and beyond STS: expert/lay controversies, secrecy and the politics of knowledge production, and risk. My ambition for this book is getting out of disciplinary silos and internal debates to favor the cross-fertilization of insights and approaches coming from different, albeit contiguous, scholarly traditions.

My choice to keep the conceptual apparatus as it appears in the present text represents an attempt to be transparent about the intellectual trajectories that inspired the interpretation and the organization of the documentary material discussed in the chapters. I am sure that most readers will agree with me that the organization of a book, even a history book, involves much more than arranging the material chronologically. There is always a "vision" of the project that inspires writers or that emerges after collecting all the material. While the organization of this book is partially the result of practical research organization, opportunities, and constraints, it is also the outcome of roads I have not taken, either willingly or inadvertently. The display of the conceptual trajectories I undertook when trying to understand the meaning of the stories I wrote, reveals just that. I hope that this effort will be as useful to the readers as it was to me.

In chapter 1 I explain that the arrival of the US Navy in 1972 did not transform La Maddalena into a typical American fleet town. The bicentennial presence of the Italian navy incorporated the archipelago into the military-industrial projects of the Italian state while forging a widespread sense of local identity around military institutions. The US Navy blended into this context without altering the fundamental traditions of the local community. The military legacy of the archipelago explains why La Maddalena was chosen for the installation of the US submarine base: the Italian government hoped that the local acquiescence to the military institutions would make the stationing of nuclear submarines just in front of the urban center of the major island less problematic. Local identity, though, explains only partially why most long-term residents and local administrators did not openly oppose the American presence.

When important sectors of the Italian scientific community, including expert radioecologists and radioprotectionists working inside regulatory agencies, protested the lack of safety measures around the base, the Italian government responded with arguments that represented nuclear submarines as nonnuclear objects or as completely safe, like inland nuclear plants. The contested nuclear status of the US base, and consequently of the archipelago, revealed the political power of nuclearity that Hecht examines in her study of uranium mining in Africa. In chapter 2, I show that the Italian government and the US Navy on one side, and antibase activists and experts on the other side, adopted different discursive registers of nuclearity, relying on the ambivalent exceptional status of nuclear technology as both safe and risky.

Combining ethnographic methods with a close reading of previously unexplored archival material, in chapters 3, 4, and 5 I examine the complementary effects of military secrecy and the haphazard organization of the Italian nuclear bureaucracy to explore the political economy of knowledge production in Cold War Italy. The material I examine in part II of this book points to the technopolitical compromises between military security and public safety that shaped the concrete implementation of scientific protocols in La Maddalena.

In chapter 3 I offer a brief introduction to the history of radioecology during the Cold War and show in detail how the field grew in Italy during the 1960s and the 1970s in connection with the development of the Italian nuclear program and its relative regulatory regime. Then the analysis shifts to the local level, describing how radioecological campaigns were conducted in La Maddalena amid public debates over the nuclear status of the US base and the difficult adaptation of standard research protocols to the unknown environment of the archipelago.

In chapter 4 I focus on the uses of secrecy and information management of US and Italian military and civilian officials about nuclear submarines' operations in La Maddalena and document how the concealment of data and restriction to its access conditioned the work of Italian expert agencies in the archipelago. The history of the radiosurveillance system of La Maddalena (which I argue is one instantiation of a typical Cold War compromise between public safety and military security) also brings to the surface the informal organization of what I call the dual system of radiosurveillance. The latter consisted of a de facto separation between civilian and military spheres of competence concerning the monitoring and the regulation of nuclear technology, with the objective of subtracting military applications from public scrutiny.

Finally, in chapter 5 I detail the complicated birth and the haphazard organization of La Maddalena's radiosurveillance laboratory. The causes of delays and malfunctions in the implementation of the environmental monitoring program were both political and organizational. I map out the complex distribution of radioprotection competences across a multitude of agencies and administrative levels. In La Maddalena's case, the difficult coordination between central expert agencies and peripheral labs added to the political conflicts between central authorities and regional and local administrations over the acceptability of risk and the financial responsibility for the laboratory's costs. These institutional short-circuits resulted in a precarious radiosurveillance system, which became partially operational only in 1986. In the last part of the chapter, I discuss the attempt by the Sardinian administration to organize a consultative referendum against the permanence of the US nuclear base in La Maddalena. The hostile reaction of the government, backed by the decision of the Italian Constitutional Court, denied Sardinians the right to express their opinions over the US installation, restating the strategic importance of the US military presence in Italy. The sentence made evident once more the exceptional status of military nuclear technology: while Italians voted to phase out nuclear power plants in the national referendum of 1987, Sardinian citizens could not express their views over the permanence of a foreign base with nuclear submarines and armaments.

Different understandings and representations of radiological risk coexisted in La Maddalena based on practical experiences (and lack thereof) and evidentiary practices of various groups living and working around the US nuclear submarines and (more or less) acquainted with nuclear things in general: US Navy personnel, Italian employees on the US base, and long-term residents, local administrators, radioecologists, and radioprotectionists working for Italian expert agencies. In chapter 6 I examine the role of material evidence in processes of signification to explain how experts and nonexperts fix, challenge, and negotiate the meaning(s) of radiological risk in sociotechnical controversies. To this end, I advance a set of methodological and substantive arguments that constitute what I call "semiotic approach to risk."

Personal interviews with retired US Navy personnel and official documentation shed light on the practices of risk control on board of submarines and around the submarine base. I give concrete examples of the misunderstandings generated by expert definitions of risk and the frustrations that these provoked among local administrators who expected clear answers from the scientific community. The transcripts of a conference on nuclear contamination held in La Maddalena in February 1975 provide insights on how decontextualized and abstract objectifications of nuclear risk by Italian experts did not allow local administrators to grasp the immediate problems of radioprotection in the archipelago. The illustration of radioprotection practices inside the US Navy provides examples of how training programs and the culture of risk control furnished radiological workers with opportunities to acquire material understandings of risk. Finally, I focus on nonexperts' guessing strategies and evidentiary practices for making invisible risks visible and explain how local activists regimented and sometimes silenced long-term residents' hypotheses about risk to construct scientized and credible arguments against the US Navy base.

In chapter 7 I explore the complex relationship between accidents and understandings of risk. A US nuclear submarine accident near La Maddalena in 2003 generated a heuristic for studying how the intervention of independent experts challenged the established radiosurveillance protocols of Italian state agencies. In this chapter I detail why national experts faced a crisis of credibility, and how the intervention of independent experts and local activists reshaped debates about the risks of the US nuclear base. My analysis moves beyond deterministic and all-encompassing explanations of public opposition to certain technologies and industrial activities as reactions to accidental "events." Instead, I show that different data interpretations that animated expert controversies after the accident derived from divergent scientific protocols and agendas shaped in the context of the Cold War and in the aftermath of the Chernobyl disaster. Italian radioecologists, most of whom worked for the state, formed their expertise during the Cold War, and adopted established methods of analysis that led them to interpret the presence of radionuclides in the archipelago as long-term consequences of atmospheric fallout from previous nuclear experiments. Other experts adopted a different method that revealed phenomena of isotopes' bioaccumulation unnoticed until then (previously undocumented). Finally, independent marine geologist Fabrizio Aumento introduced a fast alpha-particle tracing technique that produced evidence about the presence of traces of plutonium in the archipelago. Expert debates after the 2003 accident generated uncertainty in the local population, but also created a unique opportunity for local activists to openly question the efficacy of La Maddalena's radiosurveillance system. Rather than embracing a particular scientific argument, local activists demanded transparency, and organized public forums in which they invited experts with competing views to debate their positions.

SOURCES AND METHODOLOGY

This study is based on several years of multisited ethnographic and archival research in Italy. In La Maddalena, I conducted dozens of interviews with local residents, including former mayors and city councilors, Italian workers, and employees of the US Navy base, antibase activists, religious authorities, and retired US Navy personnel. Many antibase activists, local historians, and journalists shared their stories and personal archives, including films produced by local directors, photographic material, newspapers, pamphlets, and fliers used for political propaganda by parties and protest groups in the 1970s and 1980s.

I approached archival research in Sardinia, and in other parts of Italy, from an ethnographic perspective. The combination of oral and archival sources gave me the opportunity to make sense of the material and to compare the views of key actors as they appear in official documents with other perspectives emerging during interviews. This process opened new possibilities to evaluate, interpret, and reflect on the sources at my disposal. With the assistance of Italian nuclear experts, I reconstructed the history of La Maddalena's radiosurveillance system within the institutional ecology of the Italian nuclear program. When I was conducting the most intense part of my fieldwork, the main archives of the Ente Nazionale Energia e Ambiente (ENEA, Italian National Agency on Alternative Energies; previously CNEN) were dispersed or not systematically organized. Several retired radioecologists and health physicists shared personal copies of their work with me.

Thanks to the generous support of director Roberta Delfanti and radioecologist Carlo Papucci, I accessed the archive of the Center for the Study of Marine Environments of ENEA (near La Spezia, in the region of Liguria), whose personnel have five decades of collective experience in radioecological campaigns throughout the world, from the Mediterranean Sea, including La Maddalena, to Cienfuegos, Cuba, and the Arctic Pole. Beginning in the winter of 2012, Delfanti and Papucci shepherded me through the rich archival material of the center, from collections of biosamples to logbooks of data entry and notes taken during radioecological campaigns in La Maddalena, internal correspondence and preparatory research designs, the illustration of laboratory practices, and final reports. During multiple visits to the center and in extended interviews with Arrigo Cigna, one of the founders of Italian radioecology and former president of the International Union of Radioecology, I acquired detailed knowledge of the theoretical and practical steps taken by nuclear experts to assemble their knowledge of La Maddalena and of other sites across Italy. These collaborations enhanced my ethnographic work in La Maddalena and allowed me to develop a more nuanced analysis of the connections between environmental, cultural, and political dynamics that shaped both expert and nonexpert understandings of nuclear risk around the US base.

I also retrieved and assembled archival and other documentary material scattered across Italy. In Rome I consulted the libraries of ENEA and ISS. At the Istituto Superiore per la Protezione e la Ricerca Ambientale (ISPRA, National Institute for Environmental Protection) and at the ENEA research center of Casaccia I located the former archive of the Comitato Nazionale per l'Energia Nucleare, Direzione Sicurezza e Protezione (CNEN-DISP, Italian Committee for Nuclear Energy, Division of Safety and Protection). Until the 1980s, the division was responsible for the elaboration of the emergency plans for nuclear installations, including nuclear ports like La Maddalena (from 1979). Followup interviews and examination of archival sources enabled me to further detail how military secrecy forced Italian expert institutions in charge of monitoring radioactivity levels in the Archipelago to operate with incomplete data.

In La Maddalena I accessed municipal archives, which included detailed transcripts of city council debates, official administrative documentation about the US Navy base, and the correspondence between US military authorities and the local administration. At the library of the Italian Navy Command, I could access national and local newspapers articles that the Italian Navy Intelligence Office had collected over the course of thirty-five years. The collection covers every reported event directly or indirectly related to the US Navy in La Maddalena, including local and national debates from 1976 to 2008. To cover the years 1972–1976, I conducted archival research at the Public Library of Sassari, the capital of northern Sardinia. I also collected national newspapers and accessed transcripts of parliamentary debates concerning the base at the Library of the Senate in Rome. During the final months of fieldwork in La Maddalena, I found rare documentary evidence about the organization and monthly operations of the local laboratory for the measurement of environmental radioactivity. From the archives of the Province of Sassari I also retrieved allegedly lost radiometric reports and internal correspondence about the technical characteristics and bureaucratic deficiencies of the radiosurveillance system.

PART I

A Strategic Naval Outpost

History, Identity, and the Military-Industrial Complex

In the summer of 2009, I debarked on the archipelago of La Maddalena for the first time to conduct preliminary fieldwork. My research project then focused on the history of US oversea basing strategies. I wanted to understand how after World War II the US empire included Italy within its global military network. Influenced by the copious case studies of communities coping with the consequences and implications of US base politics (especially revived after September 11, 2001), in La Maddalena I expected to hear stories of local opposition to the US Navy presence. Instead, I found a widespread sense of resentment for its recent departure. My original research problem changed immediately, as I was now trying to make sense of what I considered an anomaly. La Maddalena—as I discovered throughout repeated research trips and interactions with long-term residents—was not like most fleet towns described in studies of US overseas military bases.[1]

What struck me after the first interactions with locals was the pervasiveness of narratives that connected the birth of the community with the transformation of the islands into a maritime fortress in the nineteenth century. Most of my interlocutors colored their stories with positive adjectives, sometimes even with nostalgic tones when they referred to that past, which they considered the golden era of the archipelago.

PI.1 Flag of the Glorious Battle of 1793 on the wall of the municipal hall of La Maddalena. According to the tradition, the flag was fabricated before the battle. It represents the crucifix with Santa Maria Maddalena, matron of the island, at the bottom, embracing the cross. The inscription reads: "For God, and for the King. Win or die." *Photo credit:* Fabio Presutti.

In the middle of the Tyrrhenian Sea, La Maddalena has been a strategic outpost for the control of southern Mediterranean routes, particularly since the late eighteenth century, when the Royal House of Savoy took possession of the islands and a few decades later decided to relocate there its naval fleet from the port of La Spezia, occupied by Napoleon's army.[2] Before the arrival of the Sabaudian troops, a few Corsican shepherds originally from the southern area of Bonifacio inhabited the islands during the grazing seasons.[3] The families were clustered in two communities living in the internal hilltops of La Maddalena and Caprera, protected from the incursions of pirates who often stationed inside the coves of the archipelago. With the expansion of the port of Cala Gavetta at the turn of the nineteenth century, many shepherds left their original occupations to enroll in the navy, and a few decades later some of them started their careers as either commercial patrons or entrepreneurs devoted to the extraction of granite, which they exploited on an industrial scale and exported internationally until the 1930s.[4] In the meantime, the archipelago became a destination for coral hunters, fishermen, stonemasons, and naval workers coming from different parts of the Italian peninsula. In the 1880s the Kingdom of Italy transformed the archipelago into a military fortress and established a naval arsenal, which after World War II employed on average 30 percent of the local labor force.[5] Both the Italian navy and the arsenal constituted the pillars of La Maddalena's economy, the institutions that amalgamated its composite population and forged its identity.

One of the symbols of the archipelago's history is the flag of the battle of 1793, during which the Maddalenini resisted the French assault led by the young Napoleon Bonaparte in person. This episode, together with the Sabaudian occupation of 1767 and the erection of the naval base and arsenal in the last quarter of the nineteenth century are generally recognized as constitutive moments of La Maddalena's community.

Until the mid-1990s—before the advent of a new generation of local historians who authored dozens of studies—the historiography of the archipelago consisted of a small set of books and articles describing how between the eighteenth and nineteenth centuries the islands became a strategic naval fortress.[6] Main authors included navy officers like Admiral Aristide Garelli and Captain Emilio Prasca, and scholars of the Italian Risorgimento like Raffaele Ciasca and Ersilio Michel.[7] Although they worked independently from one another, with different purposes and for different audiences, these authors conducted their research principally in the state archives of Cagliari, Turin, Genoa, and Rome, where they found documentation produced mainly by Sabaudian and Italian navy officers and administrators. Another important feature of all these works is the almost complete absence of events preceding the military occupation of 1767.[8] This explains in part the frequent overlapping and teleological structure of traditional histories of the archipelago, which equate the militarization of La Maddalena to its civilization.

Therefore, it is not surprising that representations of La Maddalena's existence (as a community) as a direct effect of the Italian navy's defensive strategies are quite popular among the Maddalenini. As I learned during multiple research trips, many interpreted the arrival of the US Navy as a "natural" continuation of their history. Let me offer some explicative examples. Maddalena Rossi, "Signora Nena," owns a copy shop and a typography inherited from her father several decades ago. The quaint shop is located near the church of Santa Maria Maddalena, in one of the central squares of Cala Gavetta. Inside that shop I spent several hours a day making copies of archival material. My interactions with Signora Nena and her family became more frequent and informal, to the point that she was introducing me to each Maddalenino who entered the door. It became common and natural for her clients and family members to ask me what I thought of the place, but after a while I realized that their inquiries were often an excuse to express their views of the "Americans" and about the history of the archipelago. What stood out to me at first was the frequency of statements like this: "La Maddalena was built by the Italian navy. We have family members and friends who work for the navy or worked for the Americans in the last thirty-five years. We do not have anything against them. How can we be against ourselves? The only ones who complain are those who did not have any economic advantage from the presence of the Americans. But there are also those who rented their houses to the Americans and still complain, only for political reasons."

I listened to these and similar claims—which connect positive attitudes toward the US Navy base with the long military history of La Maddalena—in almost every conversation I had with residents, including people who strenuously opposed the American presence, like Salvatore Sanna, one of the leaders of the local antibase front, former Partito Comunista Italiano (PCI, Italian Communist Party) activist and city councilor during the 1970s, and meticulous historian of the archipelago.[9] According to him, antimilitarism has never been part of the local struggle against the US installation; those who compared the history of La Maddalena with the oppressive presence of Italian military testing grounds and infrastructures in the rest of Sardinia "have failed to recognize that here we had quite a different experience. If we developed economically, while the rest of northern Sardinia remained a rural, impoverished place until the 1970s, it is because La Maddalena was a military fortress and the Italian navy guaranteed public services and secure stipends."[10]

The transformation of the islands into a strategic naval fortress, to which the local community has actively contributed, has shaped a common sense of identification with military institutions, which *are* a unifying factor for an otherwise diversified population. This strong presence not only created a heavy economic dependence on the military-industrial sector but also allowed linguistically, culturally, and economically divided groups, who arrived through several waves of migration, to generate a collective identity, which still holds today. The local dialect, the *isulanu*, reflects the blending of different identities (Corsican, Genoese, Gallurese, and Pontina), all amalgamated through Italian, the official language imposed by the navy both on board its ships and in the military arsenal.[11]

One of my central arguments in this book is that traditional narratives about the military vocation of the archipelago had and continue to have deep influence on local perceptions of the US Navy presence, but like in Sanna's case, this does not mean that all La Maddalena's residents unconditionally accepted living next to a base for nuclear submarines. The story is more complicated than that and reveals the complex and contested work of defining risk and making it visible within specific "regimes of perceptibility." Before starting to delve into the original controversies over the nuclear status of La Maddalena, it is crucial to understand why the arrival of the US Navy did not transform the archipelago into a typical American fleet town.

CHAPTER 1

Not an American Fleet Town

Those familiar with typical US overseas military bases and fleet towns—such as Okinawa (Japan), Rota (Spain), or the Portuguese Azores—with fast-food chain restaurants and self-contained military structures, guarded and secured from the external environment—would have found these features lacking in La Maddalena.[1] The logistics of the archipelago did not offer the same conditions of most US military outposts abroad.[2]

The US Navy tender assisting nuclear submarines during repair and refit operations was moored in front of the pier located on the eastern coast of the Island of Santo Stefano (fig. 1.1). There, on a small strip of granite and reinforced concrete, the US Navy erected a few prefabricated buildings accessible only to authorized military and civilian personnel for port operations, recreational activities, and communication services. Separated from the northern tip of Santo Stefano by a narrow water channel, the major island of La Maddalena hosted the US Navy command, which established its headquarters inside a compound, named the Mordini compound, after the family who owned the building (fig. 1.2).

Every morning, both US and Italian employees accessed the gated area. The Mordini compound—now transformed into a commercial area with a luxury resort—sits next to the Ammiragliato, the headquarters

1.1 The US Navy tender with two fast-attack nuclear submarines moored at Santo Stefano Bay, Archipelago of La Maddalena, 1988. *Photo credit:* Tatiano Maiore.

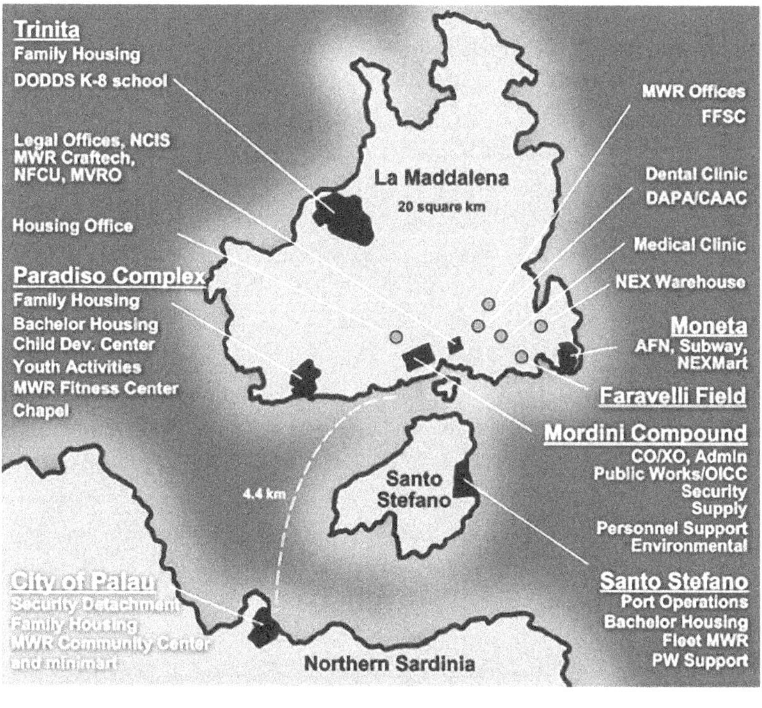

1.2 Map of the US Navy structures and compounds in La Maddalena and Santo Stefano, early 2000s. *Source:* Private archive, retired employee of Naval Support Activity La Maddalena.

1.3 A group of Marines walking on the streets of La Maddalena, 1973. *Photo credit:* Tatiano Maiore.

of Marisardegna (the Sardinian command of the Italian navy), with its beautiful façade dominating Piazza Umberto I. With the tall residences of the Italian military officers, the square symbolizes the centrality of the Italian navy in the life and history of the local community, and virtually separates the civilian residential area of Cala Gavetta from the military citadel, called Moneta, built in the southeastern corner of La Maddalena at the end of the nineteenth century to host the arsenal and the barracks for the workers and their families.

The presence of the headquarters of Marisardegna provided a formal counterweight to US military authorities in charge of the submarine base. The Italian government used the navy command to claim its ultimate authority on the archipelago, although most problems concerning the operations of the US installation were usually discussed at higher diplomatic levels. Neither the cohabitation with the Italian navy—the older patron of the community—nor the sober life of the archipelago offered the US personnel the same options and attractions of a classic "fleet town.[3] "I took all my snapshots the first week. . . . It's not like Majorca at all. There we had plenty of bars and plenty of girls," declared a young US Navy gunner's mate interviewed soon after his arrival in 1972.[4]

Correspondence between the US embassy in Rome and the State Department reveals that on multiple occasions US authorities insisted on amending, modifying, and rediscussing the terms of the Bilateral Infrastructure Agreement with the Italian government to increase the US Navy footing in the archipelago. Initially, the two parties agreed on the

1.4 The pacifist march organized by the Partito Radicale in La Maddalena on August 19, 1976. In the foreground, Julian Beck, leader of the Living Theater. *Photo credit:* Tatiano Maiore.

stationing of the US Navy tender in front of the "NATO pier" of Santo Stefano with a minimal presence on the ground, but later, the Americans required the expansion of the Naval Support Activity (NSA) structures, including the use and rental of facilities for recreational activities, housing, schooling, and so on. This put pressure on local administrators, who often showed reluctance to concede to those requests, in part because they feared that increasing the presence of the US contingent would entail a radical transformation of their community, and in part because they could bargain more favorable economic conditions by initially opposing their veto.

The first US Navy contingent debarked from the USS *Howard Gilmore*, a tender commissioned in 1944 and deployed in the Pacific during World War II. Early encounters between the US sailors and the locals were not always cordial. Cultural clashes and idiosyncrasies emerged frequently, especially at night when they crowded the same bars. The drinking excesses of American troops spurred some nocturnal confrontations on the streets of Cala Gavetta.[5] Verbal and physical disputes usually lasted no more than a few minutes, the time for the intervention of the US military police patrolling the streets of La Maddalena and the town of Palau.

More organized forms of protest, like anti-American, pacifist, and environmentalist marches against the US installation took place on November 1972 and in August 1976, but the participants were usually young activists coming from outside the archipelago. In the summer of 1976 the

1.5 The police intervene to stop young protesters who had arrived in La Maddalena for the antimilitarist march of the Partito Radicale from advancing toward the US Navy headquarters, August 19, 1976. *Photo credit:* Tatiano Maiore.

firebombing of nine American Forces in Italy cars and an antimilitarist march organized by the Radical Party seemed to foreshadow hard years to come for the US Navy in La Maddalena (figs. 1.4 and 1.5).[6] But the feared escalation of antibase sentiments did not happen because most locals remained indifferent, if not openly opposed to these "external" protests.

Due to the logistics of the archipelago, the US personnel working at the base in Santo Stefano and in the headquarters of the NSA command inevitably participated in the social life of the Maddalenini. Several US Navy personnel I interviewed told me, "Nowhere like in La Maddalena we could leave our kids playing in the central square while drinking a beer or a coffee. We spent a lot of time in Cala Gavetta because we became in some sense part of the local community."

Housing was a sore issue, though.[7] The major island of La Maddalena did not offer enough housing space for the newly arrived. The US Navy invited locals to list their rental offers even outside the archipelago, across the northern coast of Sardinia, from the towns of Palau and Arzachena to Santa Teresa di Gallura. US military personnel deployed at La Maddalena lived scattered throughout the archipelago and northern Sardinia—rather than in barracks or inside a large station.

Convinced that the average salaries of US Navy personnel were much higher than local stipends, landlords evicted Italians from apartments to make room for richer tenants.[8] To combat this, the US Navy command had to announce publicly that American sailors were not actually earning much money and that they could not afford to pay the rents proposed by the Maddalenini.[9] The housing crisis generated new opportunities for the construction industry. Brand-new apartments, in fact an entire village for American families, were constructed in an inland community of La Maddalena and arrangements were made with renters to make available some two hundred more units.[10] Adjustments were necessary: Americans found local apartments too small and lacking the comforts of TV, washing machines, and dishwashers. The charismatic priest of La Maddalena, Monsignor Salvatore Capula, appeared in all the meetings in which local authorities and the US Navy command discussed how to aid the integration of the Americans with the local community. He was a cultural broker, a spiritual guide, and the guarantor of the diplomatic relations between the US Navy, the local entrepreneurs, and the municipal administration.[11] Capula preached friendship and economic development from his pulpit while the US Navy command started a "program of intercultural relations" to acquaint its personnel with local habits.[12] All sailors and their families had to attend a mandatory crash course on Italian life and culture. To avoid surprises or unwelcome reactions spurred by cultural misunderstandings, the human resources department of the

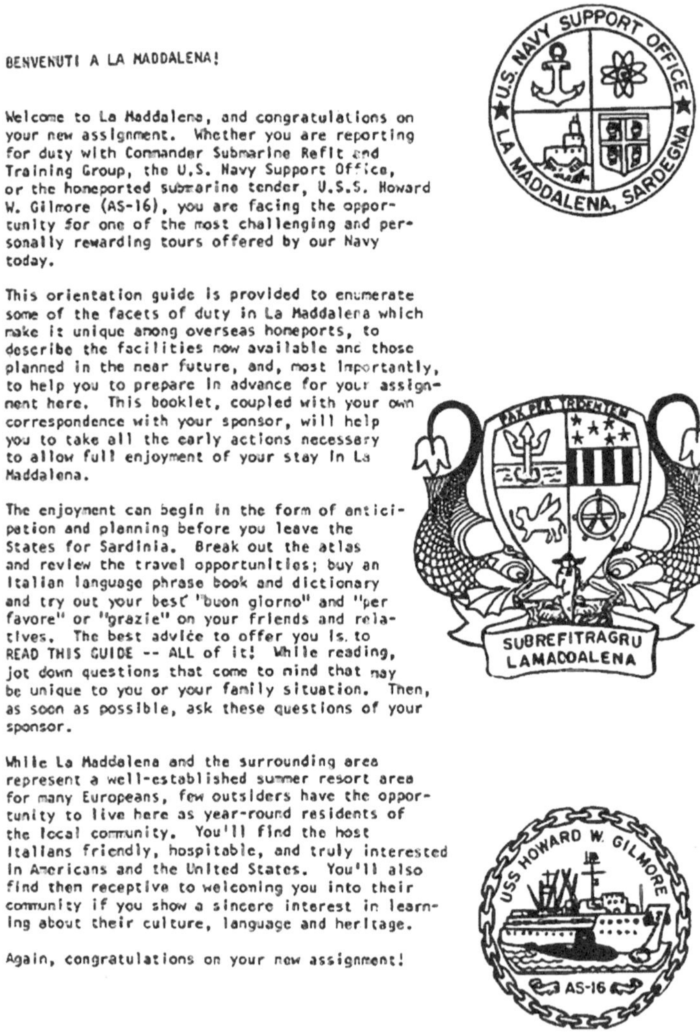

BENVENUTI A LA MADDALENA!

Welcome to La Maddalena, and congratulations on your new assignment. Whether you are reporting for duty with Commander Submarine Refit and Training Group, the U.S. Navy Support Office, or the homeported submarine tender, U.S.S. Howard W. Gilmore (AS-16), you are facing the opportunity for one of the most challenging and personally rewarding tours offered by our Navy today.

This orientation guide is provided to enumerate some of the facets of duty in La Maddalena which make it unique among overseas homeports, to describe the facilities now available and those planned in the near future, and, most importantly, to help you to prepare in advance for your assignment here. This booklet, coupled with your own correspondence with your sponsor, will help you to take all the early actions necessary to allow full enjoyment of your stay in La Maddalena.

The enjoyment can begin in the form of anticipation and planning before you leave the States for Sardinia. Break out the atlas and review the travel opportunities; buy an Italian language phrase book and dictionary and try out your best "buon giorno" and "per favore" or "grazie" on your friends and relatives. The best advice to offer you is to READ THIS GUIDE -- ALL of it! While reading, jot down questions that come to mind that may be unique to you or your family situation. Then, as soon as possible, ask these questions of your sponsor.

While La Maddalena and the surrounding area represent a well-established summer resort area for many Europeans, few outsiders have the opportunity to live here as year-round residents of the local community. You'll find the host Italians friendly, hospitable, and truly interested in Americans and the United States. You'll also find them receptive to welcoming you into their community if you show a sincere interest in learning about their culture, language and heritage.

Again, congratulations on your new assignment!

1.6 "Welcome to La Maddalena," the orientation guide distributed by the human resources department of the US Navy support office of La Maddalena to the new personnel (updated version, May 1979). *Source:* Salvatore Sanna, private archive.

1.7 A group of US Navy gunmen walking on the streets of La Maddalena, 1974. *Photo credit:* Tatiano Maiore.

US Navy gave concrete examples of how Italians might differ from Americans, deploying an arsenal of commonplaces and stereotypes about the Italian character and the exceptionally democratic ethos of Americans:

> Another point we might keep in mind is that Italy has practiced Democracy [capital D] for only about 30 years, while the United States has about 200 years of experience. Therefore, our attitudes towards individual freedom may be somewhat different (and more experienced) than the Italians. Don't be surprised to see Italians breaking into lines (i.e., at the bank or ticket office) or crowding each other (and you) in public spaces. . . . Local girls are virtually non-existent as far as personnel of visiting ships are concerned. Due to their society's rules, the local residents prefer that their daughters go out with other residents. This is not the case as far as tourist girls are concerned. The tourist girls regard visiting Americans as being in the same category as themselves. There are, however, no 'B' girls of the type normally found in the so-called "Fleet Towns." The very simple explanation for this is that La Maddalena is not and has no intention of ever becoming a "Fleet Town."[13]

Orientation guides for incoming American personnel were explicit about the uniqueness of La Maddalena: saying that the archipelago was not a "fleet town" implied that prostitution was not tolerated, but also that the local community wanted to maintain its cultural identity and did not look forward to becoming another US Navy port. According to local rumors, in exchange for his collaboration with Italian and American authorities, Capula imposed specific rules of conduct on young sailors who came without families. He welcomed prosperity for his people without moral corruption.

While La Maddalena had generally been a welcoming port, older generations of Maddalenini were not familiar with the ethnic and cultural diversity introduced with the new arrivals. Some of the elderly residents I exchanged daily conversations with remember that they feared "black soldiers," and recount retiring inside their houses early because of the "drunken" sailors lingering outside local bars.

For American wives who followed their husbands on duty in the 1970s, conditions were considerably difficult. Until the close of the decade, there were no jobs available for US Navy spouses on the base and there was little social life to make them feel they were active members of the community. Families often lived far from Cala Gavetta, which was difficult to reach without a car. While the archipelago was sunny and crowded in the summer months, the winter was desolate—and menacing winds blew through empty squares and the narrow lanes of Cala Gavetta. The US Navy exchange shop, initially located inside the base on the island of Santo Stefano, carried few items. Liberty boats—small towboats adapted for the transportation service of US personnel and families from La Maddalena to the Santo Stefano base—ran twice daily, in the morning and in the early afternoon. Medical assistance was available, but for more serious procedures the closest US Navy hospital was in Naples. Letters of complaint listed the reasons why "La Madd" or "Madd Rock," as some Americans called it, was considered a remote duty.[14] They often invited US Navy officers in Washington to visit La Maddalena to see directly what life looked like over there: "Come see us—it's just a 1½ hour plane ride—an hour bus ride—a 20-minute ferry ride from Naples and here you are."[15]

"Yes, life in La Maddalena is different, remote, and, at times, very difficult," admitted Captain R. M. Hughes, commander of the Submarine Refit and Training Group of La Maddalena. But with time, things improved.[16] Employment opportunities on the base became available also to the dependents of US enlisted personnel. The US Navy command established a recreational activities department and organized social events to promote cultural integration with the local population.

Despite the initial difficulties in accommodating the newly arrived fleet, most locals did not oppose the US base. During the annual Festa dell'Amicizia (Friendship Party) and for the Fourth of July, US personnel opened the gates of their command's headquarters and invited the Maddalenini to join the celebrations.[17] To the young locals, even those who did not have sympathies for the US government, America became

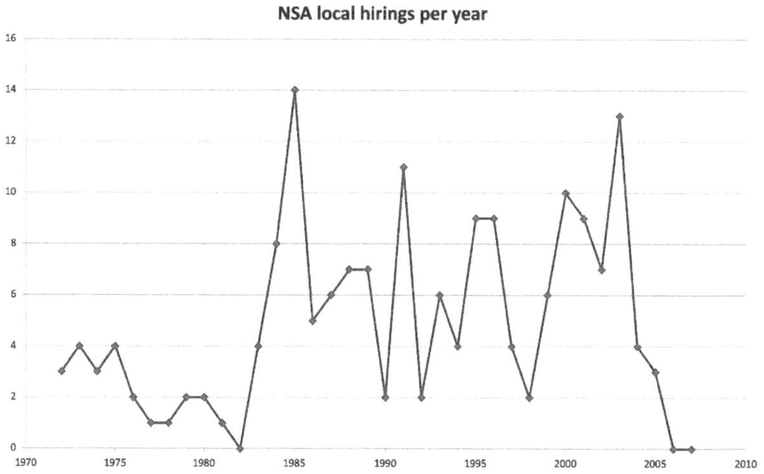

1.8 US Naval Support Activity of La Maddalena: Local Hirings of Year. Data include only the Italian personnel directly hired by the US Naval Support Office and does not take into account the personnel working for subcontractors. Source: Municipal Archive of La Maddalena. My elaboration is based on data provided by the US Navy Naval Support Activity Office and the union organizations represented inside the base. Data source: Naval Support Activity of La Maddalena. Elaboration and diagram by the author.

available at home. Cassette recorders, electric guitars, and sneakers were featured at social gatherings, fueled in part by the black market. Friendships were formed, and US sailors and local women began to marry.

The US base was undoubtedly an important source of income for the local community. At the day of decommissioning, Italian employees and workers directly hired by the US Navy totaled 175. Other activities benefited from the American business. Subcontractors inside the base employed 40 to 60 locals. The housing office of the US Naval Support Facility was one of the most crowded. It managed rentals and accommodation for hundreds of families and made sure that local housing matched, as much as possible, the living standards of US personnel at home. Second was the Morale, Welfare, and Recreation Office. The hiring system was pretty open, at least formally. Job openings were advertised, and interviews organized after candidates filled application forms and provided information of their criminal records. The hiring process was transparent, at least in comparison with stereotypical Italian standards; nevertheless, the US Navy tended to reward fidelity. It was not uncommon for two to four members of the same family (or close relatives) to obtain jobs inside the base or in the other offices of NSA (accounting, logistics, transportation, secretarial positions, and so forth).

Sometimes offers were made informally: "I was working in my parents' restaurant. One day they came [US Navy personnel] and asked me if I wanted to work for them," recounted a former US Navy Italian employee.[18] Americans did not hesitate to coopt potential adversaries if it would change their dispositions toward their presence. Some employees with leftist political ideas and union activists were hired, too. But there were limits. Only friendly unions were admitted to consultation and collective negotiations inside the base. Confederazione Generale Italiana dei Lavoratori (CGIL), for example, was officially excluded from all the US installations of Italy, in clear violation of the national law on labor employment. Discriminations and exclusions happened, but those who worked for the base proudly claim to have experienced firsthand the advantages of American meritocracy: "If you worked hard and demonstrated your desire to learn, they would give you all the opportunities you wanted. I started as a kitchen assistant at one of their restaurants and ended my career as a supply supervisor."[19]

Stories of professional success, efficiency, and personal improvements peppered interviews I conducted with former Italian employees on the base: "They paid all the courses I attended in Germany and in the US. They invested in me, and I could get better and better positions. Isn't it gratifying? You study in the US, so you know what I am talking about," Carlo, a former employee in the accounting office, told me.[20] The sense of attachment and nostalgia for the US Navy presence among former employees on the base is not surprising if one considers the working conditions of Italian public employees. The "Americans," as they were sometimes called, were a privileged and enviable group within the local community. Their salaries were higher, as were rewards and professional opportunities. Some of them described to me their sense of satisfaction with extreme excitement: "I feel like I had the luxury of being at home while going to work abroad each morning, switching language, and knowing people from which I learned a lot. What a fantastic experience!"

There were also negative sides: "After 9/11 the base was on high security levels. Also, our daily life and work routines changed. Security checks were intensified. Armed soldiers guarded the buildings where we had our offices. The atmosphere was very different from when I started working for the base." More than ever, trust was essential to security in such a sensitive workplace. Some Italians worked for the security department as well. Long-term employees were hired to assist the navy patrol's activities in Palau and La Maddalena. If conflicts between locals and Americans occurred—especially likely on weekend nights around bars and discos—Italian patrol assistants could intervene only as interpreters and cultural mediators. With time they became more integrated with the navy organization and were employed also as security personnel devoted to criminal background checks, and permit authorization for Italian workers on the base: "You know, because of our job we knew a lot about other people of the community. Being the gatekeeper is exciting but also uncomfortable sometimes. You know things but cannot say anything to anybody. I guess after a while you learn to live with that."[21]

In 2005, US secretary of defense Donald Rumsfeld announced that his department had decided to close the navy installation in La Maddalena. After the news spread the NSA command gathered all the Italian employees inside the movie theater next to the Mordini compound and officially announced that the decommissioning process would start soon and end in January 2008. Everybody was shocked. Long-term employees could not imagine La Maddalena without the base, but they were close to retirement and the closure would not affect them economically. For younger employees, the closure was more problematic. After a few months they started to mobilize. More than 150 families would have been left without a source of income: "I mean, they paid us until February 2008 even if some of us were already at home. They kept their word and assisted us in all possible ways. I was even offered a position in another base in Italy but preferred not to move."[22] Even the mayor of La Maddalena, Angelo Comiti—a former Communist Party member with no particular sympathies for the US Navy—tried to assist the future unemployed.

In 2009 the Berlusconi government signed a decree establishing that the former Italian employees on the US Navy base would be reemployed by local administrations (municipal, provincial, regional). Nevertheless, the departure of the US Navy left a profound scar on La Maddalena's economy, also because the planned reconversion that the Italian government promised after the decommissioning of Santo Stefano and of the Italian navy arsenal was abandoned quite quickly. The progressive closure of military facilities and the failure of the economic reconversion plan have entailed the dismantlement of fundamental services and a worrisome depopulation of the islands, especially by young generations in search for a job. Thirteen years later the archipelago is undergoing a slow and painful transition toward a tourist-oriented economy, without a clearly defined end in sight.

In the 1970s the US Navy personnel stationed in northern Sardinia referred to La Maddalena as "La Madd" or "The Rock." The archipelago did not offer the attractions of more preferred US military outposts abroad. In part this was the result of the vigilance with which both civilian and religious local authorities made sure that the islands would not become another US Navy port and that the community would not lose its identity.

The bicentennial occupation of La Maddalena by the Sabaudian, Royal, and Italian navies amalgamated heterogeneous groups who arrived in the archipelago through successive waves of migration from Sardinia and continental Italy in search for jobs as fishermen, coral hunters, miners, stonemasons, and naval workers inside the arsenal established in the 1880s. The Italian navy presence guaranteed stipends and infused a widespread sense of identification of the Maddalenini with military institutions. The US Navy adapted to this cultural and environmental context to the effect that, contrary to typical overseas bases—enclosed and voluminous—its facilities were scattered across the archipelago. US personnel integrated into the social life of La Maddalena, sometimes smoothly and other times with notable tensions. The local community seemed indifferent to the submarine base, exactly because it was accustomed to having the Italian navy as an integral component of its daily life. But this reading is quite simplistic. Those who opposed the submarine base focused on the problem of radiocontamination risks rather than on more abstract ideological discourses about national sovereignty and US imperialism.[23] In part, this was a strategic move, given that La Maddalena's residents were insensitive—if not openly hostile—to antimilitary slogans and ideologies.

CHAPTER 2

Becoming Nuclear

NUCLEARITY AND THE POLITICS OF SCIENCE

In 1979 the theatrical collective L'Elicriso of La Maddalena produced a film titled *Benvenutti!* (Welcome), by local screenwriters and directors Adriano Tovo and Giorgio Acciaro.[1] The movie gives a satirical account of the facts leading to the installation of the US Navy base in September 1972. The opening scene takes place on a golf course in Sardinia where the American ambassador in Rome, the Italian minister of foreign affairs, and the Italian undersecretary of defense, presumably Christian Democrat Francesco Cossiga, make an agreement about the concession of La Maddalena to the US Navy. The film shows the obsequious mayor Giuseppe Deligia and the greedy, assertive priest of the archipelago, Monsignor Salvatore Capula, plotting with the Italian government to welcome the Americans and to assuage the potential resistance of some sectors of the local community. The story contains a self-reflexive and particularly critical analysis of La Maddalena's historical acquiescence toward military and political authorities. With its hyperbolic representation of local and American characters, and of the circumstances leading to the installation of the US Navy base, *Benvenutti!* offers an accurate and polemical, albeit minoritarian, reading of traditional histories of La Maddalena, which describe the local community as naturally predisposed to live in symbiosis with military installations. In this chapter I start to uncover another side of the story and show that local military traditions and the economic advantages of the US Navy presence alone cannot explain why the opposition to the submarine installation was not widespread. I suggest that at the basis of this apparent indifference lie more complex mechanisms of risk assessment and visibility, also connected to the ambivalent nuclear status of nuclear submarines.

THE CONTROVERSIAL NUCLEAR STATUS OF LA MADDALENA

Since the 1960s the alarm for the increasing penetration of Soviet vessels into the Mediterranean Sea through the Dardanelles Strait pushed the US Navy to reconsider its strategy in the area. Particularly worrisome to the North Atlantic Treaty Organization (NATO) headquarters in the southern European flank were the agreements that the USSR signed with Syria and Egypt for the use of naval stations in Latakia, Alexandria, and Port Said.[2] As reported by *Time* magazine in June 1971, the "Soviet thrust in the Mediterranean" made the middle sea "no longer an American lake."[3] With the aid of new environmental monitoring techniques that were developed beginning in the 1960s with the contribution of NATO's scientific committees, US naval strategists decided to increase the surveillance of "choke points" and to oversee them with more units.[4]

On September 2, 1972, the main newspaper of Sardinia, *L'Unione Sarda*, reported the unusual visit of the US Navy air carrier *Kennedy* to the archipelago of La Maddalena.[5] During the same year, the air carrier was stationed in various ports of the Mediterranean: from Greece and Turkey to Spain and France. The turbulent atmosphere that characterized North Africa and the Middle East encouraged a preoccupation with Soviet submarines and pushed US strategists to embark on a "diplomatic tour" to show that the Sixth Fleet of the navy was present and vigilant.

Shortly afterward, La Maddalena became an important asset of the new Mediterranean strategy of the US government. On September 16, 1972, Lucio Manisco, the Washington correspondent for the national newspaper *Il Messaggero*, made a real scoop: unspecified sources within the Pentagon confirmed that the US Navy was going to install "a base for atomic submarines in La Maddalena."[6] The subtitle of the article mentions that the decision was a response to the increased presence of the Soviet fleet in the Mediterranean.[7] In a separate article, Manisco reported that the government of Syria had recently granted Moscow the use of two naval stations. In the following days, other national newspapers emphasized the concession of La Maddalena by the Italian government as a base for nuclear submarines of the US Navy. The little Sardinian archipelago instantly became the new front of Italian foreign policy.

MOSCA — La « Pravda » ha pubblicato nella sua edizione di mercoledì questa vignetta sulla installazione della base USA alla Maddalena. Il disegno raffigura un ammiraglio americano che offre i sommergibili atomici in una scatola di sardine con l'etichetta « Sardine di oltre Oceano ». L'Italia (raffigurata su uno scoglio con la scritta « La Maddalena ») si schermisce

2.1. Cartoon published in *Pravda* on October 25, reproduced by the Partito Comunista Italiano organ *L'Unità* on October 27, 1972.

Both in parliament and through its official newspaper *L'Unità*, the Partito Comunista Italiano (PCI, Italian Communist Party) mounted a massive campaign against the decision of the center-right administration led by the Christian Democrat prime minister Giulio Andreotti. From a political point of view, the PCI perceived the concession of La Maddalena as another sign of Italy's acquiescence to American imperialism and an act of aggression that contradicted the efforts of international détente publicly announced by the Nixon administration. At the international level, the initiative of the United States in the Mediterranean did not go unnoticed. An official visit by Andreotti to Moscow was an occasion for the Soviet authorities to express their concerns for the recent installation of the US Navy in La Maddalena. Although the agenda of Andreotti's visit primarily focused on commercial agreements, the events unfolding on the Sardinian archipelago and the Soviet reactions inevitably attracted the attention of the media.[8] On October 27, 1972, *L'Unità* reproduced a cartoon that had appeared a few days earlier on the official organ of the Soviet Politburo *Pravda*: an old American commodore offers Italy—represented as a crowned woman dressed in a classic white tunic—a bunch of nuclear submarines inside a sardine can. The woman sitting directly in front of the island of La Maddalena clearly refuses the offer (fig. 2.1).

Besides the obvious diplomatic and military implications of the US installation for Italian foreign policy, at the national level the political campaign of the antibase front focused since the beginning on nuclear safety and on the potential risks of contamination for the local population and the environment.[9] Even moderate and conservative national newspapers, generally close to the position of the Italian government, highlighted the problematic nature of the concession of La Maddalena to the US Navy. The *Corriere della Sera*, for example, published a report about the anxieties of some residents about the possible environmental contamination from the submarine reactors.[10]

Fears spread despite the government continuing to minimize the question of La Maddalena and refusing to provide any information about the terms of the agreement, the mission of the US Navy in the newly established base, and the characteristic of the armaments on board of the nuclear submarines. Only during the parliamentary debate of October 6, 1972, did the Italian public learn that the government had conceded an area of the island of Santo Stefano to the US Navy for the assistance of the nuclear propelled submarines of the Sixth Fleet. In the debate the Italian minister of foreign affairs, Giuseppe Medici, responded to the attacks of communist and socialist senators with three main arguments.[11] First, the concession of a support facility to the US Navy in La Maddalena was due to the necessity of counterbalancing the increasing menace of the Soviet activities in the Mediterranean. The act should not be surprising given that Italy was a partner of the NATO alliance since 1949. The concession of the base thus was the logical consequence of the general agreements in implementation of the defensive strategy of NATO. Second, the mission of the support facility, "not a base," was to assist and refit US nuclear submarines of the 22nd Squadron, involved in reconnaissance and surveillance activities in the Mediterranean area. A navy tender equipped with repair shops and specialized personnel would station just off the island of Santo Stefano, ready to assist the submarines. The navy tender was a "normal ship" propelled with conventional engines. It could not perform any refueling operations involving radiological work, which required conditions present only in US harbors. The nuclear reactors propelling the US submarines were not dangerous for the environment or for the residents of La Maddalena. Like reactors of civilian nuclear plants—Medici mentioned that Italy already had three, one of which was near Rome—the submarines' propellers were a safe technology. In addition, the foreign minister cited a precedent that should have convinced his opponents: "In 1964 Italy signed an agreement with the US government, allowing the use of Italian ports by a US nuclear merchant ship. Given that nuclear submarines do not differ in this regard from any surface ship, in La Maddalena we agreed to allow [the presence of the US Navy submarines] under the same general terms of eight years ago, assuring the respect of all the safety norms."[12] Finally, Medici rebutted the accusations that the US Navy base would damage the local economy: "Instead of alienating the tourists, the arrival of the US Navy personnel and their families will be an important asset to the economy of La Maddalena, similar to what happens, for example, in Gaeta and Formia [two naval bases near Naples]. Moreover, their presence will be, by itself, a demonstration of the innocuous nature of the operation."[13]

In his reply to Medici, Communist senator Ugo Pecchioli argued that the (secret) executive bilateral agreement that stipulated the concession of La Maddalena to the US Navy explicitly violated the norms of the Italian constitution, which prescribed the vote of parliament for the ratification of international agreements. Pecchioli insisted that the installation of La Maddalena was not just a support facility but "a real military base with built-in structures and a massive presence of US personnel in the archipelago." He added that, according to his sources, the base did not respond to NATO commands but operated directly under US military authorities. Therefore, the legitimacy of the concession, justified by the government as a NATO operation, was largely contestable: "With this act Italy continues to cede its national sovereignty without any parliamentary discussion. The previous installation of US bases in Gaeta, Vicenza, Naples, Livorno, and Martina Franca confirm the Italian submission to the American ally."[14] Finally, Pecchioli argued that the US nuclear base "represents an ulterior source of radioactive pollution in the hearth of the Tyrrhenian Sea, an ecological bomb that constitutes an extraordinary danger for the inhabitants of the area and for the flora and the fauna of the Mediterranean."[15]

Radiocontamination risk was already a primary issue in public debates, but it became even more pressing when important sectors of the Italian scientific community voiced their concern. Only two days before the open question time at the Senate, the Italian Society of Biophysics and Molecular Biology published a statement condemning the military applications of nuclear technology and alerting the population of northern Sardinia to the risks of contamination deriving from the presence of the US nuclear submarines.[16] Scientists promulgated several critiques to the government: 1) With the concession of La Maddalena, Italy became de facto coresponsible for US nuclear aggressions and consequently La Maddalena became a potential target for Soviet retaliation; 2) The concession of the base was an unacceptable limitation of national sovereignty; 3) The base would inevitably provoke environmental contamination due to probable discharges of radioactive substances, and the secrecy surrounding the installation would prevent national expert agencies from implementing the necessary radiosurveillance measures demanded by national and international regulations; 4) The military nature of the installation increased the risk of accidents with disastrous consequences.

On October 8, 1972, *L'Unità* published a short interview with physicist Edoardo Amaldi, who warned readers about the risks involved in the use of La Maddalena as a base for nuclear submarines.[17] A friend and close collaborator of Enrico Fermi's during the 1930s, the founder and first president of the European Organization for Nuclear Research (CERN), Amaldi was an engaged scientist who had always voiced his opposition to the military use of nuclear science and technology.[18] His opinion was certainly important, and the PCI did not miss the occasion to enroll his authoritative statement for its antibase campaign. Even prominent sectors of the cultural intelligentsia mobilized against the installation of the US base. Soon after the Italian Society of Biophysics and Molecular Biology published its document, the association Italia Nostra sent a letter of protest to Andreotti and to the secretary-general of the United Nations, denouncing the risks of an environmental disaster caused by the permanence of nuclear submarines in one of the most pristine archipelagos of the Mediterranean.[19]

The attempt by the government to minimize and gloss over these concerns failed because the PCI continued to support a strong media campaign and to back the initiatives of affiliated expert activists through its capillary organization, even within expert agencies. For example, on October 22, 1972, employees of the National Committee for Nuclear Energy adhering to the National Union of Nuclear Workers (SANN-CGIL)—a branch of the leftist union Confederazione Generale Italiana dei Lavoratori (CGIL)—took an official position against the installation of the US base. The "nuclear workers" confirmed the critiques of the Italian Society of Biophysics and Molecular Biology: "Around civilian nuclear plants safety surveillance is managed by the Istituto Superiore di Sanità [ISS], CNEN, etc. These norms, already harshly critiqued by local administrations and leftist political movements for their inadequacy and lack of democratic scrutiny, will not be implemented by the Italian authorities in La Maddalena, because everything there will be subject to the arbitrary control and secrecy of the Pentagon."[20] SANN-CGIL insisted that the problems of economic development of an area like La Maddalena, "where even the supply of running water is difficult and unemployment rampant," could not be solved with more military installations: "The center-right government, in disregard of the most elementary democratic rights, wants to subject our country to the imperialist strategy of the United States, which has already annihilated any hope for democracy and sovereignty in Turkey and Greece."[21]

Whereas political attacks addressed to the foreign policy of the Andreotti administration were part of the usual confrontation between pro-NATO and anti-NATO positions, the alarms that the Italian left and concerned experts raised about the risks of accidents and nuclear contamination required careful examination and authoritative responses. The Italian government asked the Ministry of Defense to produce a technical document to refute the alarming hypotheses advanced by the opponents to the base. With this intent, in mid-November 1972 the Centro Applicazioni Militari Energia Nucleare (CAMEN, Center for the Military Applications of Nuclear Energy) issued a preliminary safety assessment describing in very general terms the technical characteristics of the US submarine reactors and the operations that the tender would perform in La Maddalena. In addition, a confidential note from Comitato Nazionale per l'Energia Nucleare (CNEN) arrived on Medici's desk, in which CNEN president Ezio Clementel allegedly provided reassuring data on the safety of the US base. With the two documents in his hands, Medici considered the controversy virtually closed, but he was wrong.

THE NUCLEAR QUESTION EXPLODES IN LA MADDALENA

After the tense parliamentary debate of October 6, and the abundant coverage that national media gave to the "question of La Maddalena," the municipal administration had the difficult task of taking an official position.[22] The local opposition, composed of the PCI and the Partito Socialista Italiano (PSI, Italian Socialist Party), immediately mobilized against the installation of the US base through official statements, posters, and poorly attended rallies (see figs. 2.2 and 2.3). La Maddalena's residents did not, it would appear, perceive the arrival of the US Navy as an epochal change and demonstrated their willingness to accept the new presence, hoping that it would bring jobs and commercial opportunities. The mayor at the time was Giuseppe Deligia, a hardcore Christian Democrat close to the positions of the national government and a long-term civilian employee of the Italian navy arsenal. Like most conservative members of the local Democrazia Cristiana, Deligia was a devout Catholic and a disciple of Capula, whom many in town considered the real "governor of La Maddalena."[23] According to Capula's diaries and Deligia's recent "confessions," the two men knew about the arrival of the US contingent well before the official announcement of mid-September 1972. Capula was a very close friend and spiritual confidant of undersecretary of defense Francesco Cossiga, a rising star among the Sardinian Democrazia Cristiana. It was Cossiga who first revealed to the monsignor the imminent establishment of the US Navy base in La Maddalena.[24]

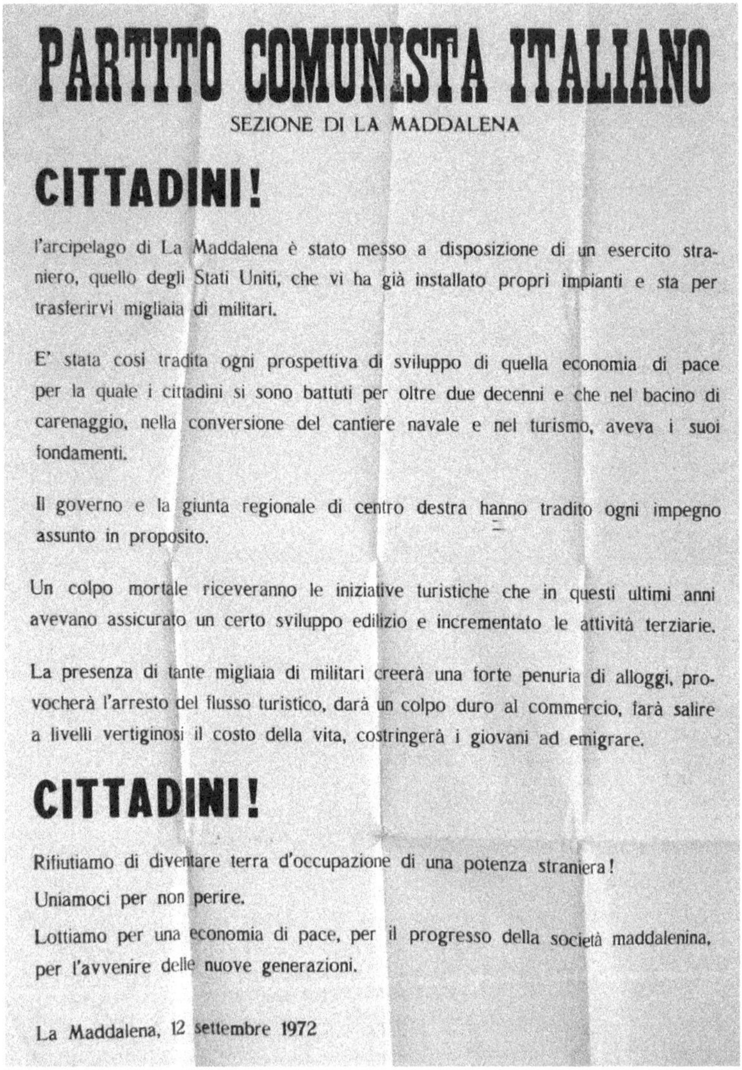

2.2. Poster of the local section of the Partito Comunista Italiano, protesting the secret agreement for the installation of the US Navy base in the archipelago, September 12, 1972.

Given the moral and political influence that the priest exercised on his community, Cossiga considered it necessary to orchestrate a strategy with him and the mayor to prepare the terrain for the new arrival, to convince the residents of the economic benefits of the base, and to reassure them about its safety.

After the news of the concession of La Maddalena to the US Navy appeared in Italian newspapers, Deligia and his majority maintained a

ITALIAN COMMUNIST PARTY
Section of La Maddalena

CITIZENS!
The archipelago of La Maddalena has been ceded to a foreign army, that of the United States, which has already installed its structures and is going to transfer thousands of soldiers.

In this way any perspective of peaceful development that our citizens fought for has been betrayed [. . .]

The government and the center-right administration of the region betrayed all the promises they made.

The touristic activities that in the past years have brought some development will disappear.

The presence of thousands of military personnel will create a lack of houses, will damage tourism and commerce, and will provoke an increase of prices, which will force the young generations to emigrate.

CITIZENS!
Let's reject the occupation of a foreign power.
Let's rejoin to resist.
Let's fight for a peaceful economy, for the progress of the community of La Maddalena and for the future of our young generations.

Translation of Figure 2.2.

strict "no comment" stance. In reality, the mayor hoped that the event would not generate much discussion among his community and that, with time, the archipelago could simply return to its normal, quiet routine. Although, as the Christian Democrats predicted, the Maddalenini did not mobilize en masse against the base, the pressure coming from the Communist and the Socialist parties and the unprecedented attention of the national media made the holding of a public debate in the city council practically unavoidable.

On October 19, 1972, the municipal assembly unanimously voted for a resolution expressing "serious concerns for the possible negative effects of the US Navy presence on the future development of the local economy, particularly on tourism, and above all for the disquieting alarms [about the risks of radioactive contamination that have] appeared in the national newspapers, which have provoked a state of profound uncertainty among

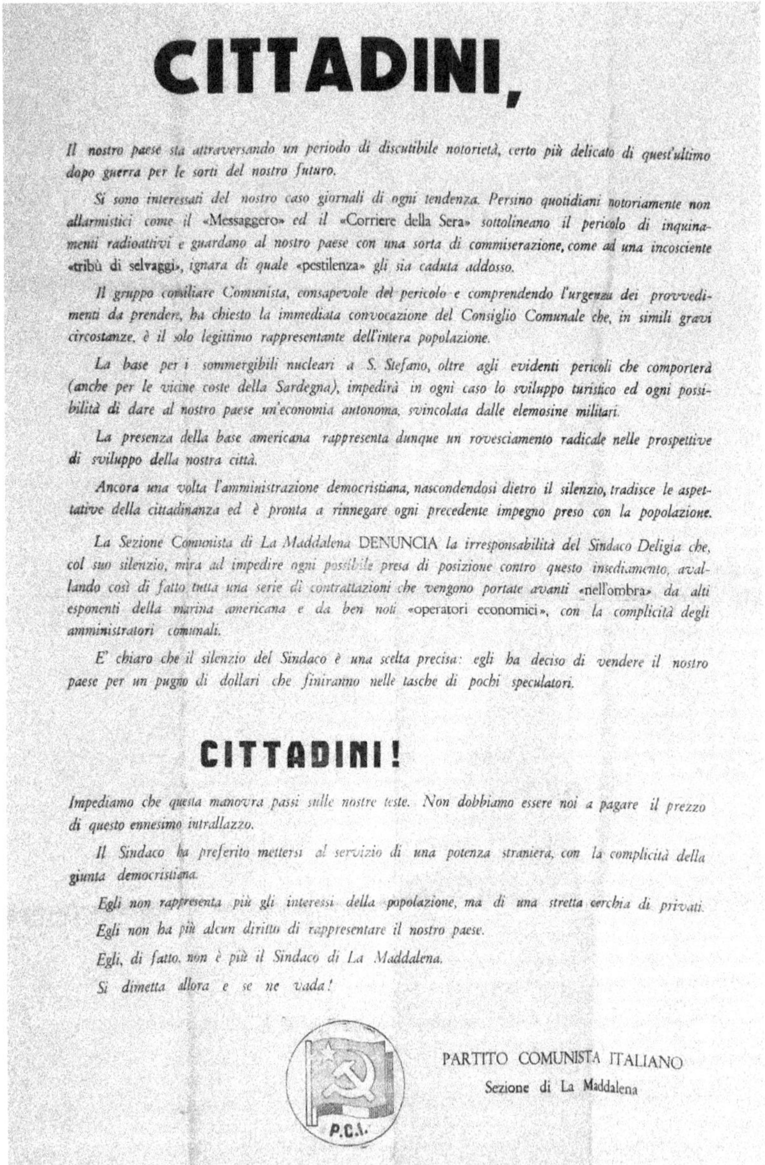

CITTADINI,

Il nostro paese sta attraversando un periodo di discutibile notorietà, certo più delicato di quest'ultimo dopo guerra per le sorti del nostro futuro.

Si sono interessati del nostro caso giornali di ogni tendenza. Persino quotidiani notoriamente non allarmistici come il «Messaggero» *ed il* «Corriere della Sera» *sottolineano il pericolo di inquinamenti radioattivi e guardano al nostro paese con una sorta di commiserazione, come ad una incosciente* «tribù di selvaggi», *ignara di quale* «pestilenza» *gli sia caduta addosso.*

Il gruppo consiliare Comunista, consapevole del pericolo e comprendendo l'urgenza dei provvedimenti da prendere, ha chiesto la immediata convocazione del Consiglio Comunale che, in simili gravi circostanze, è il solo legittimo rappresentante dell'intera popolazione.

La base per i sommergibili nucleari a S. Stefano, oltre agli evidenti pericoli che comporterà (anche per le vicine coste della Sardegna), impedirà in ogni caso lo sviluppo turistico ed ogni possibilità di dare al nostro paese un'economia autonoma, svincolata dalle elemosine militari.

La presenza della base americana rappresenta dunque un rovesciamento radicale nelle prospettive di sviluppo della nostra città.

Ancora una volta l'amministrazione democristiana, nascondendosi dietro il silenzio, tradisce le aspettative della cittadinanza ed è pronta a rinnegare ogni precedente impegno preso con la popolazione.

La Sezione Comunista di La Maddalena DENUNCIA *la irresponsabilità del Sindaco Deligia che, col suo silenzio, mira ad impedire ogni possibile presa di posizione contro questo insediamento, avallando così di fatto tutta una serie di contrattazioni che vengono portate avanti* «nell'ombra» *da alti esponenti della marina americana e da ben noti* «operatori economici», *con la complicità degli amministratori comunali.*

E' chiaro che il silenzio del Sindaco è una scelta precisa: egli ha deciso di vendere il nostro paese per un pugno di dollari che finiranno nelle tasche di pochi speculatori.

CITTADINI!

Impediamo che questa manovra passi sulle nostre teste. Non dobbiamo essere noi a pagare il prezzo di questo ennesimo intrallazzo.

Il Sindaco ha preferito mettersi al servizio di una potenza straniera, con la complicità della giunta democristiana.

Egli non rappresenta più gli interessi della popolazione, ma di una stretta cerchia di privati.

Egli non ha più alcun diritto di rappresentare il nostro paese.

Egli, di fatto, non è più il Sindaco di La Maddalena.

Si dimetta allora e se ne vada!

P.C.I.

PARTITO COMUNISTA ITALIANO
Sezione di La Maddalena

2.3. Poster of the local section of the Partito Comunista Italiano, protesting the secret agreement for the installation of the US Navy base in the archipelago, September 12, 1972.

the local population."[25] The deliberation openly asked both regional and national authorities "to give precise and unequivocal guarantees on the possible risks of environmental pollution connected to the presence and the operations of the US Navy units of the base."[26] The document was

CITIZENS,

Our community is receiving, unwillingly, a lot of attention.

National newspapers of all tendencies have paid attention to our case. Even moderate newspapers like "Il Messaggero" and "Corriere della Sera" underline the risks of radioactive contamination and look at us as a "savage tribe" ignorant of the pestilence that will fall upon us.

The Communist group in the city council, aware of the peril and understanding the urgency of the provisions to adopt, has requested the immediate meeting of the assembly, which in such a dramatic circumstance is the only legitimate institution representative of the entire community.

[. . .]

The Communist Party section of La Maddalena denounces the irresponsibility of mayor Deligia, who through his silence is trying to avoid any expression of opposition to this installation and in fact favors secret agreements conducted by the high spheres of the US Navy and well-known local contractors with the complicity of the local administration.

It is clear that the mayor wants to sell our community for a fistful of dollars that will end up in the pockets of local speculators.

CITIZENS,

[. . .]

The mayor prefers to safeguard the interest of a foreign power.

He does not represent the interests of this community anymore.

He does not have anymore the right to represent our community. He, in fact, is no more our mayor. He should resign and go home.[27]

Translation of Figure 2.3.

clearly the outcome of a political compromise between the obsequious position of Deligia's administration toward the decisions of the national government, and the firm opposition of Communists and Socialists against the base.

More importantly, from Deligia's perspective, with this tactical armistice the local Christian Democrats bought some time for organizing a common strategy with the national government. The mayor had to wear two hats. Behind the scenes, Deligia asked for government support so that he could defang the local opposition. This could be done only by dissolving the doubts about the safety of the base and by showing his community that the "sacrifice for the national interests" would be rewarded promptly. At the local level, Deligia wanted to maintain the appearance that his administration was not subservient to the interests of the national govern-

2.4. Poster produced by the Partito Comunista Italiano on La Maddalena, September 1972.

ment. In his heart, though, he knew that he could welcome the Americans only after Rome offered him technical and economic guarantees that the presence of the nuclear submarines was not dangerous but instead would bring concrete advantages. For this reason, before another extraordinary meeting of the city council fixed for November 19, the mayor of La Maddalena and his collaborators composed a document with a long list of requests to the Italian government and the Region of Sardinia.[28]

An Italian port given as a present by the Andreotti-Malagodi administration
THE ISLAND OF LA MADDALENA (SARDINIA)
TRANSFORMED INTO A SOURCE OF POLLUTION,
AN ATOMIC PERIL, AND A MILITARY TARGET
THIS IS AN ACT AGAINST DIPLOMACY AND PEACE
A SECRET AGREEMENT TACITLY STIPULATED
OUTSIDE THE PARLIAMENT
AGAINST THE WILL OF THE ITALIAN CITIZENS
Let Italian ports be ports of peace
Out of Italy the atomic bases[29]

Translation of Figure 2.3.

Articulated in nineteen points, the list included in the first place an official act of the Ministry of Health with the "unequivocal reassurance that the presence and the operations of the US naval units will not provoke any contamination. Qualified agencies will need to provide these guarantees."[30] The requests were explicit and detailed, including the estimated costs for each project: "Expansion of military arsenal school and more admissions for specialized workers . . . , the guarantee that the Italian navy schools will remain in La Maddalena," a new hospital, a more efficient water supply, a school for tourist operators, a bridge to link the archipelago with northern Sardinia, sport infrastructures and recreational areas, new houses for indigent people, a natural preservation area on the island of Caprera, the conversion of decommissioned military structures to civilian use, and so forth. The accuracy and the details of the requests contained in Deligia's list gave the appearance that the special plan had been shared in advance with the central authorities. The unanimous deliberation of October 19, expressing "serious concerns" about the negative effects of the US presence on the archipelago, was just smoke in the eyes of the local opposition.

In mid-November, just a few days before the upcoming extraordinary meeting of the city council, Medici sent Deligia a letter in which he reassured the mayor that all the requests of the local community would be examined and satisfied, in reasonable terms. He also included a technical report from CAMEN to "refute all the alarms created by interested parties."[31] In their concise safety assessment, consisting of three pages, the CAMEN experts illustrated in very general terms available information about the characteristics of the pressurized water reactors propelling the US fast attack submarines stationed in La Maddalena. The document affirmed that concerns regarding the risks of nuclear contamination were not justified: "The only peril for the population living in the proximity of a nuclear plant or nuclear ships is represented by the disposal of radioactive waste at sea. For what concerns naval reactors, the radioactive waste consists of liquid or solid material created during the activation of the cooling system. This material is stored within special tanks and is collected in particularly safe inland repositories for ulterior treatment or dispersed and diluted at sea in very high waters."[32] It's worth noting that the language used in CAMEN's reports corresponds, almost word for word, to that used by the US Navy in its annual environmental monitoring reports concerning domestic and overseas ports where US nuclear units were operating, including a description of procedures for radioactive waste management.[33] It is plausible to conclude that CAMEN's personnel did not produce any independent study, but instead accepted the US Navy's documents as valid safety assessment certification and limited themselves to providing a translation to national and local authorities.

In addition to the reassuring safety assessment of the US Navy nuclear program, the report prepared for the minister of defense explained at length that a network of radiosurveillance stations installed along the Italian coastal lines had monitored the levels of radioactivity of the seawater since 1961. The national radiometric system—according to CAMEN—had never revealed any anomalous concentration of radionuclides (especially strontium-90 and cesium-137) above the natural levels during the visits of nuclear-powered ships. Thus, military experts excluded the possibility that the US Navy routine operations in La Maddalena could be a source of radiocontamination in the future.

Backed by the reassurance of the national government and by the "scientific explanations" of CAMEN—which Deligia read aloud during the November 21 session—Democrazia Cristiana members thought that they could easily justify their proposal for a drastic revision of the deliberation of October 19:

> The City Council of La Maddalena, convened for the ordinary session of November 21, 1972, on the basis of the reassuring information received from the government, through the reliable scientific report of CAMEN, which excludes any possibility of contamination; given the substantial acceptance of the government and the Region of the Special Plan for the Development of La Maddalena proposed by the municipal administration. . . . Declares that the serious concerns expressed a month earlier by this assembly are no more subsistent, . . . and condemns the political rally organized by the Communist Party of La Maddalena [two days earlier] against our American allies.[34]

2.5. Protest march organized by the youth organizations of Partito Comunista Italiano, Partito Socialista Italiano, and Democrazia Cristiana of Gallura on November 19, 1972. *Credit:* film by Adriano Tovo, private archive.

The opposition protested in vain. The small number of Communist and Socialist councilors in the municipal assembly could not impede the approval of the resolution, and attempts to mobilize local residents on the streets were equally unsuccessful. On November 19, 1972, the PCI and PSI organized a protest march through the main squares of La Maddalena, but most of the participants came from outside the archipelago. Journalist Gino Zasso, who reported on the event for *La Nuova Sardegna*, noted that the Maddalenini "watched the march from inside their cars or through the windows of their houses, with indifference, as if what was happening did not concern them at all."[35]

The only way in which the isolated local antibase front could hope to defy the carefully orchestrated narrative of the government and the local administration was to cast doubt on the neutrality of CAMEN's assessment and put in question the authority of the technical document it produced.

Some help came from outside the archipelago. After CAMEN's safety assessment, which "convinced" Deligia and his majority to welcome the Americans unconditionally, another contested safety assessment, this time issued by CNEN, spurred a series of political reactions within civilian expert agencies.[36] Instead of closing the controversy over La Maddalena, the government's attempts to silence the opposition with the support of colluding sectors of expert agencies had the effect of shifting

2.6. Poster of local Democrazia Cristiana against the antibase march organized by the local Partito Comunista Italiano.

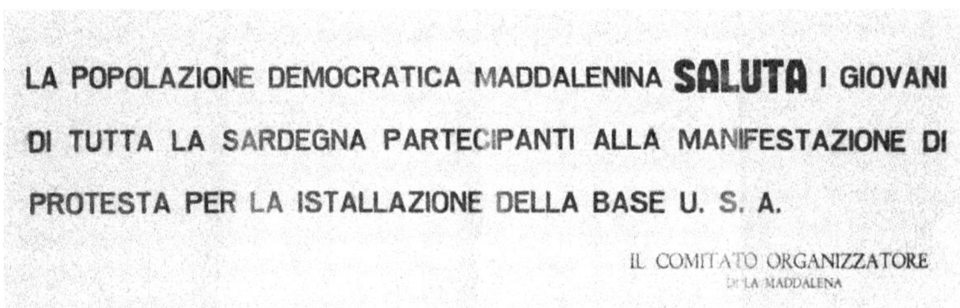

2.7. Partito Comunista Italiano poster welcoming the antibase protesters.

the terrain of conflict from international relations to nuclear safety (although the two dimensions remained, obviously, hardly separable).

ITALIAN EXPERTS SPEAK

The Italian scientific community did not remain indifferent toward the events of La Maddalena. After the Italian Society of Biophysics and Molecular Biology, the Società Italiana di Fisica (SIF, Italian Physical Society), took its official position against the US Navy installation. On October 29, 1972—just before the inauguration of the annual meeting of SIF— Eugenio Tabet, a member of the Physics Laboratory of the ISS, proposed to discuss a document about the US Navy base that he and his colleagues had redacted days earlier: "Subscribed practically by the entire ISS personnel regardless of their political orientations, with care-

THE MAJORITY
Of Citizens of La Maddalena
Pushes back against the provocative march of the communists and philo-communists against our American friends.
Go back to your own towns.
These are problems that should concern only the people of La Maddalena.
WE ARE A PEACEFUL COMMUNITY AND DO NOT WANT DISORDERS
GO AWAY

Translation of Figure 2.6.

The democratic community of La Maddalena WELCOMES the youth of Sardinia who take part in the protest march against the installation of the U.S.A. base.[37]

Translation of Figure 2.7.

fully thought technopolitical arguments, [the document] intends to submit to the scientific reflection and to the civic conscience of the physicists . . . the preoccupation of all citizens, especially those who in their institutional roles deal on a daily basis with the uses of nuclear energy, after the installation of the base."[38] The presidential committee of SIF agreed that the plenary assembly would be the appropriate place for such a debate. On November 1, 1972—at the end of the society's annual meeting held in Cagliari—the assembly, almost unanimously (only one member voted against the resolution and two abstained), decided to adhere to the protest, with only minor amendments.[39]

During the debate, different positions emerged on how to represent the society's concerns about the safety of La Maddalena. The general feeling of SIF members was that the US base presented both technical and political aspects that could hardly be separated. Thus, some insisted that "we should express our opinions not only as experts who adhere to the position of other colleagues; we should also reclaim our rights [as] citizens who have a political conscience. Therefore we should not limit our discussion to the problem of radioactive contamination."[40] In the view of the proponents of a more combative version of the document, like Tabet, the text should make clear that the US installation not only violated existing national and international norms on radioprotection, but also had a precise political meaning: "In order to be coherent with the objective reality, the document should also include a critique of the overwhelm-

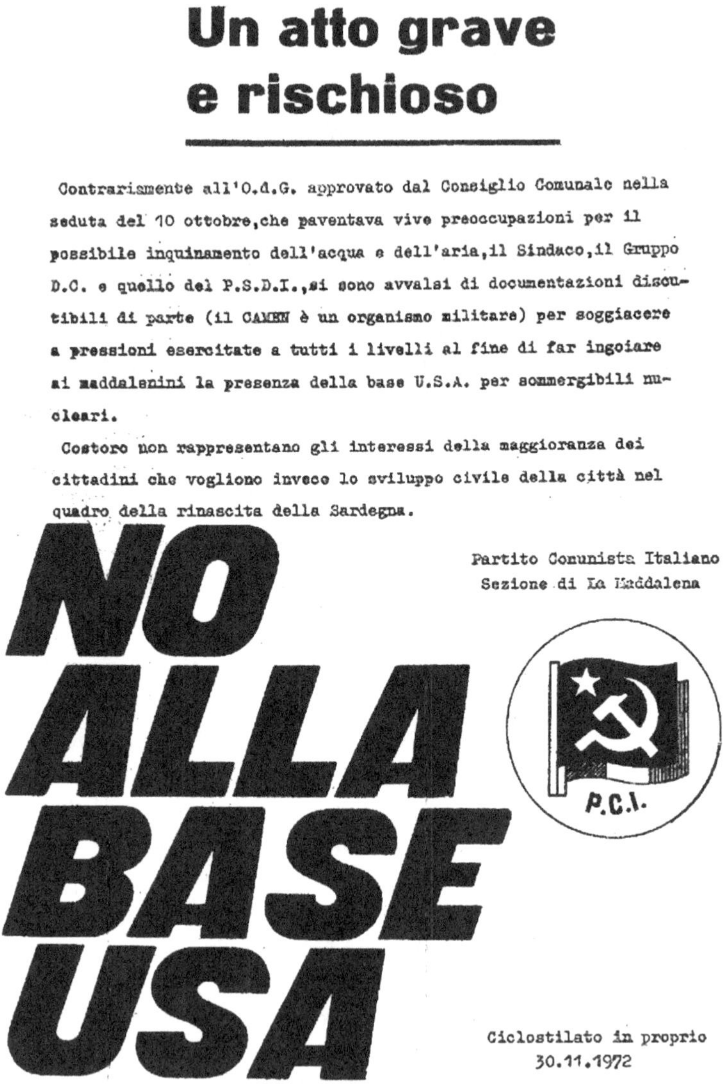

Un atto grave e rischioso

Contrariamente all'O.d.G. approvato dal Consiglio Comunale nella seduta del 10 ottobre, che paventava vive preoccupazioni per il possibile inquinamento dell'acqua e dell'aria, il Sindaco, il Gruppo D.C. e quello del P.S.D.I., si sono avvalsi di documentazioni discutibili di parte (il CAMEN è un organismo militare) per soggiacere a pressioni esercitate a tutti i livelli al fine di far ingoiare ai maddalenini la presenza della base U.S.A. per sommergibili nucleari.

Costoro non rappresentano gli interessi della maggioranza dei cittadini che vogliono invece lo sviluppo civile della città nel quadro della rinascita della Sardegna.

Partito Comunista Italiano
Sezione di La Maddalena

NO ALLA BASE USA

P.C.I.

Ciclostilato in proprio
30.11.1972

2.8. Partito Comunista Italiano flyer against the city council's deliberation of November 21, 1972. Partito Comunista Italiano, La Maddalena chapter, November 30, 1972.

ing presence of military installations in Sardinia."[41] Other participants suggested that the resolution could be more effective if written in a "politically neutral" language: "If we limit our observations to the technical aspects [of radioprotection], to which we can speak as physicists, our position could be more influential on public opinion, in Parliament, and

A serious and risky omission

In clear contradiction with the deliberation approved by the city council on October 10 [19], which expressed serious concerns for the possible contamination of water and air, the mayor, the D.C. group [Democrazia Cristiana] and the P.S.D.I. group [Social Democratic Party of Italy], have used questionable and nonobjective documentation (CAMEN is a military institute) to acquiesce to all sorts of pressures for the purpose of making the Maddalenini swallow the presence of the U.S.A. base for nuclear submarines.

They [mayor and majority in the city council] do not represent the interests of the majority of our citizens, who want instead a civil development within a general plan for the renaissance of Sardinia.

Italian Communist Party
Section of La Maddalena

NO
TO THE
US BASE[42]

Translation of Figure 2.8.

within advisory boards . . . because it will be more difficult for our adversaries to dismiss the document as the political intervention of 'a bunch of communists.'"[43]

The proposal to focus on technical arguments prevailed and—as I will detail below—became a strategic choice not only of concerned experts but also of environmentalists and antibase activists in the following years. The terrain of political confrontation was shifting toward a more technical domain. The leitmotif of the antibase front from then on was that the military nature of the US Navy base in La Maddalena could not justify the lack of safety measures routinely implemented—according to national and international laws—around other nuclear installations. Therefore, the government could not consider La Maddalena a safe site until Italian nuclear regulatory agencies completed scrupulous radioecological studies in the archipelago, installed a system of radiosurveillance, and drafted an emergency plan. Health physicists, expert activists, and participants in the antibase front alike criticized CAMEN's report for being classified as confidential and for the informal procedures through which the government requested it. To them, the fact that a military advisory body of the Ministry of Defense produced the safety assessment for the US Navy installation looked like an enormous conflict of interest.

But apparently CAMEN was not the only expert agency to issue a (preliminary) safety assessment. On November 15, 1972, the *Corriere della Sera* published some excerpts "from two studies that the government requested independently from both CAMEN and CNEN."[44] The title of the article was assertively reassuring: "'No risks of radiation from the submarines,' experts announce." The evaluations of the study attributed to CNEN—substantially overlapping with CAMEN's—provoked a political earthquake inside the civilian agency.

During a political rally on November 22 coorganized by PCI, Partito Socialista Italiano di Unità Proletaria (PSIUP, Italian Socialist Party for the Unity of the Proletariat), and SANN-CGIL, national deputy Giovanni Berlinguer—PCI's central committee member responsible for public health and scientific research policies—attacked the position of the Italian nuclear agency, which "in contrast with its regulatory mission, instead of evaluating the problem of La Maddalena in a neutral way offered [the government] a technical coverup of the entire operation."[45] Furthermore, the fact that CNEN's cited document was not signed made things even more suspect. The mysterious anonymous report became a point of contention and open political conflict within the agency. A group of CNEN technicians and laboratory researchers distanced themselves from the imprudent safety assessment: "Neither the personnel of sanitary protection and nuclear safety divisions [of CNEN] responsible for the technical evaluation of the authorization procedures for nuclear plants nor the Commissione Tecnica have been consulted about or saw any data concerning the base of La Maddalena. Therefore, we are wondering how a safety report could be legitimately produced by this agency."[46] The signatories of the accusing document sent a formal request of clarification to Clementel and to the executive committee of the agency.[47] In the meantime, the personnel from CNEN's laboratories who adhered to PCI published a seven-page document explaining why the base of La Maddalena should be considered a "nuclear installation," which required specific radiosurveillance and safety measures similar to inland nuclear plants.[48]

The insistence of expert activists on specific technical and political arguments for the recognition of the nuclear status of La Maddalena steered the debate over a discussion of the safety of nuclear submarines: How did they work? What if an accident happened? Were nuclear submarines comparable to nuclear power plants?[49] Experts and nonexperts who radically opposed the US Navy installation, or just advocated the adoption of congruent safety measures, represented La Maddalena as a nuclear site. The government instead continued to deny La Maddalena's nuclear status and to trivialize the presence of nuclear submarines in the Sardinian archipelago, representing them as completely safe. On December 9, 1972, mayor Deligia, assisted by Captain Antonio Cocco, commander of the Italian naval base of La Maddalena, and Colonel Ameli, from the Italian navy engineering department, repeated to a selected audience of journalists that the Maddalenini had nothing to fear from the nuclear submarines, as demonstrated by the technical studies of CAMEN and CNEN.[50] The polemic continued for another month, until it became clear that the government had extracted only certain passages from the preliminary safety assessment of CNEN and avoided mentioning less reassuring statements, such as those confirming the lack of information about the characteristics of the reactors installed on US submarines.[51] The mobilization of CNEN's activists against the "personal initiative" of Clementel did not have immediate repercussions on any of the protagonists (a request to discuss the episode of the "false report" in the executive committee was denied) but confirmed to the public that profound political cleavages existed inside the agency.[52]

Carlo Papucci, a former activist of CGIL-Ricerca and technician at the Laboratory for the Study of Radioactive Contamination of Marine Environment of CNEN, in retrospect sees the political divisions that animated the agency as "normal" at the time: "Inside CNEN there were fundamental divergences between those who held a technocratic view of nuclear safety, according to which experts should have exclusive competence over technical matters—especially if they involved security issues—and those who instead pushed for a more democratic approach to science, based on transparency, public discussions of data, and a closer adherence to societal needs. And then there was a third group, *i sugheri* [corks]—as I call them—those who followed the current to remain afloat."[53] *I sugheri* were researchers, laboratory personnel, and high-ranking scientific directors who preferred to maintain their neutrality or to please whoever was in power to remain in their positions. The reverberations of political tensions within the Italian scientific community could become quite evident in concomitance with national and international events. For example, in the same session in which it promoted the motion against the installation of the US Navy base in La Maddalena, SIF approved a document condemning "those physicists who had—as members of the 'Jason division'—a direct responsibility in the systemic application of scientific discoveries for the imperialist aggression of the USA in Vietnam and invites SIF to exclude from all its future initiatives all those scientists whose membership in the 'Jason program' has been proved."[54]

AMENDOLA'S CAMPAIGN

In 1973 La Maddalena did not attract the attention of the media with the intensity of the previous year, and local newspapers focused on other aspects of the American presence in the archipelago. This changed dramatically again in March 1974. The Roman judge Gianfranco Amendola was the protagonist of a new media campaign focused on the lack of safety measures around the US Navy installation. Founder of the environmentalist association Gruppo Ambiente (Environment Group), Amendola incarnated the role of the so-called *pretori d'assalto*, a group of young magistrates who—in line with the radical democratic turn of 1968 and the foundation of Magistratura Democratica—interpreted their judicial activity as service to the people and for the people, against powerful economic lobbies and industrial groups with strong political connections.[55] Amendola distinguished himself as an early advocate of ecological rights in Italy and as promoter of legislative reforms aimed at reducing industrial pollution and at safeguarding public health and the environment.[56] The involvement of the judge with La Maddalena, at least publicly, started on March 22, 1974, with an op-ed in *Il Messaggero* titled "Basi infette" (Infected bases).[57] Amendola's article brought attention to a recent scandal concerning US Navy nuclear submarines stationed in Japan. As reported by Japanese newspapers, two months earlier, during a parliamentary audit of the Japanese Diet, the secretary-general of Japan's Communist Party, Tetsuzo Fuwa, declared that the National Institute of Analytical Chemistry had forged radiometric data about the ports where the US Navy stationed its nuclear powered fleet.[58] After conducting an internal investigation, with great embarrassment and concern the Japanese government admitted to the irregularities and made other agencies responsible for the radiosurveillance program in its ports. In the meantime, the government asked the US Navy to suspend the anchorage and operations of nuclear-powered ships until a new radiometric system was put into work.

By illustrating the Japanese scandal Amendola once again called the attention of Italian public opinion to the negligence demonstrated by the government in managing the case of La Maddalena: "Is not La Maddalena exactly in the same situation of Japanese ports that host US nuclear submarines?" If Japan asked the US Navy to take its submarines away because they were potentially unsafe, insisted the judge, why did the Italian government keep denying the need of an environmental monitoring program? The rhetorical question introduced the readers to a brief review of the situation in the Sardinian archipelago. Sometime after the arrival of the US Navy, CAMEN and CNEN started to analyze samples of seawater, algae, sediments, and mollusks around the base, but the measurements were taking place every six months and were not conducted according to established standard protocols. CAMEN's data were not accessible because they were secret, whereas CNEN's monitoring campaigns were severely limited because its personnel could not access the area where the submarines were stationed. Furthermore, the results of the analyses were not directly comparable given that sampling and radiometric procedures changed from one campaign to the other. At times CNEN experts measured the concentration of radioactive elements in the algae, at other times in mollusks and seawater. The treatment of samples in radiometric laboratories also varied; thus, the results assumed different meanings according to whether the samples examined were fresh or exsiccated. Nothing was known yet about the water currents surrounding the archipelago and other atmospheric variables, which may have affected the dispersion of radionuclides in the environment. Still unknown were the characteristics of the reactors and their discharge formula. "In this situation," asked Amendola, "how can we be sure that the residents of La Maddalena and north Sardinia do not live in peril?"

Finally, the activist judge announced that preliminary data informally provided by CNEN—but not yet published—showed the presence of cobalt-60 and manganese-54 (typical reactor activation products) in biological samples, albeit below the safety threshold established by international regulations. "While this is reassuring at the moment," he warned, "we cannot have any certainty about processes of accumulation that may cause long-term genetic effects on future generations." Amendola argued that scientific authorities such as Alvin Weinberg had rejected the validity of the threshold model, admitting the impossibility, even for science, to test hypotheses about the long-term effects of low-radiation dose exposures.[59] For this reason, continued the article, those who ceded the use of La Maddalena to the US Navy were exposing La Maddalena's residents to unacceptable risks.

"Basi infette" was the first of a series of op-eds intended to break the wall of silence that Italian public authorities erected around the radiation surveillance of the Sardinian archipelago. Almost three weeks later, Amendola published another article with an equally disturbing title: "Allarme! L'onda radioattiva" (The radioactive wave).[60] Thanks to the collaboration of CNEN's laboratories, the leader of Gruppo Ambiente came into possession of the results of three ecological surveys that had been conducted around the base. The report showed that over two years the concentrations of manganese-54 and cobalt-60 had increased, although they remained well below the warning levels. For Amendola this evidence suggested that the presence of nuclear submarines provoked phenomena of accumulation, which in the long run could threaten the health of local inhabitants.

Not surprisingly, the alarming articles by the Roman judge provoked an outpouring of reactions in La Maddalena, Rome, and Washington, DC. Among the frequent notes on the evolving political scenario in Italy,

US secretary of state Henry Kissinger received detailed updates about La Maddalena. The US ambassador to Italy, John Volpe, monitored the antibase campaign of the left, including Amendola's "alarmist" articles, and suggested strategies for limiting their impact on Italian public opinion.[61] In the meantime, Italian newspapers propagated the image of La Maddalena as a contaminated place, and Communist and Socialist representatives in parliament submitted numerous requests of clarification to the government.[62]

In the archipelago, the isolated antibase groups were ready to brandish Amendola's arguments to reopen the debate about the US Navy presence.[63] On April 13, 1974, the "Antonio Gramsci" section of the Communist Party of La Maddalena distributed hundreds of fliers with copies of Amendola's latest article. The Communist group in the city council asked the mayor to immediately call a meeting of the assembly to discuss the worrisome findings of CNEN and to invite Deligia's administration to reconsider its position toward the US Navy base. A week later the city council met in an atmosphere of consternation. The Democrazia Cristiana majority unambiguously showed—maybe for the first time—the preoccupation with bearing the political responsibility of supporting the US presence in the archipelago but decided once again to take a diplomatic position. The deliberation approved by the assembly asked the Ministry of Health "to verify the reliability of the last data [CNEN's preliminary report divulged by Amendola] and to assess with more frequency the ecological situation of La Maddalena with absolute and unequivocal certainty." It also asked the government to make available to the public all the data produced so far by CAMEN and CNEN: "If the data indicates the existence of either an immediate or potential danger for local residents," warned the resolution, "this administration will ask the government to revoke its permission for the installation of the US base."[64] Some councilors of the Social Democratic Party and one from the Partito Repubblicano (who usually supported Deligia) decided to abstain during the vote. The opposition, who hoped to reach a unanimous position for the immediate removal of the base, voted against the Democrazia Cristiana document. Communists and Socialists judged Deligia's position as unacceptable and accused the majority of delaying the only possible solution to the problem: "Getting the US base out of the archipelago in light of the already alarming evidence provided by CNEN's data."[65]

News about the increased concentration of cobalt and manganese near the base pushed other sectors of the local community to voice their opposition to the American presence. Even the organizations of the Democrazia Cristiana youth of La Maddalena, Palau, and Arzachena (a little town on the northeastern coast of Sardinia) published an official statement demanding the immediate disestablishment of the US base and expressed their disappointment in the deliberation approved by Deligia's majority.[66] The students of the Istituto Tecnico Nautico (Nautical Institute) of La Maddalena took a more critical position. In a document redacted after the council's deliberation, they requested the closure of the base and asked Deligia to resign because he was "responsible for the delays with which the local administration—shamefully subservient to the political positions of the central government—decided to deal with the problem."[67]

The polarization of the debate produced visceral reactions among the opposite factions. The DC circulated a flier headed with an imperative "BASTA!" (STOP!) directed at the opposition and invited the Maddalenini to react against the "prejudiced battle of the false patriots [PCI and PSI]. The so-called paladins of freedom, who predict serious calamities for our island, are only provocateurs and disruptors of our moral basis."[68] In early May the national magazine *L'Europeo* captured the state of collective anxiety pervading the archipelago with a long reportage titled "Here is La Maddalena. Our correspondents from the radioactive island."[69] The article presented a series of interviews with Deligia's opponents and one with the mayor. Both sides admitted their disappointment in how national institutions disregarded the concerns of their community. The article quoted Deligia as saying, "I know everybody calls me 'the mayor of the Americans,' even members of my party in Cagliari, and I reply, not so ironically, that it is true. I am the mayor of the Americans because the Italians abandoned me."[70] US officials perceived that Deligia was in a difficult position when he started to veto US Navy requests with more frequency and worried that the political pressure surrounding the local administration would put the US Navy operations in the area in jeopardy. In a confidential cable to the office of the US secretary of state, Volpe admitted that

> local political developments over past several days have threatened to affect seriously US Navy's activities in town of La Maddalena. Post-divorce referendum situation and Sardinian regional elections scheduled for June 16, coupled with nuclear pollution allegations, have caused mayor and other local officials to feel that new measures must be adopted which would insure support for local administration and majority party (Christian Democratic) [*sic*]. . . . While we believe that the most pressing immediate problem will be resolved, there remains the long term one of maintaining decent relations with the mayor. In this connection we have proposed to foreign office that [the embassy office] and appropriate [Government of Italy] official, as well as the US Navy representative, visit La Maddalena in the near future to sit down with the mayor and review the entire situation with him. If this meet-

ing is held, we suspect the mayor will make certain requests of an economic nature with which we may or may not be able to comply. We will keep you informed of subsequent developments.[71]

Meanwhile in Rome, something seemed to move in the direction indicated by Amendola, as he confirmed in two subsequent editorials.[72] The mobilization of environmentalist groups and the alarming titles featured in local and national newspapers pushed the Region of Sardinia and the Italian Ministry of Health to ask expert agencies to start an environmental study of the archipelago. In late April a group of ISS and CNEN personnel arrived in La Maddalena for a preliminary survey of the site.[73] Six weeks later the minister of health, Christian Democrat Vittorino Colombo, sent a confidential note to the experts asking them to collaborate on the implementation of "a program for environmental analyses" similar to those that Italian regulatory agencies performed before releasing nuclear plants' license authorizations.[74] In a few pages the document illustrated the research design that ISS and CNEN scientists elaborated on the basis of national and international radioprotection protocols.[75] These consisted of two complementary activities. The first concerned the study of the general ecological and meteorological characteristics of La Maddalena. The second phase of the program aimed at tracking the long-term environmental effects of the US Navy operations through periodic sampling activities and radiometric analyses of marine bioindicators. The last phase consisted of the installation of continuous monitoring stations to detect in real time any alteration of the levels of radioactivity following hypothetical accidents.[76]

After two years of heated debates, the prescription of a complete radioecological program was an evident recognition of the nuclear status of La Maddalena: radioprotection protocols would be applied around the US base in analogy with other nuclear installations. The apparent closure of the dispute induced some, instrumentally or sincerely, to manifest their satisfaction, while others, including Amendola, warned that this was only a starting point: "The initiative of ISS and CNEN, with two years of delay, demonstrates that in La Maddalena the most elementary precautions have not been taken. Until these measures are fully implemented—and it will take time to do so—it is difficult to say that the population is safe."[77]

Did Amendola just have a premonition of what would happen in the following years? Or did his warning come from a profound knowledge and awareness of the complexity of the Italian bureaucratic system and of the limits imposed by Cold War military secrecy that, despite the official reassurances of the Ministry of Health, would remain largely unresolved?

In the context of the archipelago's historic acquiescence of the military presence and its positive economic benefits, debates over the US nuclear installation were only marginally inspired by anti-American or antimilitary sentiments. Instead, controversies focused on the risks related to the presence of the nuclear submarines and on the implementation of a plan for public safety. Because the local opposition was scant, outside groups usually took the lead in the antibase struggle. Between 1972 and 1974 the variegated antibase front, consisting of pacifists, radioecologists and radioprotectionists, antinuclearists, the PCI, the PSI and environmental movements, contested the absence of radioprotection measures in northern Sardinia. The Italian government and the US Navy responded with "technical" arguments that presented the submarine base as totally safe and nuclear submarines as advanced technological artifacts handled by trained experts. On the other hand, important sectors of the Italian scientific community, including experts in CNEN and ISS, insisted that the US Navy base should be treated in analogy with civilian nuclear plants, for which the Italian legislation established precise safety measures. Nuclear submarines, I argue, are not like an inland nuclear plant, with a huge reactor building that makes visible and concrete the threat of accidents and nuclear contamination. Because of their mobility and partial visibility, they were more difficult to identify as nuclear objects. Therefore, until 1974 the US Navy base's nuclear status remained a point of contention between the supporters of the US Navy presence and its opponents, who used different registers of nuclearity.

At the local level, at least immediately after the installation of the US Navy, images of nuclear technology and understandings of the risks associated with the presence of nuclear submarines were mediated by political allegiances and the observation of US personnel behavior, which shaped assumptions about the safety of the base. After Amendola led a media campaign to raise awareness about the threat of the nuclear submarines, the Italian Ministry of Health finally commissioned a series of radioecological surveys and installed an environmental monitoring system managed by expert agencies and local authorities. But this was only the first step to make radiological risk in La Maddalena visible.

PART II

Technopolitics of Risk

Secrecy, Bureaucracy, and the Production of Ignorance

During my research, I relied on the online catalogue of Agenzia Nazionale per le Nuove Tecnologie, l'Energia e lo Sviluppo Economico Sostenibile (formerly Comitato Nazionale Energia Nucleare, CNEN) to retrieve official documentation of the environmental studies that the agency conducted in La Maddalena. Among other publications, a paper titled "Uno studio dei contaminanti in mare nell'area de 'La Maddalena'" (A study of contaminants' dispersion at sea in La Maddalena), by Giuseppe Buffoni, P. Gasparini, and A. Zattera captured my attention.[1] I added the title to the list of items to borrow from the library of the ENEA research center of Casaccia (near Rome, one of the first centers built by CNEN in the mid-1950s). Fenced and guarded, the center still contains the relics of the Italian nuclear research program. Only one experimental reactor (of the three previously active) is still working and serves as a pedagogical tool for physics students.

I visited the Casaccia center in early January 2012. The personnel of the library, including the director, welcomed me warmly and invited me to make any request that would facilitate my research. All went smoothly, until I ran into Buffoni, Gasparini, and Zattera's study. The catalogue indicated that the library owned one copy, which I could borrow for up to a month. The assistant librarian looked frustrated and puzzled when he came back from his first round in the stacks: "It should be there, but I could not find it." Almost embarrassed, but determined to solve the problem, he told me, "Come back tomorrow, and in the meantime, I'll try to get this paper out. It must be here, no doubt." The next day the librarian reached me at the entry door and apologetically explained that the paper was missing. The director of the library came out of her office to meet us in the corridor and reassured me that she had already sent a request to all ENEA libraries to see if by chance they had another copy of Buffoni, Gasparini, and Zattera's study. After a few days, I received an official response: the report was missing. I interpreted the situation as just another episode of misplacement, or loss of documentation that historians sometimes deal with in the archives, so I decided, as I had done for other ENEA reports, to contact the author.

I found Buffoni through a Google search. A long list of scientific publications, together with several email addresses, appeared on the screen under his name. I tried a couple. To my surprise and excitement, Buffoni replied almost immediately and gave me his phone number. I called; he answered: "Yes, it's me. I have the report you are looking for and many others. I can send everything to you, or we can meet." We met several weeks later at the ENEA Centro Ricerche Ambiente Marino (Center for the Study of Marine Environments of ENEA), in Liguria, where Buffoni worked for almost two decades. He was still living in La Spezia, near the research center, and decided to join Carlo Papucci and me for a day of recollections and explanations of their work in La Maddalena. The day before my appointment with Buffoni, I met Papucci in his old office. We were sorting out the material from his archive when he opened a folder from which he extracted Buffoni's missing study: "Is this the paper you were looking for?"

"Yes," I replied. "Can I make a copy?"

"Hmm, better if you make just a copy of the cover and the introduction. I'll give you the rest once Giuseppe arrives here and gives his permission." For the first time since we started our collaboration, I felt that Carlo was preoccupied about something he did not want to share with me. He just mentioned that the study had "created some problems in the past" and that it was better if Buffoni himself told me about it. My curiosity now was irresistible.

The next day, Buffoni arrived at the center, welcomed as an old friend. A first-class scientist who specialized in nuclear reactor physics and calculation problems, he began working for CNEN shortly after his graduation, first on computational modeling and later, after moving to La Spezia, on numerical models of thermohydrodynamics and transport-diffusion processes. During our conversation he did not talk very much. Habituated to express concepts through formal numerical models, he tried to explain his work to me by drawing things on paper.

After responding to a preliminary set of questions about his background and scientific activity at CNEN/ENEA, he grabbed a thick folder containing a bunch of documents, including old works on La Maddalena. The missing file was finally in my hands, but far more interesting was the story behind the document. Buffoni passed on to me a series of letters: "All you need to know about this report is written there." An internal communication sheet dated November 23, 1984, was labeled, "History of the technical report *Analysis of a hypothetical nuclear accident in the Archipelago of La Maddalena* by Buffoni, Gasperini, and Zattera."[2] In it Buffoni addressed professor Pietro Metalli, a radiation pathologist who was at the time director of the radioprotection division of ENEA, to ask him why the report, completed in 1982, had not been published yet. The first thing that struck me about the 1984 letter was the title of the study. The paper as publicly catalogued by ENEA does not mention the words "nuclear accident" at all. Instead, it refers to the generic category of "contaminants." The history of the original report is an example of how the logic of secrecy became pervasively incorporated into the everyday life of expert institutions but also of the technocratic approach that CNEN/ENEA division directors and cadres adopted in cases like La Maddalena.

I continued to read avidly: "After July 1982 the work was ready [for publication] . . . I thought that the procedures for the publication as "Internal Technical Report" (RTI) had already started . . . but now [end of 1984] I come to know that they have not and that after sending the report to Metalli and Mittempergher [then director of ENEA] . . . the publication will not take place for reasons of opportunity."[3] Buffoni wondered what happened in the meantime to make the director of ENEA veto the publication after it had already received a go from his departmental director. He mentioned that, according to ENEA's regulations on publishing, the report could be classified, and its circulation restricted if necessary, but a veto on its publication was absurd "considering that its existence is known also outside the agency and that there could be requests of information. This anomaly might be noticed outside [the agency] and I am wondering if it would not be better to have at least something, even classified, rather than nothing." More than thirty years after its official publication, I was asking myself the questions that Buffoni posed to his superiors. What is the logic of hiding something that is official and purportedly public and catalogued as a contribution to ENEA's scientific production? Was it not sufficient to delete the alarming words "nuclear accident" from the title of the study? What did the officials of ENEA's censorship hope to accomplish by putting an empty box on the shelves?

Buffoni's study reappeared in official documents only in 1991, when a team of civilian and military experts collaborated for a major update of the radiosurveillance network of La Maddalena and for the relocation of the monitoring stations closer to the US Navy base. On that occasion, Papucci wrote a note to the director of the ENEA Center for the Study of Marine Environments, suggesting that to place the monitoring stations in the appropriate positions, "it would be extremely helpful to consult the dispersion models of the Santo Stefano Bay. . . . It would be of utmost importance to access, officially, the study that Buffoni, Gasperini, and Zattera elaborated at this center. The study, for various reasons, has never been published. It would be time to do so, with the agreement of the authors and the direction, so that our team could use that information for our work."[4]

In the case of Buffoni's missing file, ENEA's director and some of his colleagues used their influential position inside the agency to delete the alarming phrase "nuclear accident" from the title of the 1982 report, to considerably delay its publication, and to ultimately transform the file into a spectral archival presence. The act of depleting the document of potentially alarming signs, however, produced a multiplication of traces. With time, a secret that is no more can be left aside, forgotten. Its persistence beyond the very circumstances that justified its creation (in the eyes of the censors) makes visible the obstinate survival of secrets—and secrecy as a mental and bureaucratic form—despite the obsolescence of their content. Take the example of Buffoni's study: its title, initially containing the words "nuclear accident," has been modified to avoid public alarm, and in the end the document has been hidden within the bureaucratic belly of ENEA, only to reemerge as a spectral archival presence, devoid of its efficacy, an inert trace of routine filing practices.

Buffoni's story is revealing, in many ways, of the practices through which public institutions and state apparatuses produced and maintained secrecy by removing information that was considered sensitive, or potentially alarming. I use this example as an entry point into the object of the next three chapters: how secrecy, intended as an active removal of information, the compartmentalization of knowledge production, and the shortcomings of the Italian nuclear regulatory regime in La Maddalena shaped a system of radiosurveillance that embodied and enacted a compromise between public safety and military security.

3.1. Plastic containers used to collect samples of rainwater at the EURATOM Research Center of Ispra (Italy), Health Physics Unit. *Source:* A. Malvicini, "Le ricadute radioattive nella zona di Ispra dovute alle esplosioni nucleari durante il quinquennio 1958–1962," *Minerva Nucleare* 7, no. 7 (1963): 267–76.

CHAPTER 3

Radioecology and the Italian Nuclear Regulatory Regime

Until 1974, the Italian government denied La Maddalena the status of nuclear site. Yet, like most of Sardinia and the rest of Italy, the archipelago inevitably received atomic fallout depositions provoked by nuclear atmospheric tests that the United States, USSR, Great Britain, and France performed between the late 1940s and the early 1960s. Italian experts at the EURATOM Study Center of Ispra near the city of Varese, were able to detect the effects of the first atomic test conducted by France in the Sahara desert on February 13, 1960.[1] Italian radiometric stations installed around research centers had more or less systematically monitored the effects of atmospheric nuclear tests since the mid-1950s, including fallout depositions on the soil, rainwater, seawater, and artificial radionuclides' concentrations in edible products, like milk and meat. Specialized personnel were able to elaborate analytical models for inferring the time and place of explosions, and extrapolating data about the characteristics and origins of the bombs through radiochemical treatments of rainwater samples (fig. 3.1).[2]

Several pioneers of fallout detection in Italy conducted their first measurements in the late 1940s. Most applications were small-scale and often performed with Geiger counters borrowed from labs and comple-

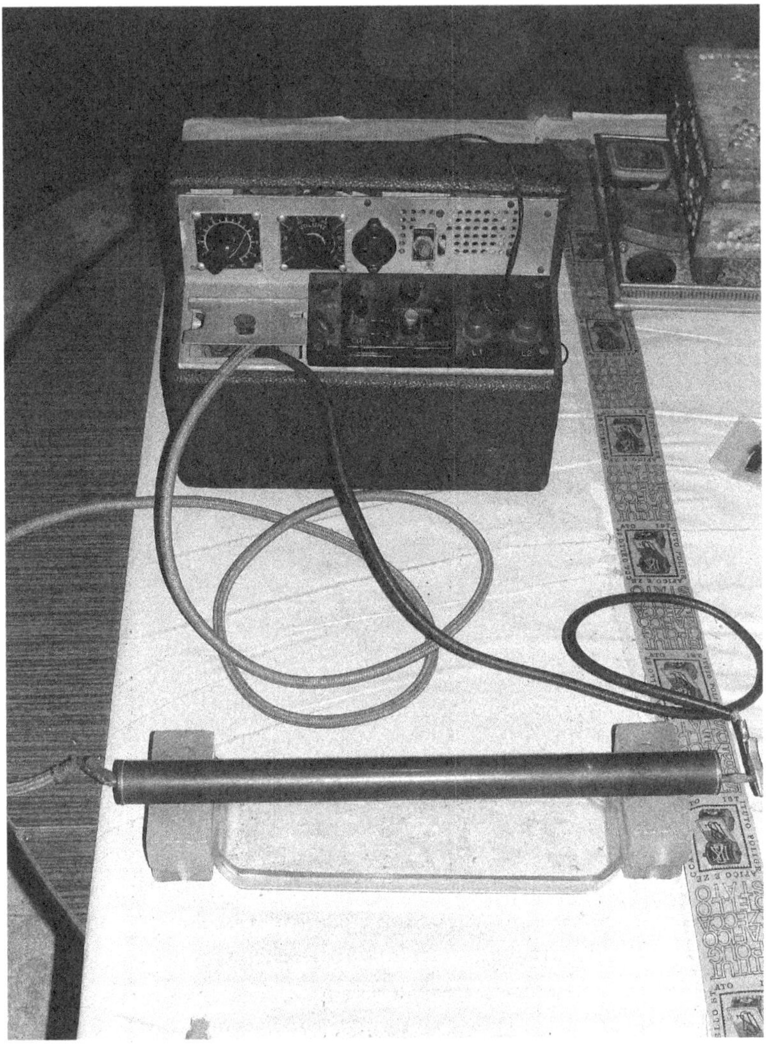

3.2. Geiger counter assembled by Arrigo Cigna in the early 1950s. Photograph by the author. Courtesy of Arrigo Cigna.

mented with variable regulated high-voltage supply. Arrigo Cigna, director of Comitato Nazionale per l'Energia Nucleare (CNEN)'s Continental Radiocontamination Laboratories from the beginning of the 1970s to the mid-1980s, and former president of the International Association of Radioprotection, conducted his early radiometric experiments in Milan with a Geiger counter that he himself built using parts of an old military radio transmitter (fig. 3.2).[3]

In 1953, Cigna presented the results of his experiments with rainwater samples at the national meeting of the Italian Geophysical Association: observations of the levels of radioactivity in the samples taken after nuclear atmospheric detonations established a correlation between the tests and the higher levels of radiocontaminants in the atmosphere detectable in Italy.[4] The name of miniscule Pacific atolls became as familiar as the Nevada desert, which the United States transformed into a nuclear testing ground. An emerging nuclear geography made visible global connections through fallout circulation tracked with new methods and instruments of calculation. Cigna's initial experiments and the more systematic environmental monitoring activities that he helped establish in Italy are part of the larger history of radioecology.

RADIOPROTECTION AND RADIOECOLOGY IN ITALY

During the early 1950s, international expert debates highlighted that the growing number of nuclear tests performed by the United States, USSR, and other countries was a clear menace to the environmental health and safety of the planet.[5] Around the same time the problem of radioactive waste disposal became increasingly controversial due to public concerns about the uncertain effects of radioactive effluent discharges and waste dumping into the ocean.[6] But nuclear contamination also generated scientific opportunities: nuclear particles spreading around the globe could be used as tracers for the study of atmospheric motions, ocean currents, and other natural phenomena.[7] In 1952 the Health and Safety Laboratory of the US Atomic Energy Commission started "a monitoring program to detect the amount of 'global fallout,' defined as radioactive debris injected into the stratosphere and widely dispersed before returning to the ground as particulate ash or in rain and snow."[8]

Radiocontamination paralleled the destiny of the bomb: it was a potential menace to public and environmental health, but also a vehicle to gather scientific knowledge of natural phenomena. As the prominent US ecologist Eugene Odum remarked, "It is generally conceded that environmental contamination with its current dangers of genetic damages, stands as the most important limiting factor in the large-scale use of atomic energy in the immediate future. This prospect is rapidly transforming ecology from a rather obscure and ill-defined member of the biological family into a more organized and coherent division, which will be expected to provide the basic answers necessary for solving practical problems."[9] Radioecology was, in short, the study of ecology in the atomic age.[10]

In the early 1960s biologists intensified their research on the processes of accumulation of radionuclides in biological organisms living in different ecological systems.[11] Many agencies, including the newly established International Atomic Energy Agency (IAEA) and EURATOM, started to fund research centers and programs to study the biological effects of atomic radiation.[12] Results of this inquiry led to important changes in the paradigms of radioprotection, which until then followed the principle of the maximum permissible dose, according to which under a certain threshold radiation exposure was not supposed to harm human beings. While the threshold principle continued to be monitored, the discovery of bioaccumulation processes revealed that organisms varied in their propensity to capture and retain radionuclides dispersed in the environment and consequently to affect human health through the food chain and water supply.[13] Thus, radiosurveillance systems had to focus not only on monitoring accidental discharges and work exposure but also on bioaccumulation processes due to controlled discharges in the environment over time.[14]

Radioecology emerged also in Italy as an interdisciplinary field of studies concerned with the biological effects of radiation and the processes of dispersion and accumulation of radiocontaminants in the environment. Between 1957 and 1958 the Consiglio Nazionale delle Ricerche Nucleari (CNRN, National Committee for Nuclear Research)—which in 1960 became CNEN—created two laboratories for the study of environmental radioactivity: the Laboratorio Contaminazione Continentale (Continental Contamination Laboratory), under the Radiation Division of the Casaccia Research Center (near Rome), and the Laboratorio per lo Studio della Contaminazione Radioattiva del Mare (Laboratory for the Study of the Radioactive Contamination of Marine Environment) in Fiascherino, located in the Gulf of Lerici (La Spezia). Both laboratories were founded thanks to the influence of geneticist Adriano Buzzati-Traverso, who in 1957 was appointed head of the Biology and Sanitary Protection Division of CNRN, and in 1959 became chair of the EURATOM biology working group.[15] Thanks to a series of association contracts with EURATOM, between 1961 and 1974 the Fiascherino laboratory could develop its research programs on the dispersion of radionuclides in marine environments.[16]

Beginning in the early 1960s both civilian and military agencies like CNEN's radioprotection divisions and Centro Applicazioni Militari Energia Nucleare (CAMEN) routinely conducted radiosurveillance campaigns, initially including four fixed monitoring stations along Italy's coastal lines—La Spezia, Venice, Taranto, and Naples—to which, in later years, more sampling sites were added.[17] These monitoring activities, though, did not entail complex radioecological surveys usually performed around nuclear installations.

Until the mid-1960s the activity of CNEN's Laboratory for the Study of Radioactive Contamination of Marine Environment at Fiascherino was oriented toward basic science: its radioecological surveys were limited to the Gulf of La Spezia and were designed mostly for training purposes and for gathering samples used in laboratory examinations. In sum, training activities and sectorial studies prevailed over field science and interdisciplinary collaborations.[18] The situation started to change in 1966, when CNEN asked the Fiascherino center to conduct a radioecological survey of the Gulf of Taranto, to assess the environmental receptivity of the site of Rotondella (Apulia), which was selected to host a nuclear fuel plant.[19] The shift toward more applied science and fieldwork experiences paralleled the growth of the Italian nuclear program and the consequent need to regulate the emerging nuclear industry.

The new Italian legislation on the civilian uses of nuclear energy of 1964 required that in addition to fallout monitoring, local radioecological studies be conducted before the start and during the operations of nuclear power plants.[20] To this end, CNEN and Istituto Superiore di Sanità (ISS)—the latter also equipped with a radiation laboratory focused on radioprotection and public health—started to conduct radioecological campaigns around civilian nuclear sites, including power plants, experimental facilities, fuel facilities, and nuclear-powered civilian merchant ships approaching Italian ports.[21]

Radioecological campaigns were either preparatory or confirmatory. Preparatory studies (like the one conducted by the Fiascherino laboratory in the Gulf of Taranto) were intended to gather information on the ecological and meteorological characteristics of a site before the installation of a nuclear plant. Their goal was to assess the so-called environmental receptivity of the site—that is, the congruence of its ecological and socioeconomic features with the presence of a nuclear installation, according to specific criteria.[22] Italian regulations required preliminary, or preparatory,[23] radioecological campaigns both for the concession of operating licenses and for the design of environmental monitoring systems and external emergency plans in case of accidents.[24]

By the beginning of the 1970s radioecology became an integrated interdisciplinary field studying bioaccumulation processes in species living in given ecological systems, meteorological and ecological parameters influencing processes of radionuclides' dispersion, and the interaction of local communities with the environment around nuclear installations.[25] Thus, radioecological campaigns included the study of "critical groups"—that is, that portion of the local population potentially more exposed to radiocontamination, either by external contact, inhalation, or by ingestion. For example, if the local economy around a nuclear plant was primarily agricultural, then farmers would be targeted for study. Socioeconomic factors—such as land use—and dietary habits were important variables included in the inquiries.

Radioecological programs followed standard protocols, but each ecosystem presented particular geomorphological and environmental characteristics affecting dispersion processes. As a consequence, each site underwent comprehensive ecological surveys before specific sampling strategies could be adapted to its distinctive features. Gathering information about radioactive discharges of nuclear plants required (and still requires) thinking about nuclear reactor technology and the surrounding environment as elements of an integrated envirotechnical system.[26] The types of radionuclides that a power plant discharges into the environment depend to a large extent on reactor designs and operations. Therefore, information about reactor technology and discharge formula becomes crucial for determining, for example, whether the radionuclides that are to be found in the proximity of a nuclear installation can be attributed to its discharges or to other sources (like radioactive fallout due to atomic atmospheric experiments). At the same time, radioecologists had to collect data to characterize the site and elaborate models to predict likely patterns of radiocontaminants environmental dispersion. This was crucial also for drafting emergency plans.

In Italy it was a specific prerequisite for siting authorization that nuclear plants licensees provide a safety report and an environmental impact assessment, along with the reactor discharge formula (that is, the type and quantity of radionuclides routinely released into the environment during operations). The Commissione Tecnica, a specialized group of experts on radioprotection and nuclear safety, was in charge of evaluating the validity and the completeness of these documents and could suggest amendments and changes to the emergency plans, elaborated by the appointed provincial authorities and prefectural offices.[27]

In addition to characterizing the receptivity of potential nuclear sites before installation, CNEN and ISS personnel performed "confirmatory" radioecological campaigns to assess the environmental impact of nuclear plants' controlled discharges over time and to provide information for emergency plans' updates.[28] As should be clear by now, radioecology was particularly relevant for its practical applications in the field of radioprotection.[29] Preliminary campaigns allowed radioecologists to select the most significant sampling points around nuclear sources in light of potential dispersion patterns, to determine which organisms should be used as bioindicators, and to gauge processes of accumulation after plants started to discharge their radioactive effluents into the environment.

A clear example of the logic underpinning the research programs of Italian radioecologists is the work of Aldo Brondi, director of CNEN's Laboratory for the Study of Radioactive Contamination of Marine Environments from 1975 to 1978, and principal investigator of several mineralogical and geological field studies in preliminary ecological surveys. A graduate of the University of Pisa with a specialization in petrography, he began his career in CNEN in 1957 as a "uranium hunter" in the Italian Alps and later collaborated with Ente Nazionale Idrocarburi (ENI, National Agency for Hydrocarbons) on various similar projects in Australia.[30] At the beginning of the 1970s he became interested in marine geology, studying the stratigraphic composition of Italian coastlines and focusing, in particular, on the processes of radioisotopes' sorption and diffusion in alluvial and marine sediments. During the 1970s Brondi joined a group of Italian geologists (mostly working at CNEN) who studied correlations between marine geomorphology, coastal morphology, sediment size, and distribution of radionuclides in marine environments. They advocated a holistic methodology—which they called a "global approach"—to study pollutants' dispersion and sedimentation in alluvial and coastal marine ecosystems. This line of inquiry presupposed a typological classification of coastlines, river basins, and marine environments according to their geomorphological and biotypical characterization. In conjunction with accurate knowledge of hydrological dynamics and of radioisotopes' behavior, typological classifications were intended to enhance the capacity of CNEN (and other regulatory authorities) to predict radiocontaminants' patterns of dispersion into the environment.[31] The final goal of this long-term research project was to create an integrated map of Italy for drafting models of contaminant dispersion on a national scale. If international debates about atmospheric fallout and nuclear waste disposal during the 1950s created the conditions for the emergence of radioecology, the development of the civilian nuclear sector required the formalization of regulatory regimes for the exploitation of this new source of energy.

THE ITALIAN NUCLEAR PROGRAM AND ITS REGULATORY REGIME

In 1962 the Italian parliament passed a general law on the peaceful uses of nuclear energy, which, with many limitations provided the regulatory frame advocated for by the *statalisti* (the proponents of state ownership and control over nuclear technology).[32] The Italian nuclear law filled a gap of almost fifteen years, during which different, and sometimes openly conflicting parties, had already started their alternative projects in an unregulated field.

After the first Atoms for Peace conference in Geneva in 1955, the possibility to acquire technology and enriched uranium from the United States made it possible for Italian competing actors to independently seek agreements with the US government and private companies. In this scenario firms like FIAT and Edison decided to circumvent CNRN's attempts to monopolize the Italian nuclear program by acquiring licenses from US companies, with the benevolent scrutiny of the Eisenhower administration.[33] This led to a multiplication of initiatives to build nuclear power plants with US technology "key in hands."[34] In 1959 the state-financed company Società Elettro-Nucleare Nazionale (SENN), in collaboration with CNRN, started the construction of the nuclear plant of Garigliano, in the southern region of Basilicata, with the economic assistance of the World Bank and the supply of a boiling water reactor designed and produced by General Electric.[35] Project Energia Nucleare per il Sud d'Italia (ENSI, Nuclear Energy for Southern Italy) embodied the spirit with which the statalisti conceived of nuclear power in Italy and was a response to the advanced design of the nuclear plant of Trino Vercellese (in the northern region of Piedmont), financed by the industrial group Edison Italia. The Enrico Fermi installation of Trino Vercellese started its activity in 1964, when the Westinghouse pressurized water reactor became critical.[36] In the meantime the company Agenzia Italiana Petrolii-Nucleare, created by the ENI, had already commissioned the first Italian nuclear plant in the vicinity of Latina (sixty miles south of Rome) with the assistance of the United Kingdom Atomic Energy Commission (UKAEC).[37] The nuclear reactor this time was designed and produced by the English Nuclear Power Plant Company, which used natural uranium as fuel and graphite as moderator. A fourth plant, commissioned in 1969, was built later between 1975 and 1981 in the Po Valley near the town of Caorso.[38]

By 1966 the variety of technological systems acquired from three different groups (two from the United States and one from Britain) made Italy the third producer of electric power from nuclear plants in the world. According to several commentators, including direct participants in the Italian nuclear takeoff, the variety of technological designs chosen for the three nuclear plants should not be interpreted as a planned effort to test the best alternatives, but as a sign of the conflicting interests and the resulting chaos that reigned in the Italian nuclear sector since its inception.[39] This is not the place to examine in depth the conflicting interpretations and historiographical theses about the development of the Italian nuclear program. Nonetheless, a brief illustration of the complex bureaucratic organization of the Italian nuclear regulatory regime is necessary for contextualizing the case of La Maddalena.

The Italian legislation on radiation protection, nuclear siting, and operator authorization, Decree of the President of the Republic no. 185 of 1964 (DPR 185, 1964) and later updates, established that the Ministry of Health was primarily responsible for the safety of workers and the public around nuclear sites. As the technical consulting branch of the Ministry of Health, the ISS oversaw the implementation of safety regulations for workers exposed to ionizing radiation hazards and monitored the levels of radioactivity in the environment. The same decree established that nuclear plant operators were responsible for the installation of local radiometric systems for monitoring the levels of radioactivity in the surrounding environment and for the control of food and water sources. CNEN, on the other hand, had the incumbency to assess the correct functioning of local monitoring systems, and the quality of data analysis, in addition to overseeing the entire cycle of nuclear plant operations, from approval of the siting proposal to the preliminary construction, including reactor safety, personnel certification, and activation tests. CNEN exercised its function of regulator through the Direzione Sicurezza e Protezione (DISP, Division of Protection and Safety), also known as CNEN-DISP, whose experts—engineers and physicists—reviewed the documentation concerning plant design, construction, and operation, and could send inspectors on site. The main office where the activities of the various agencies found coordination was the Commissione Tecnica per la Sicurezza Nucleare e la Protezione Sanitaria (Technical Committee for Nuclear Safety and Radiation Protection, also known as Commissione Tecnica) established within CNEN and consisting of fourteen (sometimes sixteen) members. The Ministries of Industry, Interior, Public Works, Labor, and Health could delegate two members each, while the president of CNEN had the power to nominate four CNEN experts to be part of the committee. The president of the committee had to be a member of the administrative board of CNEN. The last two seats were usually reserved for experts on industrial combustion, and for personnel of the Ministry of the Navy when the review process included their sphere of competence. The committee was representative of all the administrative branches involved in the operation of nuclear plants, from preliminary siting proposal authorizations to waste management.[40]

The fact that CNEN was at the same time responsible for the development of nuclear technology and safety regulator should not come as a surprise, given that in the 1960s this was quite common practice.[41] The same organization was maintained until the end of the Italian nuclear program, despite the complaints of health physicists and radioprotectionists in both ISS and CNEN who proposed to separate the two functions of nuclear promotion and safety regulation by creating two distinct agencies.[42] In Italy the role of the Commissione Tecnica has been generally underestimated in public debates about nuclear safety, especially for what concerns its ability to smooth down and amalgamate the different sensibilities and the specific objectives of agencies like CNEN (more inclined to promote nuclear technology), and ISS (more concerned with radiological safety and public health).

One of the most problematic aspects of Italian legislation on radiation protection until the late 1970s was the lack of a clear cut separation of CNEN's and ISS's roles and the lack of a technical infrastructure at the peripheral level, where radioprotection programs needed to be practically implemented. The Italian legislation put plant operators in charge of local monitoring systems, but at the local level radioprotection expertise was lacking, with the consequence that CNEN and ISS experts had to fill the gap and oversee several sites across Italy with limited resources and personnel. Thus, when in the mid-1970s the Italian government finally delegated a set of regulatory powers—including radioprotection—to the regions (established only in 1970, twenty-four years after the approval of the republican constitution), the latter could not establish monitoring systems around nuclear power plants because they lacked experts, infrastructures, and funds.[43] For what concerns the overlapping of radioprotection responsibilities of CNEN and ISS, in reality this created a latent political conflict: CNEN was run by the Ministry of Industry, while ISS was the technical branch of the Ministry of Health. The potential conflicts between CNEN's promotional goals and ISS's primary concern for radiation protection and public health found a resolution within the Commissione Tecnica, where the aspects of safety assessment and radiation protection were considered in tandem. In general, both ISS and CNEN experts I interviewed during my research confirmed that their professional interactions during radioecological campaigns and in the technical committee were almost always collaborative. Only at the ministerial level, when the logics of promotion and protection assumed political relevance, were tensions between the two agencies more explicit. This happened, for example, in 1978 when, during the discussion of the reform of the National Health System, the Ministry of Industry proposed to give CNEN an exclusive role in radioprotection. The proposal provoked a loud protest against minister Romano Prodi, who was accused of having planned to cut the ISS out of the nuclear business for its opposition to the nuclear policy of the government.[44]

In Italy two fundamental aspects of radioprotection—radiosurveillance and emergency plans—were administered and supervised at the provincial level by the Provincial Commissions on Radioprotection, and by the prefects. The Provincial Medical Officers of Health chaired the commissions consisting of two doctors (at least one of them had to be a specialist in radiology), one health physicist or suitably qualified expert, and a medical inspector nominated by the directors of the Labor Inspectorate of each region.[45] Prefectures, which, like in France, represent the authority of the state in each province, were responsible for drafting and approving the external emergency plans required for each operating nuclear installation on the national territory (both power plants and reprocessing plants, including storage facilities). With the institutionalization of the regions in 1970 and the National Health System reform of 1978, health services (including radioprotection) were delegated to the regional authorities.[46]

RADIOECOLOGICAL ROUTINES

In a little less than twenty minutes, a video documentary titled *Atomo in Mare* (Atom at Sea) describes the activity of a group of scientists from the Laboratory for the Study of Marine Contamination of CNEN in Fiascherino during one of their radiometric campaigns.[47] The film, recorded and edited between 1961 and 1962, was part of several educational initiatives undertaken by CNEN to create a "nuclear culture" in Italy. It explains to nonexperts the rationale for marine radiometric campaigns as part of the necessary measures that all countries had to take to protect their populations from the effects of radioactive fallouts caused by atmospheric tests.

The documentary starts with an animated cartoon. The H-bomb appears on the screen and the countdown begins: *nine, eight, seven, six*. . . . The initial part of the soundtrack is meant to surround the spectators with mechanical terror, as they watch cartoon images capturing the sequences of a nuclear detonation followed by the appearance of the typical mushroom. A stridor of electric sounds accompanies the dispersion of the fallout transported by yellow clouds circulating through the atmosphere far away from an unspecified ground zero. Yellow flakes of radioactive material fall into the sea, where plants and algae absorb them. After ingesting the contaminated algae, a naïvely drawn fish becomes yellow and a fisherman, unaware of what has been going on so far underwater, catches it and eats it. Now the man becomes yellow too. Conclusion: we are all at risk of radioactive contamination.

A rapid frame transition shifts the attention of the spectators from the animated cartoon to crude film footage of Japanese scientists measuring radioactivity on fishing boats in the aftermath of the infamous 1954 Castle Bravo test.[48] The narrator describes the dramatic consequences of the US experiments in the Bikini Atoll, where atmospheric fallout contaminated the waters frequented by hundreds of Japanese fishing boats thousands of miles away from the test site and months after the detonations.[49] The abstract animated cartoon is given new meaning against this powerful historical referent in the images of real events that happened on the opposite side of the globe: risk is ubiquitous.

In contrast to the general, overarching, and overwhelming threatening consequences of thermonuclear explosions, the director of *Atomo in Mare* orchestrates a rapid shift to the peaceful Mediterranean Sea, where, as the narrator's voice reassures the audience, "so far no relevant increase of the levels of radioactivity has been found in the water and in the fish." The Gulf of La Spezia metonymically represents the Mediterranean scenario where CNEN scientists worked to protect Italians from unwanted radioactive exposure caused by "fallout contamination from nuclear experiments, accidental discharges of radioactive material into the sea, and possible accidents involving nuclear propelled submarines. In all of these events CNEN experts are able to assess the concrete menace under which Italians live in the nuclear era." By showing why and how Italian scientists performed routine radioecological campaigns, *Atomo in Mare* makes nonexpert audiences aware of the risks of the nuclear era. At the same time, its script and images are intended to reassure the public. The third part of the documentary visually brings the spectators to the docks of a small harbor near Fiascherino, in the region of Cinque Terre, where, on a day flooded with sunlight, a small team of CNEN radioecologists is taking to sea for a programmed expedition. A fishing boat follows the mobile laboratory at a short distance. It will catch samples of fish that will complement the collection of specimens to be analyzed inside specialized laboratories.

The video represents radioecological campaigns as consolidated and coordinated scientific procedures transforming elements of the marine environment into samples, biological and chemical parameters that allow scientists to go back and forth from the field to the lab and vice versa through operations of reduction and amplification.[50] Sampling areas appear as fixed points on an established map guiding the routine operations of the laboratory personnel. Every operation is scheduled on a time log, scrupulously supervised and coordinated by the director of the expedition. Nothing is left to improvisation. Indeed, the site of La Spezia was a permanent station of the network of marine radiosurveillance since 1960.[51] Documentaries and reports like *Atomo in Mare* are designed to communicate facts—rather than uncertainty—and to illustrate what scientists know through coherent narratives. What lies behind the official representations of reality remains behind the scenes, retraceable in the so-called gray literature.

As is reasonable to expect, archival documents, including preparatory work, internal reports, and meeting minutes reveal that CNEN experts' monitoring practices were usually far messier than the routine radioecological campaign depicted in *Atomo in Mare*. After all, mapping and studying unknown environments is filled with uncertainty and un-

3.3. US fast-attack nuclear submarine and tender moored in the Santo Stefano Bay. *Photo credit:* Tatiano Maiore, La Maddalena, 1975.

3.4. A team from the Comitato Nazionale per l'Energia Nucleare Center for the Study of Marine Contamination of Fiascherino during a radioecological campaign on the *Odalisca*. Courtesy of Roberta Delfanti.

foreseen findings, like other practical activities of scientific knowledge production.[52]

THE RADIOECOLOGICAL SURVEY OF LA MADDALENA

According to CNEN experts involved in the radioecological study of La Maddalena, preliminary campaigns were much more demanding because they involved the multidimensional survey of a complex ecosystem from scratch. It was at the explicit request of Carlo Polvani, head of the radioprotection units of CNEN, that in October 1972 La Maddalena was included as a permanent monitoring site within the national network of radiometric stations established at the beginning of the 1960s.[53] CNEN's teams started to go to La Maddalena every six months to take samples for their measurements. Albeit limited in scope and provisory, Polvani's initiative was the first response of civilian agencies to the emergency created by the installation of the US Navy base. The comprehensive environmental study of the archipelago commissioned by the Ministry of Health in the spring of 1974 started in the summer of 1975.[54] A few months later, Brondi and his colleagues presented their preliminary results at the gathering of the Italian Geological Society. Maps of the mineralogical composition of the archipelago's marine platform, including its granulometry and sediments' deposition patterns, showed that the Santo Stefano Bay was characterized by low hydrological dynamism. This led the authors to conclude that the risks of radioactive contamination due to sediment dispersion outside the perimeter of US base was quite low. But this was only one parameter, other variables, still unknown, should have been included in the study.[55] More expeditions followed. Between the summer of 1975 and the spring of 1977, teams of marine biologists, geologists, meteorologists, and mathematicians traveled to La Maddalena to study marine fauna and flora, water currents, seasonal climatic variations, and the socioeconomic composition of the local community.

To reduce the environment of the archipelago to readable and transportable information radioecologists organized its space into a bidimensional grid of marked coordinates individuating meaningful points of data extraction, which often overlapped with local toponyms and knowledge of the place. Fishermen from La Maddalena and divers of the Nucleo Servizio Difesa Antimezzi Insidiosi (SDAI) of the Italian navy command supported CNEN's personnel in various phases of their campaigns. Instruments and tracing materials made the invisible visible.[56] Aerostatic balloons with anemometers were used to measure atmospheric currents; rhodamine B, fluorometers, and current meters were used to understand the dynamism of the archipelago's marine currents; microscopes, chemical separations, and spectrometric analyses were used to assess the amount and the types of radionuclides absorbed in collected samples. All these measuring techniques amplified La Maddalena's environment by transforming samples of fish, microorganisms, water, and sediments into quantifiable and standardized parameters, whose meanings could be immediately intelligible to the scientific community.

This involved selecting organisms as bioindicators to monitor potential processes of radioactive accumulation in the archipelago. The prin-

3.5. Experiments with current meters, 1976. *Photo credit:* Carlo Papucci.

cipal candidate for the role of "sentinel" was *Pinna nobilis*, a bivalve mollusk whose feeding method made it particularly suited to the monitoring activities (fig. 3.6). In fact, these big-shelled mollusks (commonly known in the archipelago as *nacchere* or *gnacchere*, or castanets, because of their shape) filter water to capture micro-organisms, and in so doing retain other particles, including radioactive ones, which tend to accumulate in their

3.6. Exemplar of *Pinna nobilis*, a typical bivalve mollusk of the archipelago of La Maddalena, now a protected species. *Photo credit:* Antonio Outis Colacino, La Maddalena.

3.7. Exemplar of *Sabella spallanzanii*, a polychaetae worm living on sandy bottoms in the Mediterranean Sea. *Photo credit:* Antonio Outis Colacino, La Maddalena.

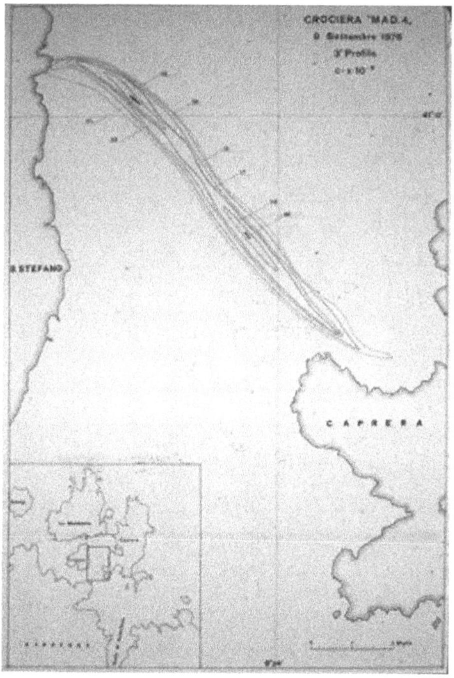

3.8. Tracks of rhodamine B dispersion in the Santo Stefano Bay after experiments, September 9, 1976. *Source:* Carlo Papucci, private archive.

3.9. Comitato Nazionale per l'Energia Nucleare radioecologists and Italian navy divers taking samples near the US Navy tender unit (in the background) in La Maddalena (1980s). Photographer unknown. Courtesy of Roberta Delfanti.

tissues.[57] When at the end of the 1970s *Pinna nobilis* was declared a protected species, though, CNEN radioecologists resorted to using a marine worm: *Sabella spallanzanii* (fig. 3.7).[58]

Knowledge of the environment in which measuring devices are to be located is essential for designing a network of fixed radiometric stations. For example, current-metric analyses and atmospheric tests help experts determine the most probable patterns of radionuclide dispersion in the case of accident, like the arrow of a wind vane. Based on contaminants' dispersion models, the choice of the location of measuring devices was strategically tailored to the environmental characteristics of the site, including the proximity of urban centers, and of other sensitive targets (fig. 3.8).

In many ways this is like choosing sampling sites according to their significance as probable spots of radioactive particle deposition. That is what Brondi and CNEN's team did preliminarily when studying new sites. Once the sampling points were established, confirmatory radioecological campaigns could take place routinely: CNEN and ISS personnel monitored La Maddalena from 1975 to the beginning of the 1990s, when the Region of Sardinia took charge of the radiosurveillance system (fig. 3.9).

RADIOECOLOGY WITH MISSING DATA

During his intervention at a national conference titled "US Nuclear Bases, Local Populations, and the Environment: The Case of La Maddalena," organized by Gruppo Ambiente in September 1975, Carlo Papucci, a CNEN radioecologist and activist in CGIL-Ricerca (Confederazione Generale Italiana dei Lavoratori-Ricerca), called the audience's attention to various organizational problems that were slowing down the study of the archipelago. "The first aspect concerns the current lack of expertise on physical oceanography, which we should try to acquire with the collaboration of other research centers. . . . The second problem regards the concrete possibility to implement this ambitious radioecological program within a reasonable amount of time. Some of our units cannot respond adequately within such a tight schedule considering that we lack means and materials in key sectors."[59]

Originally planned as a yearlong cycle of seasonal campaigns, starting in May 1975, the radioecological survey was delayed for months. As Papucci declared, the scientific program laid out by the Ministry of Health required expertise and equipment that the CNEN Center for the Study of Marine Contamination alone could not provide. For this reason, in May 1976 CNEN signed a collaboration agreement with the Venice Oceanography Research Center of the Consiglio Nazionale delle Ricerche (CNR, National Research Committee).[60] Documents concerning the preparatory phase of the radioecological campaign for the summer of 1976 reveal that the scale of the program was extraordinarily complex and vast in comparison to the operations routinely performed by the two research centers. Current meters were scarce, to the point that they had to be rented or borrowed from other institutions. CNEN personnel adapted a fishing boat, called *Odalisca*, into an efficient floating lab but it needed a licensed crew, which was finally recruited after several attempts and months of delay.[61] After two years of uncertainties and tense political debates about whether La Maddalena should be considered a nuclear site, Papucci and his colleagues were still struggling with bureaucratic and organizational obstacles that limited their capacity to do their job.

The coordination of specialized units (marine biology, oceanography, radioprotection, meteorology, geology) within CNEN and with other agencies—primarily CNR and ISS—resulted in exhausting delays that did not allow Italian experts to respond adequately to the urgent need of monitoring the archipelago's environment. From their first interventions in public debates CNEN and ISS radioecologists and radioprotectionists described La Maddalena as an emergency: the situation, they explained, was anomalous in comparison to other Italian nuclear sites, where preliminary radiosurveillance measures were taken before nuclear plants and facilities started their operations.

While reading radioecological reports from other sites, such as Garigliano and Caorso, I realized that La Maddalena, in fact, was not an exception. Nor was dealing with uncertainty and scarcity of environmental data a novelty for CNEN radioecologists.[62] Official CNEN reports of radioecological campaigns at other Italian nuclear sites frequently denounced the lack of preliminary studies and the difficulties of arriving at a precise determination of radiocontaminants' dispersion models. For example, in the case of Latina—a few kilometers south of Rome—CNEN personnel suggested that: "During the programming phase of the sampling design, the scarcity of meteorological data did not allow the precise individuation of the prevalent directions of winds in the area, and for this reason a 'dense geometric' criterion for the choice of the sampling sites has been followed for a radius of 2 kilometers around the nuclear plant, privileging points near farms, rural houses, and small urban centers."[63]

While this is in part not surprising (as explained above, radioecological surveys are not instantaneous pictures of the ecological status of a given ecosystem, but dynamic representations that require continuous updates and integrations because both demographic and ecological conditions around a nuclear site change over time), the systematic lack of information emerging from CNEN radioecological reports seems to indicate the presence of a form of "institutionalized ignorance" that has to do with the organizational shortcomings of Italian regulatory agencies.[64] Polvani, the head of CNEN Radioprotection Unit, illustrated one aspect of this structural problem when he explained the lack of a uniform national siting policy: "The Italian legislation . . . reflects the status quo of the 1960s, when siting requests were extremely sporadic. The law leaves to the entrepreneur the initiative to propose a site for the construction of a nuclear plant and requests that public authorities evaluate its appropriateness. But this procedure does not allow for choosing the best sites from an environmental point of view. . . . What is worse is that siting choices are made independently from one another without a global view of the problem, on the basis of shortsighted evaluations."[65] Complaining about the lack of a centralized siting procedures, Polvani denounced what had become accepted praxis in Italy: the construction of nuclear plants and facilities in places selected on the basis of political and economic considerations, rather than after detailed preparatory ecological campaigns.[66] Gloria Campos Venuti, director of the Radiation Laboratory of the ISS, and her collaborators stressed the limits of the Italian legislation on nuclear siting on multiple occasions. When the regulation on siting authorization and radioprotection was issued in 1964 (DPR 185, 1964) three nuclear power plants and several nuclear research facilities were already in place or under construction; therefore, the procedures established in the new legislation could be applied only for sites selected after the mid-1960s. But even after 1964, siting procedures continued to follow the same irregular path, as suggested by Cigna during a personal interview in February 2012: "When the installations were ready to start, all the radiosurveillance systems were up and running with, I want to say, great accuracy. But if you ask me whether the choices of the sites were based on previous studies of ecological receptivity, the answer is that only a few of them were performed before construction. One case that I remember well is Montalto di Castro at the end of the 1970s, which, as you may know, was never activated due to strong local opposition and by virtue of the referendum of 1987 after the Chernobyl accident."[67]

Eugenio Tabet, for a long time a member of the Radiation Laboratory of the ISS, offered the same response as Cigna when I asked him about his experience with safety assessment reports around Italian nuclear installations. For example, referring to the nuclear power plant of Caorso, completed in 1981, he affirmed: "The decision to build the plant in Caorso was taken by CIPE [Comitato Interministeriale per la Programmazione Economica] without any preliminary study of the environmental receptivity of the site. We did our safety assessment after the construction was completed."[68]

In sum, the organizational problems impinging upon the work of ISS and CNEN radioecologists in La Maddalena were not new or unique. The biggest anomaly instead rested in the regime of secrecy that US and especially Italian authorities maintained over the environmental data produced between 1972 and 1975 and about the technical characteristics of the nuclear reactors propelling the submarines.

At the beginning of the 1960s Italy was among the most active countries in adopting and developing nuclear technology for energy production and other civilian applications. With the construction of nuclear power plants and the spread of auxiliary facilities across the country, the need to regulate safety and security aspects of nuclear technology became crucial. Radioprotection and radioecology developed in Italy beginning in the early 1950s in conversation with international research and debates about the health and environmental effects of radioactivity. In the late 1950s specific units and research centers were established within the Consiglio Nazionale delle Ricerche Nucleari, later denominated Comitato Nazionale per l'Energia Nucleare. While the Cold War enhanced the work of radioecologists globally, in La Maddalena the presence of a US Navy installation for atomic submarines posed severe limitations to Italian experts in charge of the radiosurveillance program. When they started their studies in 1975, Italian radioecologists had to adapt their routine protocols to the distinctive political and environmental features of the archipelago. The Fiascherino Laboratory for the Study of Radioactive Contamination in Marine Environments, established at the end of the 1950s, was not equipped for such a large and complex environmental survey, and bureaucratic delays impeded the implementation of the radioecological study designed in 1974. In addition, the US Navy and the Italian government refused to provide Italian scientists with information about the technical characteristics of the reactors propelling the nuclear submarines. To compensate for the lack of technical information about potential radiocontaminants, CNEN and ISS experts had to multiply their monitoring efforts by enlarging the scope of their research and in so doing they incorporated elements of uncertainty into their protocols. Since Italy lacked a specific regulation for nuclear ports, Italian experts (CNEN and ISS) treated the US Navy base in analogy with civilian nuclear plants.[69] CNEN and ISS experts used the argument about the lack of preliminary environmental surveys in La Maddalena to reach a substantial political goal: ensuring the implementation of independent radiosurveillance measures by Italian civilian authorities and opening a breach into the wall of secrecy erected by Italian and US military apparatuses. Their strategy was only partially successful.

CHAPTER 4

A Technopolitical Compromise

THE DUAL SYSTEM OF RADIOSURVEILLANCE IN ITALY

On July 18, 1975, the *Odalisca*—the ship laboratory that Comitato Nazionale per l'Energia Nucleare (CNEN) used for its marine radioecological campaigns—finally arrived in La Maddalena. Everybody expected to receive an update from the team of CNEN and Istituto Superiore di Sanità (ISS) radioecologists, who instead delivered a laconic "no comment" and invited newspaper correspondents to submit their questions to the press offices of their agencies.[1] The reticence of both US and Italian authorities contributed to undermining the credibility of the radiosurveillance program in the eyes of environmental activists like Gruppo Ambiente and local commentators as well.[2] For example, Gian Carlo Tusceri, the correspondent for *La Nuova Sardegna*, wrote: "It is clear that within CNEN there are two groups, 'the workers' (those who, like [Aldo] Brondi, are here to collect data for the radioecological survey) and 'the bosses' (the elements of a pyramid that reaches the highest levels of political authority where science ceases to be science and becomes subject to considerations guided by the reason of state). . . . [I]n this complicated situation the public will never know how things really are, at least from the voice of the bosses."[3] As I will document below, Tusceri's impression, albeit partial, was quite true.

INFORMATION MANAGEMENT AND THE VISIBILITY OF RISK

The belated response of Italian institutions to the "problem of La Maddalena" was, in their admission, a reaction to the public campaign that a variegated antibase front organized between 1972 and 1974. The alarming tones of judge Gianfranco Amendola's articles and the growing anxieties among La Maddalena's administrators pushed the Italian Ministry of Health to take the initiative. The first step consisted in asking the Ministry of Defense and the Ministry of Foreign Affairs to intercede with US authorities to obtain information about the consequences of potential accidents on submarines' reactors, the specification of maintenance activities performed by the US Navy tender, a list of operations that submarines conducted when in port and during navigation in the archipelago, and data of all radioecological surveys that the US Navy did in the harbor until then. In addition—maybe provocatively—the Ministry of Health proposed that the US Navy finance the installation of a network of continuous monitoring stations around the base.[4] The US Department of State at first did not receive these requests favorably and rebutted the argument that La Maddalena's base should be treated like an inland nuclear plant:

> If the quote domestic and international criteria unquote being referred to in this question are what is normally used in determining acceptability of locating a land-based reactor, such criteria are obviously not applicable to mobile reactors. In no country that has nuclear ships are such criteria applied because there are many differences (e.g., ship's reactor is mobile and can be removed from the site; when it is in port it does not produce anywhere near full power in contrast to a land-based reactor). As to the kinds of investigations and surveys, GOI [Government of Italy] may state that on [the] US side everything that would have been done to establish nuclear submarine support facility in CONUS [the continental United States] was also done at La Maddalena. This includes preoperational and periodic environmental monitoring surveys, results of which have been given to GOI. Latter have shown no increase above natural background levels in La Maddalena harbor environment as a result of US naval nuclear-powered ship operations. (2) There are no "legal norms and procedures" which require safety analysis of warships, regardless of their means of propulsion. Further such a safety analysis would require access to US military security information which the US is constrained by law from releasing to other governments. Such information has not been released by US to any other government in conjunction with port entry. It has not been necessary because GOI has received official USG [US Government] assurances that reactor safety aspects of design, crew

training and operating procedures of nuclear propulsion plants of United States nuclear-powered warships are reviewed by United States Atomic Energy Commission and the statutory advisory committee on reactor safeguards, and are as defined in official approved manuals.[5]

In sum, while characterizing the demands of the Italian Ministry of Health as technically groundless and politically unrealistic, US authorities restated that all safety measures were already in place and their assurance should be accepted as a certificate of guarantee. The Department of State explicitly reminded the US Embassy Office in Rome that the Italian government already possessed the radiometric data for La Maddalena. This seems plausible and raises an important question: who, within the Italian government, had access to the data and why was the Ministry of Health not aware of them?

It's worth remembering that in the United States, at least since the 1970s, the Environmental Protection Agency and other civilian authorities oversaw the radioecological and radioprotection protocols of the US Navy, which in turn issued annual reports with environmental monitoring results from all naval nuclear facilities at home and overseas. La Maddalena was not an exception.[6] A document dated 1990 and other reports that I found in the Municipal Archive of La Maddalena indicate that the US Navy had been taking ecological samples around the base since June 1972—that is, months before the arrival of the first nuclear submarines in the archipelago: "A survey of the harbor near La Maddalena was conducted by the U.S. Navy in June 1972 prior to the arrival of the Naval nuclear powered ships and reported in U.S. Navy Report NT-72-12 dated November 1972. Environmental monitoring by the U.S. Navy in La Maddalena for the period of June to December 1972 and calendar years 1973 through 1988 is discussed in seventeen previous U.S. Navy reports."[7] Radiometric results excluded the presence of radiocontamination due to submarines' operations. Thus, one could ask: why was the existence of this reassuring data never discussed in public? The answer is twofold. First, the US administration wished to avoid "an undesirable precedent. Preparation of meaningful reports from sampling results is difficult and expensive. Requirement or preparation of quarterly as opposed to annual reports would overtax US capabilities, particularly if other countries heard we were providing quarterly reports to Italy and requested similar reports for their ports. We can, of course, reiterate commitment made previously to Italian [representatives] that we will advise them asap should any of our monitoring data indicate that a significant increase in La Maddalena harbor background radioactivity has occurred."[8]

In sum, US authorities were not willing to supply their environmental reports on a regular basis, but they also wanted to avoid the allegation of infringing upon the independence of Italian state agencies by passing on data that were not produced by neutral experts. US officials used American environmental reports when they needed to reassure their allies that all radiological safety measures were in place but understood that public disclosure of data supply could expose the Italian government to accusations of collusion with the US Navy. Secondly, while US Navy environmental reports were probably available to deputed sectors of the Italian government—such as the Ministry of Defense—the latter, through their military expert agencies (like Centro Applicazioni Militari Energia Nucleare, CAMEN), adopted a strategy of reticence toward the public. A possible explanation resides in the policy of concealment regarding nuclear military applications that the Italian authorities implemented beginning in the 1950s. This coincided with the US nuclear sharing proposal and the intensification of US-Italy defense integration through the establishment of the Southern European Task Force (SETAF). While Democrazia Cristiana leaders and military elites pursued collaboration with the United States as a way to obtain, albeit indirectly, nuclear status, they explicitly requested to downplay, omit, and sometimes erase the "nuclear" component of bilateral and NATO agreements to avoid negative public reactions.[9] As William Arkin and Richard Fieldhouse have amply documented, "the standard device to avoid dealing with such dissatisfaction among allies is to neither confirm nor deny the presence of nuclear weapons" in foreign territories.[10]

In addition to the impossibility of conducting a preliminary environmental survey of the archipelago prior to the US Navy arrival, and to the reticence of Italian military agencies concerning US radioecological data, CNEN and ISS personnel had to deal with restricted information about the nuclear reactors propelling the US submarines. This was considered classified material to which only the Italian Ministry of Defense and CAMEN would have (indirect) access.[11] Admitting that the situation at La Maddalena presented unprecedented challenges to their work, in the summer of 1974 the Ministry of Health sent a note to ISS and CNEN scientists reassuring them that "a request was forwarded to the Ministry of Defense in order to acquire all the data and useful information that it already possesses or will collect about the technical characteristics of the submarines' reactors and their modalities of operation."[12]

The transmission of technical data, however, never took place. Both the Italian Ministry of Defense and the Ministry of Foreign Affairs tried to dilate the time of response as much as they could in the hope that the pressure from the Ministry of Health would diminish eventually. The US embassy in Rome received constant updates from Italian diplomatic officials. In a cable dated July 15, 1974, from US ambassador John Volpe to the State Department and to the Command of the 8th Submarine Group in Naples, the Rome office made clear that

> 1. [. . .] According to [Eric] Da Rin, Health Ministry had been pressing Italian Ministry of Defense for past several months. [. . .]
>
> 2. Da Rin said that the Ministry of defense had attempted to turn off the Ministry of Health, but its delaying tactics proved unsuccessful and [the Ministry of Defense] decided to refer the entire problem to the Ministry of Foreign Affairs for coordination with us.
>
> 3. EMBOFF [embassy office] reminded Da Rin of the discussions they had had during the last two years on these aspects of NPW [nuclear-powered warships] operations and the fact that US is severely constrained by the security requirements for technical information on nuclear powered warships. [. . .] Da Rin acknowledged all this, but observed that, given the Ministry of Health's general public health mandate for La Maddalena, the Foreign Office had no choice but to pass the request on to us. He said Health's requests were "maximum" in nature and they would settle for a lot less. He asked that we be as forthcoming as possible. EMBOFF emphasized he could give no reassurances on any of the requests and, indeed, was very pessimistic about what we could do. He asked Da Rin to send a memorandum to the embassy outlining the requests. On July 13, director general for political affairs Ducci made same point as Da Rin, i.e., we know there are limits as to what you can provide but do the best you can to make it look good.
>
> 4. Comment: We fully appreciate how these requests will be received in Washington. Nevertheless, we believe we must do that we can to assist the Foreign Office and Defense in its dealings with the Ministry of Health. We would hope to be able to provide unclassified technical material as is in any way relevant. (Sheer quantity would not be unhelpful in this exercise).[13]

Two months later, Volpe reiterated his view that nonclassified material, including US Navy environmental monitoring reports, could be made available "to help MFA [Ministry of Foreign Affairs] and [Ministry of Defense] keep the Ministry of Health in line.[14] US diplomats used a variety of strategies to deflect requests they could not satisfy. At times they were adamant that classified information could not be shared, without exceptions, but were willing to provide unclassified data to help Italian officials shield the US Navy installation from domestic "politically inspired attacks."[15] US diplomatic personnel did not limit themselves to seeking collaboration with trusted Italian military officers, diplomats, and politicians. Their communication strategy to rebuff allegations of lack of emergency plans and proper safety measures in La Maddalena involved organizing ad hoc press conferences, providing data to complacent journalists who could write positive stories on the US Navy base, asking sympathetic personnel within expert agencies to pass on information about the results of (yet to be published) environmental surveys, or influencing decisions about the content and final publication of official radioecological reports. Events and news related to the US Navy presence in Sardinia—such as the conference that took place in La Maddalena on February 24, 1975, or the initiative of Amendola's Gruppo Ambiente in September of the same year—were closely monitored and countermeasures evaluated in view of their potential negative influence on public opinion. On March 31, 1975, for example, US secretary of state Henry Kissinger wrote:

> 1. We are concerned with substance of draft CNEN article as contained in reftel particularly because such an article will appear to be official views of the Italian agency that is the recognized authority in radiological matters. If published, article will assist those elements attempting to show US NPW represent an unacceptable health hazard. Article clearly does not put radiological aspects of US NPW operations in perspective and leaves overall impression that these operations could result in a growing public health hazard of undefined [pro]portions. Accordingly, embassy should make presentation to CNEN in attempt to have publication of the article cancelled. Since we find the article objectionable in virtually all aspects, there [are] no "minor changes" we can suggest that would adequately correct it. Further, if article is published as currently drafted, we want to be dissociated with it, as to be in position to publicly refute the distorted implications if necessary.
> . . .
> 4. Should CNEN be unwilling to cancel or completely revise the article, at its discretion, embassy may wish to inform appropriate Italian MFA officials of concern proposed article could have on La Maddalena basing.[16]

In his missive, Kissinger expressed a certain degree of confidence that within CNEN somebody could be willing and able to inhibit or substantially neutralize the publication of a negative (for the US Navy) report upon US officials' request. In this specific case the pressure of the US embassy in Rome seemed to be successful, as confirmed by Volpe in August: "1. Results of CNEN radiological [monitoring] at La Maddalena were published in the June 1975 issue of house organ Notiziario (copies pouched). The report on the February 24 ecology conference did not appear."[17] While the identity of those in the Italian nuclear agency who worked against the publication of the problematic article is uncertain, other telegrams reveal that Achille Albonetti, director of the external relations office of CNEN, was a frequent and friendly interlocutor of US authorities. The Italian diplomat—a fervent and active anticommunist—provided US representatives with CNEN's radiological reports at La Maddalena months before their publication and offered his views on possible options for the presentation of results and the choice of wording "to the effect that there was no scientific basis for publicly expressed worries."[18]

US diplomats and navy officials calibrated their power—intended as the ability to influence others' behavior—according to the level (local, national, international) of their interlocutors, and to the matter of concern. Secrecy, the active removal of knowledge and data sequestration, was just one component of a variegated repertoire that US and Italian officials employed to screen, select, and release information concerning nuclear safety at La Maddalena. Their strategy included supplying and exchanging data, showcasing the efficiency and safety of US military technologies and infrastructures, inviting "not hostile or prejudiced" journalists to write reassuring cover stories on La Maddalena, and monitoring the activities of the opposition to thwart their public resonance. Inside the Italian government, privileged and trusted interlocutors of the US embassy mediated with those sectors of the state that appeared less inclined to succumb to Cold War realpolitik and unwilling to take US Navy safety reassurances as valid without further (and independent) evaluations. In assisting the Italian government with countering polemical attacks and concerned voices from opposition parties and unsympathetic experts, US authorities had to navigate the endemic instability of the Italian political system and to patiently assess their limits and possibilities for suggestions and interventions behind the scenes. But what concrete effects did the information management strategy of US and Italian authorities have on ISS and CNEN experts' work in La Maddalena?

NUCLEAR SAFETY AND MILITARY SECURITY

The initiative of the Ministry of Health to direct a radioecological study of the archipelago activated bureaucratic procedures that CNEN and ISS routinely followed for the redaction of emergency plans around civilian nuclear installations. Upon the request of the minister of interior affairs, the Ministry of Defense consulted CAMEN for technical support. Given the impossibility to access classified data on US Navy submarines reactors, CAMEN sent a "synthesis" of the hypotheses adopted for the redaction of emergency plans when US nuclear vessels were stationed at Italian naval bases:

> In the nearly absolute absence of documentation for the individuation and the evaluation of the [maximum credible accident] for [military units propelled with nuclear reactors], it was deemed reasonable to assume, in analogy with merchant nuclear ships, that the reactors considered here are of the type PWR [pressurized water reactor] and that the maximum credible accident is the release of primary cooling elements with the consequent melting of the reactor core and the dispersion of fission products.
>
> As for the power of the reactor taken as baseline for the calculations, in the absence of established data, it was considered a power of 130 [megawatts thermal], which is comparable to that of reactors [installed on] bigger units and approximately doubling that of a submarine.[19]

During their research activities, Italian radioecologists had to constantly navigate the halo of secrecy surrounding US Navy nuclear reactors.[20] Consider the recollections of Eugenio Tabet, one of the ISS experts involved in the initial phase of the radiosurveillance program and member of the Commissione Tecnica: "At the beginning we faced a situation that recalled Ionesco's theatrical work. We asked CAMEN to give us the technical details of the US submarines' reactors, but their response was never direct and clear. To get started we had to assume that the reactors propelling the submarines were pressurized water reactors with a power of seventy megawatts or so. CAMEN's answer was that we were probably not too far from the truth."[21]

When designing the radiosurveillance system for La Maddalena, Italian experts had to make assumptions about crucial aspects of the US Navy submarines' operations. Based on the reactors' characteristics described by Tabet, ISS and CNEN identified some typical radionuclides that in routine discharges and in anomalous circumstances, like accidents, would be dispersed in the surrounding environment. In analogy with what would have happened with a typical pressurized water reactor plant like the one installed in 1964 at Trino Vercellese (in Piedmont, Italy), the radioactive elements of concern could be categorized as fission products, namely iodine-131 and cesium-137, or activation products, in particular cobalt-60 and manganese-54.[22] Thus, the main goal of the radiosurveillance program was to detect the presence of those elements in order to assess the environmental and sanitary effects of the US Navy presence and operations in the archipelago.

The task was complicated. Given the lack of environmental data prior to the US Navy installation in the archipelago, Tabet and colleagues were not able to calculate and identify the possible patterns of radiocontaminants' dispersion in case of an accident: direction, strength, and intensity of winds and marine currents, levels of natural background radioactivity, presence of significant bioaccumulators, and the socioeconomic characteristics of the site were all blank spots that needed to be filled. To obviate this problem, CNEN and ISS personnel decided to adopt an overdimensioned survey strategy with many sampling points. In the technical report they sent to the Minister of Health, ISS experts warned that the lack of basic information "makes the radioecological study particularly urgent in order to assess the environmental and sanitary impact of the installation as soon as possible."[23]

Before the official initiative of the Ministry of Health, CNEN had already included La Maddalena among the sites forming the national network of marine radiosurveillance. Thanks to this zealous initiative of the Continental Contamination Laboratory of Casaccia, in November 1972 CNEN's personnel began to measure the levels of radioactivity of various samples coming from the archipelago. CNEN, ISS, and CAMEN repeated their analyses every six months, although they did not apply identical sampling methods and radiometric procedures.[24] This monitoring system guaranteed only the minimum standard of radioecological surveillance.

During his intervention at the first national conference titled "Basi nucleari U.S.A., popolazione, e ambiente: il caso de La Maddalena" (US nuclear bases, local communities, and the environment: the case of La Maddalena), organized by Gruppo Ambiente in September 1975, Tabet denounced the serious deficiencies of the radiosurveillance program.[25] Not only did the lack of basic technical information regarding the US reactors force ISS and CNEN to design a very large network of sampling points; even more problematic was the scarcity of radioecological expertise and research infrastructures in Sardinia, which meant that the central radioprotection agencies had to provide technical equipment, methodological assistance, and radiometric measurements in Rome. Specimens collected in La Maddalena had to be sent to Rome, prepared, and treated in ISS and CNEN laboratories, and finally examined. The entire process took months, and the numerically limited personnel devoted to this task could not continue to supply the service ad infinitum, as clearly stated in the initial guidelines prepared for the Minister of Health in 1974.[26]

Thus, the time separating samples' collection from their final radiometric analyses was too long for the data to be significant in case of an accidental discharge of radiocontaminants. Taking samples every six months, the so-called second-level network, was, in fact, intended to trace the processes of radionuclide accumulation in the environment over an extended period, but was not designed to identify and respond effectively to an accidental discharge. For this reason, ISS and CNEN technicians

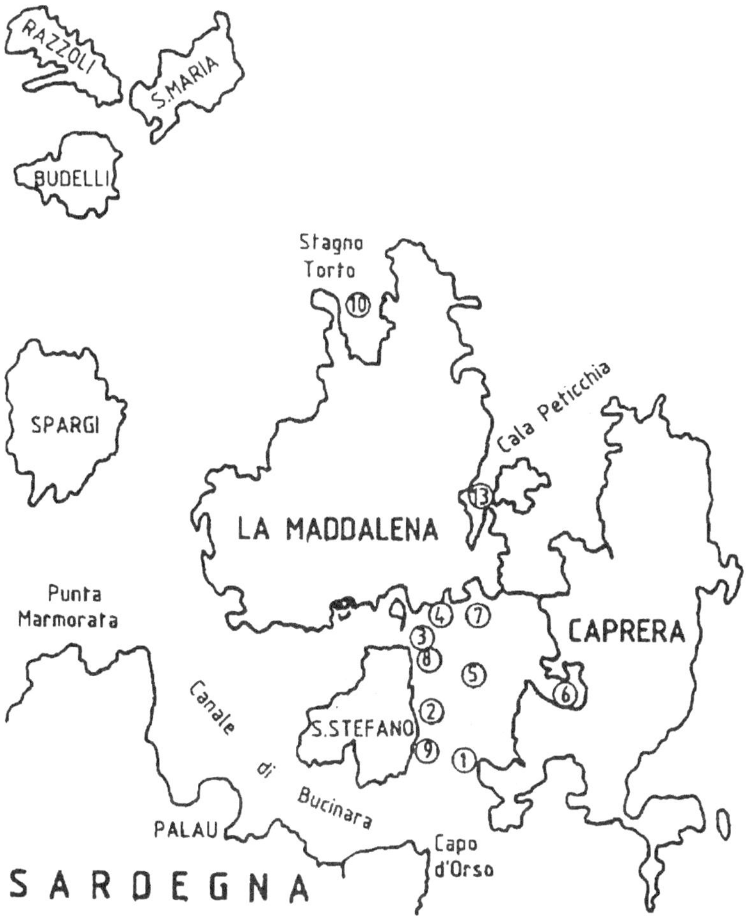

4.1. Comitato Nazionale per l'Energia Nucleare and Istituto Superiore di Sanità radiosurveillance system in La Maddalena: sampling sites. (In 1974, point 10, the so-called point zero, was added to the network). *Source:* Carlo Papucci, "Indagini radiometriche a lungo periodo," in *La salvaguardia dell'ecosistema marino costiero de La Maddalena: Vent'anni di sorveglianza radiometrica e di indagini ecologiche*, conference proceedings, Associazione Italiana di Protezione contro le Radiazioni, La Maddalena, April 13, 1991.

designed a so-called first-level network that would allow measurements for twenty-four hours a day thanks to the installation of automatic radiometric stations connected to monitors for the immediate transmission and visualization of data. In case of anomalous signals, local and regional authorities would be alerted and, if required, start emergency protocols. To complement the monitoring system, qualified personnel would conduct monthly radiometric measurements through gamma spectrometry on specimens collected and analyzed in a laboratory that would be built in La Maddalena. According to the original plan, the Region of Sardinia should complete and manage the last part of the radiosurveillance system with its personnel, but the lack of expertise and resources at the regional and provincial levels delayed the implementation of the last phase of the program for eight more years.[27]

Following the intervention of Tabet, Mario Mittempergher, then director of the Radiation Department of CNEN, lamented the state of uncertainty in which radioecologists and radioprotectionists had to work in La Maddalena. They had to adapt to the unusual situation by selecting many sampling sites and using a remote point situated on the northwestern corner of the major island (Stagno Torto) as a control case, the so-called point zero (point 10 on the map in fig. 4.1), where US submarines supposedly would not transit and no currents coming from the bay of Santo Stefano (points 2 and 9) could transport radionuclides.[28]

Tabet and Mittempergher detailed the contradictions and the obstacles that ISS and CNEN experts encountered in La Maddalena due to the special circumstances of the US Navy installation. While they tried to preserve, to different degrees, their independence from external pressures and constraints, Italian experts could not avoid taking into consideration the obvious political and technical consequences that military secrecy had on the design of the radiometric network and on their sampling activities.

In 1976, the Ministry of Interior asked the Commissione Tecnica for assistance with the elaboration of an emergency plan for the archipelago, given its role of consulting body on matters of radiological protection. On February 2, the members of the committee met to discuss the *presupposti tecnici* (base technical data) of the emergency plan for the US Navy base.[29] The presupposti tecnici include the characteristics of the nuclear reactors under examination necessary to assess the consequences of hypothetical accidents. In normal circumstances, the Italian legislation required operators of nuclear reactors to send a detailed technical report of the characteristics of the technology in use, but La Maddalena was not business as usual. "After examining the available documentation," began Maiani of the ISS,

> we can formally deduce the following: 1) The [Commissione Tecnica] has been asked to examine a problem with a very unusual procedure; 2) In particular there are no technical data available, official or not, provided by the operator (the US Navy); 3) The documentation provided by CAMEN cannot be considered a valid substitute for data that in normal circumstances the operator should provide. Furthermore, from a technical point of view, the

> documentation of CAMEN does not satisfy the requirements of art. 144 of the [Decree of the President of the Republic no. 185, 1964] for the elaboration of an emergency plan. In fact, in the document the calculations about the release [of radionuclides] after the hypothetical accident are based on technical hypotheses formulated for nuclear merchant ships. . . . This situation, nonetheless, must be considered in light of the pressing need to produce an effective emergency plan for the safety of the local community as soon as possible.[30]

Thus, despite the anomalous situation, Italian radioprotection experts decided to move forward and to take the hypotheses formulated by CAMEN—and confirmed by CNEN-Direzione Sicurezza e Protezione (DISP)'s preliminary assessment—as valid on two conditions. First, the US Navy should provide guarantees that the containment barriers around the reactors installed on the submarines could reduce the release of radionuclides as CAMEN's document assumed. Secondly, the committee members prescribed that a US Navy or Italian navy vessel be always ready to tow away a submarine within one hour after a hypothetical accident to remove and isolate the source of potential contamination and limit the exposure of the local community.[31] The Commissione Tecnica made the same recommendations in December 1977 when its members met to evaluate the emergency plan proposed by the office of the prefect of Sassari. On that occasion the committee acknowledged that the Minister of Interior Affairs sent a confidential note stating that the US Navy confirmed the hypotheses made by CAMEN; therefore, the two conditions posed in the previous meeting were satisfied.[32] In addition, the members of the committee raised the point that the emergency plan, as drafted, took into consideration only the release of radioactive contaminants after a hypothetical accident in the bay of Santo Stefano, but no hypotheses were made about accidents that could occur to submarines during navigation. In subsequent meetings between February and April 1978 the Commissione Tecnica approved the changes based on its earlier observations and exhorted responsible authorities to refer to the results of the radioecological campaigns that CNEN personnel conducted in the archipelago between 1975 and 1977.[33] The entire radiosurveillance system was based on a fragile compromise between security needs (military secrecy of the US Navy) and safety protocols.

THE DUAL SYSTEM OF RADIOSURVEILLANCE IN ITALY

The Italian legislation on the "peaceful uses of nuclear energy" and radioprotection (DPR 185, 1964) did not contain explicit norms for the regulation of the military applications of nuclear technology. Italy has never had concrete possibilities to acquire or develop nuclear technologies in the military sector, apart from its participation in the short-lived and inconclusive plan for the construction of a "European" atomic bomb with France and Germany.[34] Nevertheless, similarly to other European countries, in Italy CNEN and CAMEN officially managed a radiosurveillance system to monitor the effects of nuclear fallout from military atmospheric experiments and industrial emissions. The role of military institutions in this field was tacitly acknowledged but never formally regulated. For example, the Italian legislation lacked general norms about safety measures at military nuclear ports—that is, those ports where the anchorage and the traffic of nuclear ships were allowed.[35] Italian authorities prepared ad hoc emergency plans for specific situations, like the arrival of the US nuclear ship *Savannah*, in analogy with the prescriptions followed for inland nuclear plants.[36] This was not the case for military nuclear ports. In Augusta (Sicily), La Spezia (Liguria), and Gaeta (Latium)—where the circulation and anchorage of US Navy nuclear ships was quite frequent—it would have been impractical and unrealistic to follow routine authorization procedures adopted for civilian nuclear ships.

The effectiveness of US military strategies depended crucially on secrecy and time. Consequently, the Italian military authorities had minimal information about the movement of US military nuclear vessels and it was related to access and mooring authorizations. In such circumstances the Italian navy commands were required to share the information with the prefects—in charge of the emergency plans at the provincial level.[37] The personnel of civilian radioprotection agencies were not allowed to enter the security zone established around US nuclear ships moored at Italian ports. Only CAMEN and/or Italian navy specialized personnel could take radiometric measurements around the nuclear vessels.[38] The presence of US forces in Italy was regulated through a secret bilateral infrastructure agreement signed by the Italian and the US government in 1954, and further amendments related to site-specific arrangements. It is likely that the use of Italian military ports became part of the agreement with later integrations, through a *una tantum* authorization to conduct nuclear ships in Italian territorial waters, as indicated in a telegram exchange between the US State Department and the US embassy in Rome in the spring of 1974: "The United States Government has also certified that all safety precautions and procedures followed in connection with operations in United States ports will be strictly observed in foreign ports. *The GOI [Government of Italy] has thus followed an approach consistent with other countries in permitting nuclear powered warships to visit its ports*. In addition, the 19-year accident-free record of US nuclear powered warships is demonstrable proof that this certification is valid."[39]

For this reason, CAMEN included military ports in the national radiosurveillance network and conducted periodic radiometric measurements that guaranteed the application of basic radioprotection standards without interfering with US military strategies.[40] This arrangement, however, was neither formally stated nor publicly acknowledged. Exclusive military authority over radioprotection and radiosurveillance at nuclear military ports was the result of the state of anomie in which Italian legislation left the military applications of nuclear technology. Only in 1979, based on the experience acquired with the case of La Maddalena, the Italian government, in consultation with CNEN, decided to fill this normative gap.[41] Thus, La Maddalena was a point of reference for the extension of nuclear safety norms to all nuclear ports and military nuclear ships.

The new regulation, however, did not substantially alter the technopolitical compromise between military security and public safety in place in La Maddalena (and presumably in the rest of Italy). Rather, the evidence I gathered suggests that even after the start of the radiosurveillance program in the archipelago, CAMEN (and later, its successor Centro Interforze Studi per le Applicazioni Militari, CISAM) maintained a degree of control over the radiometric system, independently from civilian agencies, by installing its own radiometric network, with monitoring stations directly managed by its personnel and by the Italian navy. A de facto separation between civilian and military spheres in radiosurveillance programs structured what I call Italy's dual nuclear regulatory regime. The presence of the dual system of radiosurveillance is yet another element of the technopolitical compromise between military security and safety protocols that Italy assumed as a US ally during the Cold War.

For example, after the accident of the submarine *Ray* in September 1977, the Italian minister of defense declared that the Italian navy monitored the level of radioactivity in La Maddalena with its radiometric system.[42] Given the ambiguity with which the actors involved in the environmental monitoring of La Maddalena often described it, assessing what military authorities exactly meant by "radiosurveillance system" (monthly sample analyses—first level; semestral sample analyses—second level; or continuous radiometric monitoring—zero level?) requires a close and contextual reading of the available documentary material. ISS and CNEN documents confirm that CAMEN collaborated with the two civilian expert institutions since 1972, when they began conducting twice-yearly sample analyses of specimens from the archipelago. But did the Italian navy install its own continuous monitoring system independently from civilian expert authorities?

I could find only indirect, and sometimes contradictory, traces of its existence. In one report issued in June 1980 Gian Nicola Cabizza, the first director of the local laboratory of La Maddalena, mentioned that the false alarms signaled by the aerial monitoring station of La Maddalena were also appearing on the monitors of the Italian navy.[43] Another report redacted by the Marisardegna Command of La Maddalena dated September 28, 1989, detailed the results of a meeting between the representatives of the Italian navy, Centro Ricerche e Studi Applicazioni Militari (CRESAM, formerly CAMEN), ENEA (formerly CNEN), and the directors of the local laboratory Floriana Manca and Nicolino (Nicolò) Erre. The object of the discussion was the "civilian monitoring network" of La Maddalena and the intercomparison of monitoring results between it and the "homologous military network."[44] In 2004, during the audit organized by the Environmental Committee of the Italian Senate after the *Hartford* accident, Armando Benedetti, a technician of Centro Interforze Studi per le Applicazioni Militari (CISAM, Interforce Center for the Study of Military Applications, previously CAMEN), explained that the Italian navy controlled a network of three monitoring stations, designed and managed by military authorities, but he did not specify when they were installed. He only mentioned that one of the stations was "positioned directly on the pier where the submarines are moored."[45] More recently I received confirmation from a former Italian navy officer that the monitoring system of the Italian navy existed and was activated between September 1986 and September 1988. According to the officer, the navy decided to put its own radiosurveillance stations in place because it did not want to deal with the uncertainties and the bureaucratic complexities of the Province of Sassari. He remembers that CISAM personnel went periodically to La Maddalena for maintenance operations: "They usually asked our workers [of the arsenal] for assistance and for technical equipment. . . . I can tell you that one monitoring station was certainly installed at the US Navy pier in Santo Stefano, and another one in Stagnali, on the island of Caprera, just in front of the US base. I do not know if there were other stations, though."[46]

The dates suggested by the Italian navy officer correspond with those I found in other documents. In November 1985 the Office of Environmental Affairs of the Province of Sassari sent a fifteen-page report to the president of the province with a detailed description of the state of the art of the radiosurveillance service of La Maddalena: "The radiometric stations [purchased in 1979] delivered to the laboratory of La Maddalena in 1982 have not been installed yet due to technical problems (delays in the connections with electric and phone cables) and due to the veto of military authorities on their installation near the US Navy pier, which the experts deem crucial for the effectiveness of the radiosurveillance system."[47]

A year later, the radiometric stations in the proximity of the US Navy base were still inactive. During a meeting held in La Maddalena in November 1986, twenty-six participants, representing military, civilian, and expert institutions from the local to the governmental level, decided to solve once and for all the "technical problems" impinging on the activation of the monitoring system.[48] The report of the event, titled *Rete di monitoraggio civile Arcipelago di La Maddalena* (Civilian monitoring network of the archipelago of La Maddalena), implicitly indicates a distinction from noncivilian monitoring systems.[49] Indeed, as the document clarifies, military authorities gave civilian agencies permission to install three monitoring stations near the US Navy pier (but much further away from the original points indicated by CNEN and ISS experts) with the condition that military personnel maintain exclusive control over them. Thus, the personnel of CISAM and the Italian navy obtained their own radiosurveillance system (of three stations), nested inside the larger radiometric network managed by the Province of Sassari.

Among ISS and CNEN experts the impression was widespread that the unofficial duplication of radiosurveillance systems was strongly upheld by military authorities. On November 1977, after the USS *Ray* accident, military and civilian authorities organized the umpteenth conference to discuss the delays and problems in the implementation of the radiosurveillance program.[50] Carlo Papucci, who attended the meeting, wrote a report for CGIL-Ricerca (Confederazione Generale Italiana dei Lavoratori-Ricerca) in which he lamented the tendency of military authorities to keep an exclusive control over military nuclear affairs: "the emergency plan has been made by the [Ministry of] Defense, the radiosurveillance system is managed by the military. ISS and CNEN have been working for two years only to cover up the problems during the difficult times"—that is, controversial debates during which local institutions and the public needed reassurance about the efficacy of the radioprotection program.[51] Papucci concluded that what happened in La Maddalena was not new (because it happened elsewhere), but it was in open contradiction with official legislation. This was even more evident for radioprotection programs at nuclear ports, where military authorities had a real conflict of interest as they were simultaneously the controller and the controlled: "This raises serious doubts as to whether the protection of civilians is really among their [military authorities'] priorities."[52] While CNEN maintained its authority over the radiosurveillance of civilian nuclear ships approaching and mooring at Italian ports, for Papucci and colleagues nuclear-powered military ships continued to be off limits.[53] This did not exclude collaboration between the personnel of CAMEN/CISAM and those of ISS and CNEN. In fact, even in La Maddalena the exchange of information during data comparisons, sampling, and specimen collection was frequent. The collaboration, though, was limited to specific areas of the research. As Papucci suggested, the opacity of the radiosurveillance system of La Maddalena was not unprecedented. Parallel systems of radiosurveillance (even if not necessarily structured like that of La Maddalena) existed in all those situations in which military authorities perceived that national defense interests prevailed over the respect of protocols established by the Italian law.

The Italian nuclear law left the military field virtually unregulated. Civilian agencies (CNEN and ISS) could intervene in La Maddalena only by establishing an analogy with inland nuclear plants, but this was an informal, ad hoc arrangement. In Italy military agencies could appropriate an unofficial role over nuclear activities of military relevance exactly for the absence of clear general regulations of military uses of nuclear energy and of safety protocols at nuclear ports, more specifically.[54] Until 1978 the Italian legislation did not assign CAMEN any explicit authority in the field of radiosurveillance.[55] In fact, contrary to what happened for CNEN and ISS, CAMEN was not even instituted by a law approved by the parliament.[56] The center started its activity around the mid-1950s with a small number of specialized personnel working under the supervision of nuclear physicist Tito Franzini, within the naval academy of Livorno.[57] In 1961 the construction of an open pool experimental reactor (the RTS-1, more commonly known as Galileo) in the new site of San Piero a Grado (between Pisa and Livorno, in Tuscany) made the center an important incubator for research projects and experiments in the fields of material testing, radiocontamination detection and defense, and nuclear propulsion.[58] Until recently, the Ministry of Defense had exclusive regulatory authority over the activities and organization of CAMEN and apart from rare exceptions it has done so through ministerial decrees, which have "protected" the military institute from the scrutiny of the parliament and the "interferences" of civilian agencies.[59] CAMEN, thus, responded only to the Ministry of Defense and to military commands. The first parliamentary law addressing the center was passed in 1962 (without even mentioning CAMEN) to authorize the Ministry of Defense to hire external personnel with nuclear expertise.[60]

The ambiguous status of CISAM persists today, even though more recent legislation explicitly requires that military agencies and research centers conform to European Union directives on environmental radiosurveillance and radioprotection.[61] Citizens and elected representatives continue to evidence problems of transparency regarding CISAM's activities and lack of compliance with national regulations. For example, in 2013 members of the Italian parliament and environmentalist movements in Tuscany denounced CISAM's reticence about the decommissioning of its experimental reactor (following the 1987 referendum) and the procedures for the disposal of radioactive waste it produced.[62] In her official response to the question time act, Minister of Defense Roberta Pinotti (Partito Democratico), confirmed that other ministries "have not been involved in the authorization process . . . because the special norms applicable to the sites of the Ministry of Defense did not require it, as stated in the decree of the Prime Minister n. 183 of June 24, 2005."[63] During the decommissioning of CISAM's reactor, civilian authorities intervened only marginally.

The legislative anomaly concerning CAMEN/CISAM is twofold. On one hand, the administrative (in the sense of extraparliamentary) nature of the Italian regulation of military nuclear activities protected the agency from public scrutiny and guaranteed a de facto compartmentalization of military and civilian nuclear sectors. On the other hand, the legislative vacuum concerning safety regulations at nuclear ports (at least until 1979) nominally left this field under the responsibility of civilian agencies, who could intervene in analogy with inland nuclear plants, but substantially allowed CAMEN/CISAM to keep its exclusive control over it. The case of La Maddalena sheds light on the existence of two parallel radiosurveillance systems in Italy, which rarely intersected: a civilian system, officially regulated and under public scrutiny, and a military system, operating without formal legislative mandate thanks to a special regulatory status supervised directly and exclusively by the Ministry of Defense and the military personnel.

La Maddalena also sheds light on an additional aspect of the technopolitical compromise between public safety and military security during the Cold War. The exceptional status of radiosurveillance at nuclear ports (until 1979) and of the military applications of nuclear technology in Italy created an unregulated space over which military authorities could exercise an exclusive control. Thanks to the lack of official acts, protocols, and parliamentary scrutiny, this invisible radiosurveillance system could work in the shadow, fulfilling its mission discreetly, in collaboration with US objectives.

There were two dimensions of the effectiveness of the shadow radiosurveillance organization. First, maintaining the dual system guaranteed the independence of military authorities from the administrative supervision of CNEN and ISS. Second, the ambiguity surrounding the shadow system helped shoulder its authority, in the sense that its existence could be mentioned in times of crisis, like after the *Ray* accident in 1977, to assuage public anxieties about the possible environmental and health consequences of unforeseen events. Within expert circles its existence was a public secret, something that everybody knew but was not able to speak about. It revealed itself at times of crisis or during emergency situations, and then withdrew again from public attention.

The information management strategy of both US officials and sectors of the Italian government interested in keeping the US Navy base of La Maddalena out of public scrutiny consisted of various tools, not just concealment and removal of existing data: supplying nonclassified information to allied Italian officials, inviting favorable reports and coverage by nonhostile journalists and newspapers, monitoring the activities of the opposition, and so on. The calibration of this complex strategy happened by taking into account the general Italian political context and more local circumstances. Military secrecy and the production of ignorance more generally is a multifaceted phenomenon involving silence, deception, duplicity, opacity, ambiguity, and the proliferation of bureaucratic apparatuses. On multiple occasions, both US and Italian officials used the tactic of delaying the time of their responses to deny radioecology experts and the Ministry of Health access to restricted data or to smokescreen operations and sensitive information that could allegedly create alarm among the public.

La Maddalena's case suggests that ad hoc regulatory and diplomatic arrangements allowed the deployment of US nuclear ships and weapons in Italy to operate outside the supervision of civilian radioprotection agencies. Thanks to legislative gaps, the Ministry of Defense de facto maintained exclusive control in this field. The result was the tacit and informal establishment of a dual radiosurveillance system, one military and one civilian, sometimes overlapping in idiosyncratic ways. In the archipelago the existence of a radiosurveillance network directly managed by the military authorities, in addition to the civilian one, emerged only in times of crisis or accidents to reassure the local population that everything was under control. Instead of making the radiosurveillance service more efficient and reliable, the invisible military network created conflicts with the civilian monitoring system and wasteful overlappings. In addition, the radiosurveillance network controlled by civilian authorities was always limited by the vetoes of military authorities, who for security reasons precluded the installation of monitoring stations in the proximity of the US Navy base.

The shadow radiosurveillance network that the Italian navy and CAMEN built in La Maddalena exemplifies the duplicity that the Italian state, whose logic resided in the possibility of remaining hidden (and therefore autonomous), but capable of being brought into the open in times of crisis and distress to assuage public anxieties.

CHAPTER 5

Italian Bureaucracy and the Production of Ignorance

The Region of Sardinia was nominally responsible for the radiosurveillance program of La Maddalena, but for reasons of geographic proximity and political opportunity, regional authorities decided to delegate the management and supervision of the service to the Province of Sassari.[1] As discussed in chapter 2, in 1974 the political campaign that judge Gianfranco Amendola and various antibase groups carried out and broadcast in local and national media provoked heated debates over the risks of radiocontamination in northern Sardinia. In June of the same year, the Ministry of Health requested that experts from Istituto Superiore di Sanità (ISS) and Comitato Nazionale per l'Energia Nucleare (CNEN) design a radiosurveillance program for La Maddalena.[2]

On May 28, 1977, the City Council of La Maddalena petitioned the Region of Sardinia to build a radiosurveillance laboratory in the archipelago. Through a fund made available by the Ministry of Health, the Province of Sassari would cover the costs of the new building and the purchase of technical equipment. In addition, the province would be responsible for hiring the laboratory personnel.[3] A year earlier, the provincial council had approved a detailed draft of the budget and of the contract for the installation of an air monitoring station in partial

Table 5.1. The organization of the radiosurveillance system in the archipelago of La Maddalena

Level	Administration	Agencies	Activities	Dates of implementation
National	Ministry of Public Health	**CNEN-ENEA** Division of Marine Radiosurveillance Division of Continental Radiosurveillance Radioprotection Division (DISP) **ISS** Radiation Laboratory **CNR** Marine Division	Preliminary radioecological campaigns (CNEN-ENEA Radiosurveillance Divisions and CNR Marine Division) Design of radiosurveillance system (DISP and other divisions) First level (monthly) Second level (biannual) Zero level (continuous) Design of Emergency Plan (DISP)	1975–1977 1975–1979 Starts 1978 Starts 1972 Starts 1986 1975–1979
Regional	Assessor of Public Health	**Regional Health System** - Radioprotection division	Administration of radiosurveillance system - Technical advisory board (only administrative, but no direct involvement with radiosurveillance management until 1990)	1975 and thereafter
Local	Provincial administration	**Laboratory of Hygiene and Prophylaxis Province of Sassari**	Management of local radiosurveillance (local laboratory and personnel) First level (only specimen collection) First level (Geiger counters and aerial monitoring station) Zero level (continuous radiosurveillance monitoring system)	1979 1975 1978 1986
Local	Prefecture	**Prefecture of Sassari**	Public safety - Approval and implementation of emergency plan	1979 and further revisions

compliance with the radiosurveillance plan.[4] Between 1975 and 1977 the Province of Sassari—delegated by the Region of Sardinia's department of health and environmental protection—and a group of experts from CNEN and ISS—delegated by the Ministry of Health—managed the programming phase of the plan.

The ad hoc committee was chaired by the president of the province (or one delegate) and included two experts from the Provincial Laboratory of Hygiene and Prophylaxis, a secretary from the Province of Sassari, three experts—one ISS and the others delegated by CNEN—and a delegate of the Ministry of Health.[5] The rationale behind the diverse composition of the committee was to include all the administrative levels involved in the implementation of the plan. According to the original design, five beta/gamma aerial monitoring stations should have been installed in the following points: one in La Maddalena, two in the island of Caprera (Stagnali and Sailing Center) in the area facing the US Navy installation across the bay of Santo Stefano, one in Palau, and one on the pier of Santo Stefano, in the proximity of the US Navy tender.[6] In addition to the beta/gamma aerial stations, two fixed gamma spectrometers should have been installed on the US Navy pier of Santo Stefano for the continuous monitoring of the seawater. The stations would pump water into a spectrometric chamber with filters capable of retaining and detecting radionuclides. A cable transmitter linked to the local laboratory would send the data to a monitor, which would enable the personnel of the laboratory to visualize the results and have an almost immediate measurement of the levels of radiation. As described in chapter 4, the two monitoring stations for the seawater were installed only in 1987 because the Italian navy, probably after the veto of the US Navy, denied access to the area. Three levels of radioactivity concentration were set up to determine the corresponding levels of alert in case of accidental releases from the base of Santo Stefano. If detected by the monitoring instruments, radioactivity concentrations above the natural background would activate acoustic signals for an immediate pre-alarm procedure, and an emergency alert would be issued in case of serious accidents, for which an emergency plan would be implemented.

THE LOGIC OF EMERGENCY

The construction of the local laboratory and the beginning of its activities took more than four years (1975–1979). During this time several events contributed to exacerbate the disputes between local, regional, and national authorities over the completion and the efficiency of the radiation surveillance system. After the local elections of February 1976, for the first time the Democrazia Cristiana (DC) and the Partito Socialista Italiano (PSI) formed a political coalition with the ambition of transforming the administration of La Maddalena. The new mayor, Gavino Canopoli, was close to the Christian Democrats but contrary to his predecessors he revealed himself to be quite independent from the strong control that his party and some sectors of the local church traditionally exercised on the municipio. A lawyer and retired navy officer, Canopoli was considered a suitable person to lead La Maddalena toward a more promising economic future, which in his view should not be exclusively dependent on the Italian navy presence. Like the rest of Sardinia, the archipelago hosted many military facilities overseen by various commands and authorities who had direct jurisdiction over land and other resources. This dense military occupation of the territory made it difficult for locals to think of and develop other economic activities, like tourism, for example. In this context, the US installation not only added to the constraints of the historical military presence in the archipelago but also generated alarms for its potential health and environmental consequences.

Canopoli was the first mayor to openly put the US Navy base and all its problematic aspects at the top of his political agenda. As he stated on March 10, 1976, before the newly elected city council:

> As administrators of La Maddalena we reserve the right to express our opinion about all the decisions concerning the permanence of the US Navy installation, especially in relation to its economic, political, and social consequences and to solicit the responsible authorities to proceed without further delays with the construction and the management of the program of environmental protection. . . . The most urgent problem that we face is the lack of a radiosurveillance network and an emergency plan. . . . On this occasion and without receiving any formal communication [from the central political and expert institutions] we affirm that if the implementation of the radiosurveillance service will be delayed any further this administration will adopt all the possible measures to guarantee the safety of this community. . . . We agree with the recent position expressed by the Region of Sardinia: if the central administration of the state continues to be unresponsive to this problem, it will be necessary to adopt energic political actions to remove the US installation for nuclear submarines.[7]

The direct participation of the PSI in the local administration was a tangible sign of a real change afoot in the strategy vis-à-vis the military authorities. Canopoli's position with regard to the US Navy was also the result of a general state of agitation that mounted after February 1976, when ISS and CNEN announced the discontinuation of their monthly radiometric analyses, the so-called first level surveillance.[8] The news seemed to take everybody by surprise, but the plan submitted to the Ministry of Health in 1974 already prefigured the suspension of the service after two years in order to give local health authorities full control over the system.[9] The Region of Sardinia, which, according to the protocol, should have managed the radiosurveillance service starting in 1976, did not prepare the necessary plan and lacked organization, technical equipment, and adequate expertise. More importantly, at the origin of this situation was the unsettled question of who, among central and local institutions, should finance and oversee the radioprotection measures around the US base. Regional deputy and secretary of the Sardinian Communist

Party Mario Birardi, from La Maddalena, made clear that "the problem is not of economic nature. . . . The issue has political relevance: the central government cannot ask the Region to pay for a decision that it made unilaterally [i.e., allowing the United States to install its base in La Maddalena]. The Region of Sardinia was and still is completely opposed to the USA installation for nuclear submarines; therefore, it cannot assume responsibility for the surveillance of radioactive contamination but should pursue the objective of getting rid of the base."[10]

A few weeks after his election, Mayor Canopoli found himself in the crosshairs of this dispute, which risked leaving his community without protection. His protest was loud and appeared in major national newspapers: "This situation is absurd. . . . Frankly, we don't care whose responsibility this is, but if there are no guarantees for the safety of this community the base must go."[11] The US Navy was an interested observer of this bureaucratic quagmire and when possible, deployed its diplomatic repertoire to deflect attacks from local and national opponents of the base, influence the press, and bring the mayor to more "reasonable positions." During a conference hosted on May 10 on the *Howard Gilmore* (the US Navy tender ship anchored in Santo Stefano), and on the aircraft carrier *Saratoga* (temporarily operating in the Mediterranean), Captain Edward Burkhalter and other US Navy officers illustrated to Italian journalists the "scrupulous" systems of radiosurveillance present at all ports visited by the US nuclear vessels around the world. With more than twenty years of experience, the commodore assured the journalists and their readers, "our operations have never caused even the smallest accident."[12] In a cable to the office of US secretary of state Henry Kissinger at the end of May, US ambassador to Rome John Volpe confirmed his satisfaction with the diplomatic strategy orchestrated to obtain a favorable coverage of the US Navy operations in Sardinia: "Mission considers both goals to have been fully attained. The coverage on La Maddalena exceeded our best expectations; In the future it will be difficult for Italian journalists to sustain a charge of nuclear pollutions or secret strategic weapons at that site. Moreover, we now have on public record a volume of favorable material for possible use if and when the advent of NPSS Nimitz, California, or South Carolina begin to generate controversy in Italian media."[13]

The only negative note concerned the local press, which Volpe accused of being politically prejudiced: "For local political reasons two of the three Sardinian newspapermen wrote . . . snide or inaccurate stories, including one accusation of nuclear missiles. However, articles have not been picked up by mainland press, probably due to the small size of the papers concerned and to their remoteness."[14]

Despite these diplomatic efforts, 1976 was a complicated year for the US Navy command of La Maddalena. Distrust toward public authorities and doubts about their objectivity were often reported in the media. Three weeks after the press conference that the US embassy organized on the *Howard Gilmore*, the newspaper *L'Unione Sarda* published the story of three babies born with cranioschisis in La Maddalena only during the past seven months.[15] The families who experienced such traumatic loss timidly advanced the hypothesis that the cause of the malformations could be related to the presence of the US nuclear submarines. This episode obviously generated a debate involving expert and public authorities, from CNEN radioecologists to local doctors, microbiologists, and local administrators. If the new political climate and the controversial status of the radiosurveillance system increased the anxieties about the environmental risks of the nuclear submarines' operations, the first accident officially reported by the US Navy command of La Maddalena rekindled the sort of political debates that had been in vogue two years earlier, when Amendola launched his campaign against the base. On September 20, 1977, the nuclear submarine USS *Ray* ran aground in an unspecified area south of Cagliari during underwater operations. The news appeared on local and national papers only three days after the damaged submarine approached the US Navy pier of Santo Stefano.[16] The US Submarine Group Command in Naples revealed what happened without previous consultation with Italian authorities and after the arrival of USS *Ray* in La Maddalena denied the commander of the Italian navy in port the possibility of inspecting the damaged unit. The new US ambassador to Italy, Richard Gardner, alerted the secretary of state that

> the accident suffered by the USS *Ray*, its subsequent visit to La Maddalena and the following, uncoordinate[d] press release, created furor in the press and among Italian officialdom that has only begun to subside. The incident has given renewed impetus and focus to the claims of opponents to our presence at La Maddalena that the NPW [nuclear-powered warships] support facility constitutes a danger to the citizens of the archipelago. The reaction to the incident has placed in jeopardy both the Italian emergency nuclear disaster plan (present lack of which is an issue seized by environmentalists) and the completion of a protocol to the bilateral infrastructure agreement, painfully negotiated over many months, that would legitimize the presence of the shore support facilities and personnel.[17]

For the first time, the Maddalenini had tangible proof that an accident could happen, regardless of the meticulous measures that both the US Navy and the Italian authorities promised to take. Despite reassurance by military and health authorities that nothing serious had happened to the "nuclear parts" of the submarine, public distrust grew to unprecedented levels. The Communists and the PSI organized a petition against the permanence of the US Navy base and requested that the mayor call an extraordinary session of the city council to discuss the evident lack of progress that expert agencies and public health authorities made on the radiosurveillance program.[18]

The grounding of the USS *Ray* inevitably had serious repercussions on Canopoli's coalition. On October 18, 1977, the city council convened in the *municipio* after the reiterated requests of Partito Comunista Italiano (PCI) and PSI.[19] The two parties presented distinct but convergent motions inviting the assembly to vote against the permanence of the US Navy in the archipelago and to petition the central government for the immediate removal of the base. The Democrazia Cristiana (DC), instead, opted for a more diplomatic solution, proposing the creation of an ad hoc parliamentary committee to study the case. Despite a general agreement on the intolerable rebounding of responsibilities between central and regional institutions and the incredible delays of the radiosurveillance program, the vote on the three motions presented in the council produced an irrecuperable fracture between pro-Americans and anti-Americans.[20] The *Ray* accident shook the alliance between the local DC and the PSI, and in December the socialists decided to resign from their administrative positions.

Canopoli acknowledged the political crisis opened by the PSI and decided to offer his resignation. After three difficult months, during which the DC attempted to recuperate the political alliance with the PSI, the city council formally accepted the mayor's resignation and approved the formation of a new alliance between Christian Democrats and the Partito Social Democratico Italiano (PSDI, Social Democratic Party).[21] The different positions emerged in the assembly after the accident redefined political alliances and pushed the clock of the archipelago back to when Socialists and Communists fought together against the US Navy presence and the Christian Democrats governed La Maddalena with the support of a conservative coalition. If at the national level the PCI and DC were taking the first steps toward a "historic compromise," in La Maddalena the dispute over the US Navy presence kept the classic Cold War ideological divide alive and well.

THE LOCAL LABORATORY

By the end of 1977 the Laboratory of Hygiene and Prophylaxis of the Province of Sassari became more involved in the radiosurveillance operations in La Maddalena. With funds from the Ministry of Health, it acquired basic instruments for the chemical treatment of the biological samples to be sent to Rome for the monthly spectrographic analysis (always conducted by ISS and CNEN). The agreement between the provincial administration of Sassari and the Ministry of Health was valid for only two years (1977–1979); after that time the radiosurveillance service would continue, pending the stipulation of a new agreement between the two institutions.

The first disbursement of funds could only cover the installation of one aerial monitoring station in the urban center of La Maddalena and the purchase of a couple of medium-intensity mobile radiometers. Thus, the first monitoring system put in place in the archipelago corresponded only partially to the original plan.[22] At the end of 1978 the province issued a call for applications to select the personnel for the laboratory of La Maddalena, which started working in 1979 under the direction of the Provincial Laboratory of Hygiene and Prophylaxis. In late December, after stipulating a new contract with the Ministry of Health, the Province of Sassari approved the budget for the second phase of the radiosurveillance program, but more time went by before it became operative.[23] After almost eight years since the establishment of US base, La Maddalena was formally radiosurveilled, but one year later funding problems, bureaucratic complications, and political idiosyncrasies put the entire plan at risk.

A ZEALOUS DIRECTOR

In 1978 Gian Nicola Cabizza was teaching mathematics and physics in high schools in Sassari. He had recently graduated with an experimental thesis on the spectroscopy of solid matter. One day he came across an advertisement published in Sardinian newspapers saying that the Province of Sassari was hiring technical personnel for the new radiosurveillance laboratory of La Maddalena. The position was appealing to a young physicist with a temporary teaching contract. The job did not seem complicated; it required routine operations and standard laboratory procedures that would be easy for somebody with Cabizza's background to perform. The advertisement also specified that the selected personnel would be offered qualified training and assistance by CNEN and ISS.[24]

In February 1979 Cabizza signed the contract, but the situation in La Maddalena soon appeared more complicated than expected.[25] After receiving the funds from the Ministry of Health, the Province of Sassari built the local lab and equipped it with a monitor linked to a beta/gamma filtration chamber and a couple of portable gamma spectrometers that Cabizza and the electrotechnician, a Mr. Spanu, could use to measure the levels of radioactivity around the archipelago. It was an itinerant job. Spectrometric surveys were to be conducted, at least in principle, from a truck and a motor-powered boat purchased by the province with the rest of the equipment. In the event, the boat was inoperable and Cabizza had to request the assistance of the Italian navy for the sampling procedures at sea. The young physicist started working in February 1979, only after testing and calibrating the instruments, which took several weeks. Cabizza's early reports suggest that the Province of Sassari and the Provincial Laboratory of Hygiene and Prophylaxis were not able to supply him with the necessary material: the equipment was faulty, and the laboratory needed constant assistance from ISS and CNEN experts. The only continuous monitoring stations in place at this time consisted of a mobile chamber for measuring the intensity of atmospheric exposition installed on the rooftop of the laboratory and an aerial pump for the filtration of atmospheric particles for total beta spectroscopy. The portable gamma spectrometer did not work properly, "to the point that even during the first tests we had to replace certain parts. . . . Later, in collaboration with the technician Spanu, we could identify the faulty parts but could not intervene because we did not have new ones."[26]

One year and a half later, in June 1980, the activity of the laboratory was still limited "due to the technical problems with the installation of the instruments and the lack of the necessary equipment for our daily activities." The biological samples that Cabizza and colleagues collected in the archipelago for the gamma spectrometry still had to be sent to Rome, where ISS and CNEN experts could perform the analyses with the appropriate instruments. Moreover, the only working monitor in the lab was experiencing electric power oscillations, which caused faulty emergency alarms, and other instruments seemed to have frequent calibration problems.[27] Despite the pressing requests coming from La Maddalena's administration and Cabizza's intention to make his monthly reports available to the public, the provincial authorities continued to be reticent and maintain a certain reservation about divulging the results. This was probably one of the main causes of attrition between the young director, his colleagues at the provincial Laboratory of Hygiene and Prophylaxis, and the provincial public health assessor, Giovanni Desini.[28] This opacity generated a state of uncertainty around the accuracy of the analyses. Local newspapers often reported minor technical malfunctions with alarming tones and local politicians cast doubts on the efficiency of the laboratory on the basis of unspecified or anonymous sources.[29] Conscious of the crucial importance of providing information to the public, Cabizza organized guided tours with local students to explain the activity of the laboratory.[30] In his official reports to the Province of Sassari, and in more informal correspondence with ISS and CNEN experts, he often lamented the scarcity of resources and the lack of proper instrumentation, for which he suggested possible solutions.

As mentioned above, the agreement between the Ministry of Health and the Province of Sassari stated quite clearly that the funds available for the daily operations of the laboratory were limited and contingent upon the successful implementation of a biannual plan of radiosurveillance in view of the incoming installation of fixed monitoring stations around the US Navy base. The agreement would be renewed if both the Ministry of Health and the Province of Sassari verified that the laboratory achieved the expected results.

The temporary arrangement that allowed Cabizza to start his job revealed its fragility within the year, when the Ministry of Health and the Province of Sassari had problems renewing their agreement. A series of staggering bureaucratic slippages underscored the precarious position of the laboratory personnel and put the continuation of the radiosurveillance service in question. The local correspondent for *La Nuova Sardegna* described the absurdity of this delay as an example of the Italian "bureaucratic grotesque."[31]

In March 1980 the president of the Province of Sassari, Giovanni Maria Cherchi, a regional PCI leader, sent a note to the Ministry of Health asking for new funds to guarantee the continuation of the radiosurveillance service in the archipelago. After a few weeks the ministry officially responded that the central administration noticed some fiscal anomalies and did not receive any report of the radiosurveillance activities conducted by the province.[32] The ministry made clear that, in light of the current situation, it could not issue another disbursement. The province replied, explaining that an oversight of the treasury office prevented the administration from using previous funds destined to the radiosurveillance service.[33] The purchase of the electronic monitors and the rest of the equipment amounted to 395,000,000 lire (of the 400,000,000 disbursed by the ministry), which was fronted by the provincial administration. Cherchi asked the ministry to issue a new payment so that the province could guarantee the installation of the recently purchased equipment and the routine operations of the laboratory for the following two years.[34] The bureaucratic impasse persisted. The Ministry of Health maintained its position, lamenting that, in contravention of the agreement in 1978, "the provincial Laboratory of Hygiene and Prophylaxis has not sent yet any report concerning the radiosurveillance activities in La Maddalena."[35] In the meantime, CNEN experts advanced doubts about the management of the radiosurveillance system assembled by the provincial Laboratory of Hygiene and Prophylaxis providing additional motives of distrust to the already strained relations with the ministry.[36]

After further requests met no response, Cherchi attempted to break the stall by announcing publicly that the province could not guarantee the radiosurveillance of La Maddalena anymore and that the laboratory would interrupt its activities.[37] On May 30, 1980, he sent a brief letter to Cabizza and colleagues communicating the decision to suspend the activities of the local laboratory. Exonerating himself from further responsibilities, he wrote that closing the laboratory was the only thing he could do. Cherchi had even the audacity to appeal to the good conscience of the laboratory personnel: "I leave to you the decision whether or not to continue your work, without any obligation toward this administration."[38] That a well-known regional PCI member like Cherchi showed such an inflexible conduct should not come as a surprise, after all. As PCI's regional secretary Mario Birardi explained three years earlier, the Communist Party considered the central government responsible for the establishment of the US nuclear base in La Maddalena. The PCI strategically used the skirmishes between the provincial administration and the government to demonstrate once again the problematic consequences of that decision. The goal, in sum, was to remove the US Navy base, not its slow incorporation into the archipelago's life. One may speculate that, paradoxically, in the long run the PCI had more to gain from leaving the radioprotection problem unresolved so that the choice of the Italian government would reveal itself to be unsustainable.[39]

Thus, less than two years since he took charge of the local radiosurveillance service, Cabizza found himself in the middle of this tragicomic quarrel between central and peripheral authorities, with little certainty about his position and the future of the laboratory. Cabizza and his colleagues decided to keep the laboratory open, hopeful that the Ministry of Health, the Region of Sardinia, and the Province of Sassari would soon find a reasonable solution.[40] But the local elections of June 8 put the provincial administration on standby until the new president and staff took office. Three months passed by without results. At the end of August an exasperated Cabizza decided to hold a press conference to denounce the ineptitude of the Province of Sassari, where he proposed that the Region of Sardinia assume direct control of the radiosurveillance system and that the local laboratory expand its activities by becoming a center for the environmental monitoring of northern Sardinia.[41]

Without pay since early May and left without a clue about the destiny of his laboratory, at the end of the summer Cabizza returned to his teaching position in Sassari but kept traveling to La Maddalena twice a week to check on the laboratory activities. Instead of obtaining the effect he desired Cabizza's words triggered the hostile reaction of local authorities who perceived him as a "troublemaker" whom they were gladly getting rid of. The June elections changed the composition of the provincial administration, now led by a coalition between DC and PSDI.

The provincial assembly elected a new president: Giovanni Desini, the former assessor of public health and main target of Cabizza's critiques over the past two years. As soon as he took office, Desini made clear that Cabizza's zealous initiatives were not in line with the objectives of the new administration and used a pretext to get rid of the unwelcome physicist.

When he realized that he could not expect a positive response from the Province of Sassari, in a last, desperate attempt to mobilize the local community and to garner the attention of national expert agencies, Cabizza sent a letter of resignation (mostly symbolic, given that his contract had already ended in May) to the Region of Sardinia, ISS, CNEN, and the Ministry of Health. His decision was motivated, he wrote, with his "profound disagreement with the ways in which the province [had] mismanaged the funds for the radiosurveillance program, . . . by not providing the necessary equipment for the correct functioning of the laboratory, [and by] not divulging the results of the analyses to the local authorities."[42] His appeal to change the radiosurveillance system remained inconsequential and came too late. The Province of Sassari probably had already decided to hire someone else in Cabizza's place (maybe less critical of his/her employer). A few days after Cabizza's symbolic resignation, the new provincial assessor of public health, former mayor of La Maddalena Giuseppe Deligia, declared, "Not only is the laboratory not going to close but, according to our program, it will be improved and developed, and the first step in this direction is to have full-time employees instead of personnel who work at the laboratory in their spare time while holding teaching positions. . . . The laboratory will have to become fully independent and capable of doing all the analyses, instead of limiting itself to taking samples and calibrating the instruments as it has so far."[43]

With a letter to the director of the school district of Sassari dated December 12, 1980, the new assembly president Desini requested information about the teaching positions held by Cabizza from February 1979 (when he started his job as director of the laboratory) to the present.[44] Once the funds that the Ministry of Health refused to disburse for almost a year finally became available in late November, the provincial administration used Cabizza's presumed unavailability as an excuse to deny him the renewal of his contract, asserting that his teaching activities were incompatible with the directorship of the laboratory. By the end of November 1980, the Province of Sassari, with the funds from the Ministry of Health, had already hired Angelo Parodo as the new director of the laboratory.[45]

POLITICAL STABILIZATION AND ROUTINE MALFUNCTIONS

A health physicist and faculty member of the University of Sassari, Professor Parodo was a *barone*, a term used in the Italian academy to denote someone with power and influence. Thanks to the temporary contract signed with the provincial administration, Parodo asked to have two of his collaborators, Floriana Manca and Nicolino (Nicolò) Erre, temporarily employed as assistant researchers in the laboratory of La Maddalena.[46] This time Desini and Deligia showed a degree of flexibility they did not use with Cabizza. Both Manca and Erre, recent physics graduates, collaborated with Parodo at the University of Sassari. Like the former director of the lab, they held teaching positions in high schools, where they continued to work even after their collaboration with the province started. Due to their jobs and academic commitments, they did not reside in La Maddalena. From the documentation concerning their paychecks, it is evident that Parodo's assistants received periodic reimbursements for their "research trips."[47] On average they went to La Maddalena six times every two months. The province probably did not deem their daily presence at the laboratory necessary because in the meantime other personnel were hired for routine operations and maintenance.

In a letter sent to Carlo Papucci at the CNEN Laboratory for the Study of Marine Contamination, Cabizza bitterly commented that it was curious that "the Province was not able to renew the contract of one physicist [Cabizza] and now it hires three! Manca and Erre also applied for the job [when Cabizza got it in 1979] and arrived second and third, behind me. Parodo instead never applied. All of them participated in the environmental study of the Province, amply criticized by national experts."[48] In reality, the province did not hire Manca and Erre. The temporary contract (one year of duration, renewable) signed with Parodo explicitly mentioned them as research assistants without direct remuneration. From this point of view, the new contract presented the same elements of uncertainty and precariousness that put an end to Cabizza's experience. This time, however, the relationship between the provincial administration and the laboratory personnel seemed much improved—at least, this is how Manca and Erre perceived it.

According to Erre, working in La Maddalena was difficult because the archipelago was isolated from the rest of Sardinia. Back then, Sassari was the nearest city with research centers, a university, and qualified personnel who could supervise the activities of the laboratory.[49] Reaching Palau by car took up to two hours and then the ferryboat to La Maddalena could take from twenty to forty-five more minutes, depending on the frequency of the runs. In sum, the job in the archipelago was not appealing, but

Erre and Manca continued to collaborate with Parodo until 1987, when the old professor decided to not renew his contract. That year, Manca became the primary person responsible of the radiosurveillance service and the province temporarily hired five more people for routine maintenance of the laboratory.[50] By the end of 1986 the installation of the electronic monitoring stations purchased by the province in 1979 (and idly stored outside the laboratory since 1981) was finally completed.[51] Local newspapers celebrated the "event" with relief. [52] Until 1989 the province remained directly involved in the management of the radiosurveillance program, when another reform of the Italian National Health System delegated the administration and organization of public health services to ad hoc regional institutions—the so-called Unità Sanitaria Locale (USL, Local Sanitary Units). At that point in time, the radiosurveillance laboratory fell under the direct administration of the USL of Sassari.[53] Finally, in 1993 Manca and Erre were hired as health physicists within the regional health system and their jobs at the laboratory were regulated with standard contracts for public employees.[54]

After Cabizza's experience, the radiosurveillance system of La Maddalena was not substantially improved. The new hires worked with the same "obsolete and fragile equipment" at their disposal, whereas the network of monitoring stations became operative only after the Chernobyl accident (six years later). [55] Erre recalls: "When the radioactive cloud hit Italy, we were totally unprepared. In fact, we did not detect it until a few days later, when we checked the filters of the aerial station installed on the laboratory's rooftop. If we had the electronic network in place, we would have known immediately, because the alarm would have gone off."

As mentioned earlier, the installation of the automatic monitoring stations was not easy. Complex bureaucratic procedures, endless meetings, and vetoes from Italian and US military authorities impeded the timely implementation of the radiosurveillance service. But even after its completion, the network presented numerous deficiencies.[56] First, when the electronic system was finally ready to work, the software was obsolete and required extensive updates. After buying the sophisticated equipment in 1979, the province stored it outside the laboratory without proper precautions. The deterioration of the material required substantial refitting and repair operations. Second, the calibration of the instruments was laborious; it needed to keep into account the baseline of the natural background for the stations to reliably measure significant variations in the levels of radioactivity. Finally, La Maddalena proved to be a very hostile environment for electronic devices. High temperatures in the summer made the natural radioactivity background rise significantly, and this required seasonal recalibrations. Strong winds, humidity, and the corrosive salinity of the seawater hit the monitoring stations quite hard. During the winter frequent sea storms and the violent washing of the archipelago's coasts submerged completely the water pumps (and their electronic apparatuses) of the radiometric stations installed around the US base, provoking false alarms and damages whose repair necessitated lengthy suspensions of the monitoring activities.[57] Some technical reports, issued between 1989 and the beginning of 1990, listed in detail the number and types of malfunctioning episodes registered monthly. In February 1989, for example, some of the stations failed to transmit signals for forty-four hours; in March the station installed inside the Italian military arsenal did not transmit for almost eighty hours. Other frequent malfunctions were related to failures of battery recharge systems, high temperatures, water pump failures, and electric energy interruptions (very common in the rest of La Maddalena).[58] In sum, saying that La Maddalena was monitored sounds like a very optimistic account of the radiosurveillance system in place in the archipelago. Between 1987 and 1988 local newspapers reported cases of malfunctions and denounced the disorganization of the laboratory numerous times, putting into question the reliability of radiometric data.[59] On October 26, 1988, *L'Unione Sarda* published a reportage in which journalist Alberto Testa described the widespread alarm circulating among the Maddalenini.[60] The article featured the stories of two fishermen who allegedly saw US Navy personnel working around a damaged submarine trying to contain a major leak of chemical substances: "Since then, the seafloor of that polluted area has been completely covered with a white powder that has killed all the marine vegetation, as if a bulldozer had passed over there." The two men further suggested that the episodes of birth defects that "many local families experienced" could be explained only by the presence of the nuclear submarines. According to a local family physician, it was "inconceivable that the regional health authorities had not yet investigated the causes of the extraordinary rates of cancer on this island."

For Erre, public alarms were exaggerated and unjustified, contributing to an "unfair image of the laboratory, which some Maddalenini already regarded with suspicion."[61] Very often, he explained, the whistleblowers were members of the laboratory personnel (especially locals) frustrated with their working conditions: "Most of the time they got paid late and wanted to be hired permanently by the public administration. When complaining about their working conditions with family, friends, and journalists, they did not fail to mention those episodes of malfunction, which the newspapers reported with great emphasis. This is how it worked."

The first network of monitoring stations installed in 1987 was a fragile system exposed to the merciless attacks of the natural elements and subject to the organizational fallacies of local and national institutions. But the laboratory also became a target of public outcry when the political tension over the US Navy presence escalated. One example is the battle for the regional referendum that antinuclearists, pacifists, and a large part of the left organized against the submarine base. During those moments of tension, even small malfunctions, or bureaucratic delays, reported with emphasis by the frustrated personnel of the lab, became sources of alarming rumors. A false alarm due to a connection failure, for example, could be easily interpreted as a radioactivity increase. When the directors of the lab tried to explain that nothing was wrong, local journalists replied with understandable skepticism. In turn, this reinforced experts' prejudice against the public. As Erre recalls: "I remember one time when Professor Parodo sent a letter to *La Nuova Sardegna* trying to explain that false alarms should not be interpreted as real ones, but the journalist the next day changed his words and made up the story that the network was substantially inefficient. After that episode Parodo decided to not respond anymore because he thought it was just a waste of time."[62]

Parodo's response to the uncontrolled rumors and interpretations of the malfunctions of the local laboratory points to the clear difficulties that expert institutions had in communicating with the local community. True, sometimes journalists and interested Maddalenini alike, for various reasons, mischaracterized and dramatized even small problems, but in the past, apart from Cabizza, nobody had campaigned to inform the public about the role and the work of the laboratory. Rumors proliferated in a context in which secrecy and the innumerable contradictions that surrounded the radiosurveillance system made it difficult to understand exactly how the monitoring stations worked (or did not work properly). The laboratory directed by Manca and Erre was out of sight for most Maddalenini. The edifice was in a scarcely frequented part of the main island. Furthermore, the directors of the lab sent their monthly radiometric reports only to central authorities (Ministry of Health, CNEN, ISS, Province of Sassari, Region of Sardinia, and the municipal administration). Despite sporadic requests and repeated declarations of good intentions, local authorities never established transparent procedures for divulging the data. Even after repeated alarms, like the birth defects of 1976 and the USS *Ray* accident in 1977, municipal administrators limited themselves to denouncing delays in the installation of radiosurveillance network, but ultimately did not take concrete initiatives to change the status quo, like instituting an ad hoc commission of the city council, for example.

The systematic exclusion—sometimes self-exclusion—of local institutions from the decision-making process concerning the radiosurveillance system of La Maddalena and restricted access to radiometric data were the result of inertial bureaucratic responses in which delegation of responsibilities and forms of technocratic mentality matched and accommodated each other. The apparent lack of concern about risk, and the failure of public authorities to respond in a timely fashion to an urgent problem of public safety, created a myopic view of risk management in the archipelago and a culture of skeptical response to alerts that would have a long-lasting impact on the future structure and routine operations of the radiosurveillance program. But the fact that many long-term residents of La Maddalena did not voice their concerns openly does not mean that they were not ready to mobilize against the US base when given the opportunity.

ANTINUCLEAR ESCALATION AND THE MISSED REFERENDUM

In the early 1990s the Italian government announced the end of the national nuclear program. The decision followed the result of a referendum held in 1987, a few months after the Chernobyl disaster. Environmentalist movements, and large parts of the left (including the PCI), campaigned to abrogate three laws concerning nuclear plant siting and the activity of the public electric company Ente Nazionale Energia Elettrica (ENEL) in the nuclear sector.[63]

So far, the literature on the 1987 Italian "nuclear" referendum has been dominated by communication studies focused on the media coverage of the Chernobyl disaster and on survey data preceding and following the vote.[64] Among Italian nuclear experts and advocates of the nuclear industry the opinion is widespread that the post-Chernobyl referendum determined the phasing-out of the national nuclear program. This thesis is quite simplistic and does not consider the long-term economic, political, and organizational deficits that impinged upon the functioning of the nuclear program, as documented in the present study.[65]

Nevertheless, the governmental decision to phase out the four nuclear plants operating on the national territory inspired a wide antinuclear front in Sardinia to promote a regional referendum against the permanence of the US Navy base of La Maddalena.[66] Regional referendums—according to Sardinia's statute—could only be consultative and as such could not have any immediate legislative or administrative effects. Even so, the idea that Sardinians could express their political orientation toward the US nuclear base acted as a political catalyst: backed by the regional administration, antinuclearists, environmentalists, Communists, Sardinian autonomists, and younger generations in the Socialist

Party and the Christian Democrats formed a bipartisan alliance to promote the consultation.[67] In June 1988 the organizers commented enthusiastically on the first results of their campaign. In La Maddalena alone they collected more than a thousand signatures of the ten thousand required for the admission of the referendum.[68] At the end of the summer, the regional administration announced that the consultation would take place in December.[69] "The regional law that introduced [in Sardinia] the possibility to celebrate consultative referendums was primarily conceived for the problem of La Maddalena, which has become an issue of national and international importance. This is the peak of a battle for peace and for the protection of the environment," said one of the promoters.[70]

The referendum specifically targeted the US Navy base but quickly surged to a symbol of Sardinian identity and right to self-determination, which sectors of regional elites wanted to reassert in opposition to the history of colonial subordination to the Sabaudian state and the Italian Republic.[71] In the mid-1980s the president of the region was Mario Melis, a long-term exponent of the Partito Sardo D'Azione (Sardinian Action Party). The party was a branch of the Partito D'Azione, a small republican formation led by young intellectuals particularly active during the liberation movement against the Fascist dictatorship. Contrary to what happened in the rest of Italy, after the liberation the party maintained a significant presence (also from an electoral point of view) in Sardinia especially due to its autonomist positions (but not independentist). The election of Melis as president of Sardinia was directly supported by the PCI, with the external votes of the PSI and the Partito Repubblicano Italiano.

The referendum catapulted La Maddalena once again to the epicenter of Sardinian and national politics. The archipelago became a target of international campaigns organized by ecopacifist movements like Greenpeace.[72] The Italian government, guided by Christian Democrat Ciriaco De Mita, seemed aware of the potential political consequences of the initiative (and the spillover effect that it could have on Sardinian politics and Italian diplomacy) and decided to appeal the regional decree in the Constitutional Court. According to Rome, the consultation could not take place because a regional constituency did not have the right to vote over the status of a military installation that was legally established through a bilateral agreement.[73] The Constitutional Court case provoked a profound fracture within the Sardinian administration. Socialists and Christian Democrats aligned with their national leaders and distanced themselves from the Communists and the Partito Sardo D'Azione's proposal to fight the government in court.[74] Melis accused the socialist allies of acquiescing and subordinating to the colonial logic of the Italian government: "This is not about foreign policy. This is about the right of the Sardinians to express their will. It is . . . a battle against the restrictive interpretation of Sardinian autonomy and of the Italian constitution."[75]

The legal dispute placed the celebration of the referendum on hold until the Constitutional Court released its verdict, but the political conflict escalated. In December *L'Unione Sarda* published the results of an opinion poll commissioned by the agency Abacus. The results were striking: if the regional referendum were to be held, 68 percent of Sardinians, on average, would have voted against the permanence of the US nuclear base of La Maddalena.[76] After numerous deferrals, in early March 1989 the Constitutional Court ruled in favor of the Italian government and decided that the regional referendum could not be admitted because it violated the principle of "exclusive authority of the State over foreign affairs and national defense."[77]

By preventing Sardinians from democratically expressing their will about the presence of the US base, the sentence reaffirmed one hierarchical order of the Italian state: national interests came before the interests of a single region. This was not new to Sardinia, where the Italian state had already imposed the presence of 60 percent of its military installations and training camps, a colonialist tradition that continues today.[78] With this decision the Constitutional Court reestablished Sardinia's subordination to the exploitative and hegemonic policies of the Italian state. The decision of the court appears to be quite contradictory if one reflects on the fact that in 1987 the residents of La Maddalena voted to determine the destiny of Italy's civilian nuclear program, but in 1988 they could not have a say about living alongside a foreign base for nuclear submarines. The formalistic explanation and objection that the national referendum of 1987 was far more decisive than the one concerning the fate of the US Navy base of La Maddalena because the first had direct abrogative affects while the second was only consultative is not convincing. First, if the Italian government wanted to maintain the civilian nuclear program, even after the results of the national referendum, it could do so by simply proposing some legislative changes and integrations. For example, in 2010 the Berlusconi government launched the idea of a national nuclear renaissance by changing old and proposing new laws in parliament.[79] Second, if the referendum in Sardinia could not have any direct legislative effect, why impede its celebration? These questions are obviously rhetorical ones, considering the documental evidence discussed so far.

With the missed referendum about La Maddalena, the contradictions of Italy's relationship with nuclear power during the Cold War reached their peak. As historian Leopoldo Nuti explains, while Italy renounced the use and development of nuclear power for military purposes, its pro-NATO political elites perceived the American deployment of nuclear weapons on Italian territory as a guarantee for national defense and higher international status.[80] If Italy could not achieve military nuclear capabilities directly, it could serve, with substantial returns, as a logistical platform for the control of the Mediterranean area. La Maddalena was an integral and crucial component of Italy's strategic alliance with the United States and its status could not be challenged.

Scott Frickel and Michelle Edwards argue that ignorance is not necessarily the product of secrecy or of purposefully deviant, or bad, science.[81] Rather, scientific protocols, standards, and epistemic traditions produce areas of "undone science," which determine which risks are visible and which will remain invisible. In this chapter I discussed various examples of how the production of ignorance about the environmental conditions of the archipelago was intrinsic in the radiosurveillance program commissioned and implemented after 1974. I did so by looking simultaneously at the institutional arrangements of the Italian radiosurveillance system and the concrete process of its implementation in the archipelago, including the effects that limited access to information and frequent technical and bureaucratic malfunctions had on local politics and administrative practices.

The bureaucratic organization of the radiosurveillance program also suffered tremendously from the lack of coordination between different administrative levels (central, regional, provincial, municipal). The Italian nuclear bureaucracy was a very centralized administration organized around specialized research centers (continental contamination laboratories were in Rome, while marine contamination laboratories were in La Spezia, Liguria). Teams of radioecologists and radioprotectionists from CNEN and ISS conducted radiosurveillance campaigns around Italy. Radioecological expertise was highly centralized and not easily available in peripheral areas. Regions like Sardinia started to have their own research centers only at the close of the 1980s.[82] In the meantime, CNEN and ISS had to supply expertise, instruments, and radiometric measurements. In a context such as the archipelago, the time separating the collections of biological samples and the laboratory analyses in Rome or in Liguria was considerable (sometimes more than a month) and the results were always out of synchrony with the real ecological situation of La Maddalena.

The design of the radiometric network followed standard procedures. Two problems, however, remained unresolved. The monitoring system did not consider the impact that the environmental conditions of the archipelago would have on the instruments. The radiometric stations were purchased in 1979 but were not installed until 1986. After seven years the equipment was already obsolete and needed major updates and repairs. Even after that, the false alarms and the malfunctions of the instruments did not allow continuous radiometric measurements.

In La Maddalena frequent malfunctions, delays, bureaucratic inefficiency, and lack of expert communication about the radiosurveillance system, with time, produced a sense of hopelessness and resignation about the legitimate expectations for a ready and efficient response of public institutions among both the Maddalenini and local administrators. This sense of induced hopelessness produced a de facto incapacity of the local community to remain vigilant and actively involved in the decision-making process to effectively make public institutions and individual officials accountable for their actions or inactions. In other words, one of the main characteristics of bureaucracies—distributed and impersonal agency—made it difficult to attribute specific political responsibilities for recurrent malfunctions and delays. In a rich study on the politics of waiting, sociologist Javier Auyero has documented the strategic use of time by state officials and bureaucrats to describe the "temporal processes in and through which political subordination is reproduced."[83] As Auyero calls it, this "tempography of subordination" explains, at least in part, the incredible delay in the implementation of radioprotection measures in La Maddalena. Time transformed individual and collective rights into favors, establishing asymmetrical power relations and networks of patronage.

The exceptional status of the nuclear base of La Maddalena was sanctioned once more in 1988 when the Italian government, backed by a ruling of the Constitutional Court, blocked the regional referendum on the US Navy base, arguing that a regional constituency could not have a say on matters concerning national security and defense. The permanence of a base for nuclear submarines in Italy once again made evident the contradictory nature of the Italian nuclear regulatory regime and the unstable nuclear ontology of the US submarines. For clear political and strategic reasons, Italian authorities considered the nuclear base of La Maddalena different from civilian nuclear plants.

PART III

Risk, Accidents, and Political Mobilization

In 1977, Gloria Campos Venuti, director of the Radiation Laboratory of the Istituto Superiore di Sanità (ISS), proposed her own sociological explanation for why local administrators seemed to be skeptical about the effectiveness of the radioprotection program designed by Comitato Nazionale Energia Nucleare (CNEN) and ISS experts:

> I cannot definitively judge the primary cause for the extreme slowness and only partial implementation of the entire program [of radiosurveillance], despite the efforts of the two institutions [ISS and CNEN]. An accurate analysis of the situation could reveal the political and bureaucratic difficulties encountered in financing the entire plan, and the uncertainty of the Region of Sardinia, on the one hand, and the Province of Sassari, on the other hand, to allocate the necessary structures, personnel, and expertise. I think above and beyond these obstacles, at the local level there is a general distrust about the real guarantees that the entire program could provide for the protection and safety of the public. There is, perhaps, a difficulty in understanding the real health dimension of the problem and the concrete actions that can minimize the risks, regardless of the ideological struggles that different parties conduct around it.[1]

Campos Venuti's comment laid open the frustration of ISS and CNEN experts when they tried to communicate their analyses of risk to both local administrators and national politicians, but also provides insights about why most Maddalenini seemed to be uninterested in the radiosurveillance system, a question I was trying to answer thirty-five years later.

In the fall of 2012, I was conducting fieldwork in La Maddalena. After several attempts to access the recently relocated municipal archive I was finally able to consult it. I had been working for several days in the *municipio*, taking pictures, and checking out folders. Something seemed off. For a place that had been dealing with radiological risk for over thirty years I found it strange that among the copious documentation archived in the municipal palace I could not retrieve any radiometric report issued by the local laboratory. I knew for a fact—after interviewing three directors of the lab—that the mayors received radiometric reports quite regularly. And yet, apart from recorded debates of the city council I could not find any trace of radioactivity. One day, unexpectedly, one of the secretaries inside the municipio called me into her office and showed me a folder: "Maybe this can be interesting to you. Inside here are a bunch of documents about radioactivity. I saw them by chance, and I thought that they could be important, so I put them inside my drawer. You know, we had that laboratory there, because of the base. Let me know if you want to look at it." Inside the folder I found only the radiometric reports dating from the end of 1989 to the beginning of 1990, but it was enough to give me some understanding of how the laboratory worked and of the malfunctions that it recorded each month.

My excitement about that serendipitous discovery was equal to my puzzlement. When I interviewed former mayors of La Maddalena they mentioned those reports without much emphasis and only after I explicitly asked them, as if that data was just part of an empty bureaucratic ritual without substantial meaning. After reading the reports, I had the impression that the documents were written with the exact same template. Very concise, almost brutal, impersonal, and with the same happy ending: "The levels of radioactivity registered during this month are not relevant from a sanitary point of view. Best regards." After all, that was the only thing that mattered to both the laboratory personnel and the mayors. But that mechanical editing of the reports and the superficial readings allowed by the succinct formulas used to communicate the result—what mattered—come across almost as a tacit agreement, maybe unconscious, between experts and local administrators: the first would keep doing their job without interference, and the second would keep receiving updates without asking many questions about the why and how as far as the results continued to be irrelevant from a sanitary point of view.

Floriana Manca, director of La Maddalena's laboratory from 1987 to 2008, also testified to the fact that the municipal administration did not pay much attention to the monthly radiometric reports. She told me that after the USS *Hartford* accident at the end of 2003—with the consequent controversy about its possible environmental effects—"I suggested that we display the radiometric data outside the office of the mayor so that everybody could see them. They answered: 'What data?'"[2] I objected that the laboratory personnel could take the initiative to make the data available to the public, but Manca explained, "I worked for the region, not for the municipal administration. I had to respond only to the authorities and my job was to give the data to a predefined list of institutions."

Several decades later, Campos Venuti's words resonated with me. In the next two chapters, I analyze how communities of experts, practitioners, and nonexperts developed, fixed, exchanged, and contested definitions and representations of radiological risk around the US Navy base. In the mid-1970s expert communication about risk generated misunderstandings and frustration among the local administrators, who were expecting conclusive answers about the presence or absence of radiocontamination in the archipelago. Established radioprotection and risk control practices shaped understandings of radiological risk among US Navy personnel, while citizens without knowledge of radiation hazards tried to understand what risks were involved in the US Navy operations in the archipelago by interpreting changes and continuities in the surrounding environment as well as allegedly unprecedented events—such as birth defects—as evidence of nuclear contamination, or lack thereof.

The radiosurveillance system in La Maddalena not only was the response to a technical problem; it was the result of expert radioecological protocols, practical adaptations to environmental challenges, and organizational and bureaucratic instantiations of regulatory regimes. As detailed in earlier chapters, the assemblage of these components stabilized the radiosurveillance system as a sociotechnical system that held together until the power relations that configured it remained fundamentally unquestioned.

In October 2003, an accident occurred to a US Navy submarine in the waters of the archipelago. The event did not have environmental or health effects, but the uncertainties that emerged from scientific and political debates that unfolded in its aftermath provoked a crisis of credibility of the institutions in charge of monitoring environmental radioactivity in the archipelago. A historical analysis of the reactions to the *Hartford* accident reveals how state-sponsored agencies, and, later, independent experts, activists, and residents constructed and challenged technoscientific practices and narratives of safety and failure of the radiosurveillance system.

CHAPTER 6

Material Semiotics of Risk

On February 24, 1975, a conference on "pollution" (the adjective *nuclear* was purposefully omitted by the organizers) took place in La Maddalena. It was the best-attended meeting on the problem of the US Navy presence and its ecological implications since the installation of the submarine base. The Region of Sardinia organized the event and invited all the relevant actors involved in the radiosurveillance program: political authorities and representatives of all administrative levels, like the Province of Sassari, the mayors of several towns of northern Sardinia, and, most importantly, some key experts from Comitato Nazionale per l'Energia Nucleare (CNEN) and Istituto Superiore di Sanità (ISS) working on the radioecological survey and on the elaboration of the emergency plan.

The transcripts of the conference yield rich insights into the divergent understandings of risk connected to the operations of the US Navy base.[1] Because it was organized by public authorities and was held in the city hall of La Maddalena, the conference required that scientists and local politicians prepare their contributions in advance. Crucially, both CNEN and ISS experts came to La Maddalena after a long media campaign focused on the lack of transparency and credibility of their agencies. The public knew that the obstructionist position of the Italian government, military authorities, and the US Navy precluded the access of nuclear experts to technical information necessary to their studies. Further, in 1972, the president of CNEN, Ezio Clementel, attempted to assuage the public with a document that denied the presence of health concerns pertaining to the US base. Several CNEN laboratory technicians openly opposed this document.[2] This and similar episodes of internal conflict damaged the reputation of the agency and strengthened suspicions that public authorities, including expert institutions, downplayed the risks of contamination and accidents due to political pressures.

ISS and CNEN experts at the 1975 conference had specific goals. The first was pedagogical. They started their contributions by illustrating the rationale of radioecological surveys, the meaning of the available data, and the future steps required to draft an emergency plan. The second objective was to demonstrate their scientific integrity, despite the difficulties of working around a nuclear submarine base in which military secrecy and lack of knowledge about the local environment aggravated the already complicated task of the Italian regulatory agencies.

The political authorities who organized the event, primarily the Sardinian assessor of ecology, Alessandro Ghinami of the Partito Social Democratico Italiano (PSDI, Social Democratic Party), took the conference as an occasion to show that the region valued the safety of the Maddalenini above all. In his opening remarks, Ghinami condemned the "absence of the state" and called the conference "an act of democracy"—a moment of truth in which the experts were invited to speak openly about the results and the future measures to be taken in La Maddalena.[3] ISS and CNEN representatives took Ghinami's invitation seriously and displayed the knowledge gaps present in their preliminary assessments. Their openness created even more alarm and misunderstandings.

La Maddalena's administrators—especially among anxious US Navy supporters—expected to know whether the risks of nuclear contamination in the archipelago were real or overemphasized by the opposition for political reasons. In his short welcome speech, mayor Giuseppe Deligia reminded listeners that "we expect a clear and unequivocal answer, because if there were doubts or indecisions, our position could change. . . . After all, the source of potential pollution is a boat, so if we get rid of it, we will solve the problem."[4] City councilors, and citizens in general, were uncomfortable with the idea that "science" could not provide immediate answers to their questions. For the local Christian Democrat majority, who accepted the presence of the US Navy and portrayed it as a source of income, the political and moral responsibility for a decision that could have negative effects on their community was unbearable. Deligia and colleagues hoped (and maybe believed) that transforming a political issue into a technical one would place all the responsibility on the experts from whom they expected an unequivocal and positive verdict about the safety of the US Navy presence in the archipelago.

The first part of the conference consisted of long presentations in which CNEN, ISS, and Sardinian experts explained how they were preparing for the radioecological study that was supposed to start in the spring. While reassuring the audience that the data collected until then did not indicate any anomaly, the panel nonetheless underlined the limits of the radioecological program and introduced unfamiliar concepts that spurred more questions. Arrigo Cigna of CNEN explained that over the past two years his laboratory had found traces of radionuclides deriving from the activation products of the submarine reactors, but their concentration in mollusks and aquatic plants was so low as to be irrelevant to human health.[5] In the following contribution Gloria Campos Venuti, director of the ISS Radiation Laboratory, exposed the uncertainties that Italian nuclear experts had to face in La Maddalena: "The radioecological survey that we are doing now should have been done earlier [before the arrival of the US Navy], although it is never too late to move in the right direction."[6] According to Campos Venuti, Italy lacked regulations concerning the circulation and approach of nuclear ships at national ports; therefore, Italian nuclear agencies decided to adopt protocols established for civilian nuclear power plants. Osvaldo Ilari, a radioprotection expert from CNEN, probably talked for half an hour (his presentation is ten pages long) introducing general radioecological principles before focusing on the specific challenges that his colleagues faced in La Maddalena. Italian nuclear agencies, he explained, did not know the characteristics of the submarine reactors and their discharge formula during routine operations. For this reason, experts had to adopt a wider spectrum analysis that could compensate for the lack of those data.

Most of the questions from the audience focused on the uncertainties and knowledge gaps that experts laid open when illustrating the unresolved aspects of their study. In the words of Partito Comunista Italiano (PCI) regional secretary and Sardinian deputy Mario Birardi, "If this radioecological study presents such shortcomings from the start, how can it generate reliable data about the ecological situation around the US base?" Franca Careddu, also a PCI Sardinian deputy, added, "If the current data show that the levels of radioactivity are too low to be a menace for the local community now, could you exclude that the accumulation of low quantities over time will cause health effects in the future? And if not, when could the safety threshold be reached?"[7] Family physicians echoed similar concerns. Dr. Bentivenga, the health officer of the nearby town of Palau, emphasized: "In thirty years we will probably be able to assess whether these low levels of radioactivity are dangerous or not. We will see the possible effects not on ourselves, but on our children. I have to say that I am not convinced by this kind of radioecological study. They are useful but cannot answer these questions."[8] Questions from the audience multiplied and the frustration of municipal representatives mounted. Francesco Bardanzellu, a city councilor from Movimento Sociale Italiano (MSI) addressed the panel of experts and representatives of the Region of Sardinia:

> Has this illustrious committee of politicians, administrators, technical experts, and scientists been assembled with the goal of forecasting inherently possible accidental events or for studying the effects of radioactivity coming from the ship [the US Navy tender]? I ask because it looks like we are becoming a center for the study of radioactivity; that is, we are playing the role of guinea pigs for your experiments. There are still too many questions, among which the biggest is . . . that they [the experts] cannot give us any certainty because they do not know the characteristics of the reactors. Earlier, Ghinami told us that this information has been requested from the ministry of foreign relations. I am wondering when it will be delivered.[9]

When questions from the audience became increasingly polemical and alarmed, regional assessor Ghinami perceived that the conference—originally organized to reassure the local community—risked becoming a backlash: "We invited the experts here to illustrate with clarity and transparency their work because we do not want people to think that we are trying to hide something. Politicians cannot answer technical questions. For those we have the experts, who are keeping the situation under control. If one day they tell us that the data are worrisome, we will take the necessary measures."

But when the experts spoke about technical matters, the political dimension of the US Navy presence became even more evident. Professor Luciano Maiani, an ISS expert on reactor safety exposed these links with alacrity:

> Earlier somebody asked if we know the characteristics of the rectors. The answer is that we do not know anything about them. This is a different situation from those we usually face in civilian nuclear plants. . . . That information would be very useful because it would allow us to calculate the hypothetical consequences of an accident and predispose a proper emergency plan. . . . The other question that I wanted to answer concerns the risks connected to low-dose exposures. Currently, scientists are oriented toward the idea that a safety threshold below which humans are completely safe does not exist. That is, there is a certain probability of damage at each level. In this case [La

> Maddalena], this is very small, but it exists, like for any nuclear plant, and more generally, I would say, for any industrial activity. For this reason, we talk about risks and benefits for the local communities living around industrial installations. In general, for civilian nuclear plants we can calculate the risks in a very reliable way, exactly because we know the characteristics of the plant. We can tell you what our opinion is, but honestly, we cannot give you the reassurance that you are looking for.[10]

The experts who gathered in La Maddalena found it difficult to explain to local politicians that risk is not a 0–1 variable, but a probabilistic function: the product of the probability that a certain type of accident occurs, multiplied by the foreseeable consequences of that accident. While Deligia and his colleagues expected experts to give the final word about the safety of the archipelago, Campos Venuti, Ilari, Maiani, and their fellow scientists responded that they needed more information about the US Navy submarines and that even the most accurate monitoring techniques could not exclude the possibility of an accident or of increasing radioactive discharges in the future. In sum, Deligia and colleagues had to learn to deal with risk. This was why ISS and CNEN were designing a radiosurveillance system to constantly monitor the situation. Toward the end of the conference a restless Deligia could not hide his frustration and disappointment:

> The experts from the region, CNEN, and ISS expressed their individual opinions; It is my idea instead that we should invite them to express a collegial assessment—that is, [the expert committee should give us] a coordinated answer, because today, it is my impression that Professor [Mario] Ladu gives us one set of results, but Professor Cigna has very different ones.[11] They [the experts] have gone through a list of technical details: risk, danger, potential risk, data showing a slight increase of radioactivity, numbers, as Professor Ladu calls them. On the other hand, Professor Cigna says that there is an increase and then a decrease [of radioactivity]. I'll tell you what, gentlemen: you have to gather together, as experts of CNEN, ISS, and the region; take a look at the numbers written in your charts, decide whether there is a tendency in radioactivity increase or decrease, or if, even at a very slow pace, we are going to reach the safety threshold in, say, one hundred years. I ask the president [Ghinami] that the committee [of experts appointed to study the case of La Maddalena] collegially evaluates the numbers and gives us an answer that leaves no doubts![12]

Physicist Mario Ladu—director of the physics laboratory at the University of Cagliari and consultant of the Region of Sardinia—intervened to demarcate once more the "different roles of experts and politicians.

. . . We are speaking two different languages, but now we are asked precise questions that we probably cannot answer satisfactorily. When you ask: 'Is there a risk or not?' . . . Well, this question posed in those terms cannot be received."[13] He went on to say, "[You ask] 'should we agree on having these submarines here?' The answer is a political one. We, as experts, can only provide data and numbers and monitor the levels of radioactivity. The answer to your question is political in nature and you will have to discuss the solution."[14]

The "democratic experiment" orchestrated by Ghinami ended with local representatives, like Deligia, asking experts to retreat, do their job in isolation, and communicate their findings only after they had reached an internal consensus. The paradox is that after three years of requests for more transparent communication, local representatives blamed the experts for being too open about the uncertainties (created by the sequestration of data of Italian government and the US Navy) that characterized the environmental monitoring program at this stage.

The conference of 1975 illustrates the existence of different understandings of risk at the beginning of the radiosurveillance program planned by ISS and CNEN. The transcripts that I synthetically analyzed above offer a static representation of the conflicting conceptions of radiological risk present at that moment of the conference. But objectifications and representations of risk are never static, immutable, or essentially attached to one's social or cultural identity. Instead, they change over time in relation to environmental, epistemological, social, and political circumstances. Reciprocal misunderstandings among experts and nonexperts, as I will show below, are neither manifestations of irreconcilable lifeworlds nor expressions of fixed, identitarian, and (consequently) insurmountable epistemic divides. Instead, interactions between expert and nonexpert (sometimes through forms of collaboration) shaped objectifications and representations of risk, and how the latter changed over time in connection to two important factors: the material circumstances of processes of signification, and their social regimentation, which I call "politics of coherence." In the remaining sections of this chapter, I describe how established radioprotection and risk control practices shaped understandings of radiological risk among US Navy personnel and how residents of La Maddalena without practical knowledge of radiation made hypotheses about the presence of radiocontamination (or lack thereof) based on their daily observations of the environment and others' behavior.

THE CULTURE OF RISK CONTROL IN THE US NAVY

Santiago was only eight years old when he left the Philippines to relocate in the Los Angeles area with his family. Like many other US sailors, he decided to join the navy because he needed the money to pay for his education.[15] Santiago did not want to enroll in the army, and at that time the air force did not often admit non-US citizens. He enrolled in the navy thanks to a diplomatic agreement signed on March 14, 1947, between the Philippines and the US government according to which the US Navy could keep its bases in that country, in exchange for admitting a quota of voluntarily enlisted Filipino citizens every year.[16]

Santiago arrived in La Maddalena in April 1974; it was his first assignment out of boot camp. He flew from San Diego, California, to Norfolk, Virginia, for a two-week firefighting training program, before continuing to Rome and then to La Maddalena. "I remember a military bus was waiting for us. The airport was literally in the middle of grass fields. The road from Olbia to Palau was curvy and desolated, almost no constructions along the way. When I arrived here, I thought that my first duty was tough."[17] La Maddalena was considered an "isolated duty," a fair representation of the conditions that US sailors faced in the archipelago in the mid-1970s. The *Fulton*, the navy tender moored in the bay of Santo Stefano, was the only place where US personnel could purchase American products, cigarettes, some alcohol, and other comfort goods.[18]

As Santiago remembers: "La Maddalena was also difficult because of the nuclear stuff. There was a lot of politics around it and sometimes we could feel the tensions. But after all, the economic benefits outweighed fears."[19] Americans consumed cases of beer, spending money in bars and discos on the weekend. The navy tried to minimize conflicts with locals. The shore patrol policed the streets every night and was known for repressing any improper behavior of US personnel with alacrity. Under the aegis of the Intercultural Relations Program, the Naval Support Office hosted a series of events to create an atmosphere of cordiality and mutual respect. "We shared holidays, like the Fourth of July, and organized the 'Festa dell'Amicizia' (Friendship Party) every year, consuming tons of beer, hamburgers, and all kind of food with the rest of the population."

Santiago's first journey to "La Madd"—the shorthand used by fellow sailors for the submarine base in La Maddalena—was brief. By September 1974 he was back in San Diego to attend a training school for machinists. Upon his return to active duty, he stayed in La Madd for only one year before moving to Holy Loch, Scotland, where in 1961 the US Navy had installed a strategic base for fleet ballistic missile (FBM) submarines, much bigger, heavier, and slower than the fast attack submarines stationed in Sardinia.[20] There Santiago worked as a steam generator repair specialist, but the job involved traveling a lot without knowing the destinations: "I had my toolbox and they took me whenever a submarine needed a repair guy on board. So, planes and helicopters took me to subs on patrol. I stayed on board until the problem was fixed, sometimes even two weeks, then they would reemerge and leave me wherever they could." Sometimes Santiago was lucky enough to be landed on a US base from which he could be sent back to Scotland by airplane. Other times, returns were more complicated. Because he was still a Filipino citizen, in Holy Loch he needed a top-secret clearance to go on board FBM subs. After his return to La Maddalena, a year later, Santiago's status was downgraded to confidential: "You know, in La Madd I was mostly working on the tender. I could still go on board if I needed to do a repair when the submarines were anchored next to the ship."

Submarines are highly compartmentalized environments, in which only a commanding knowledge of signs allows the personnel to map out the intricate ensemble of machines, buttons, tubes, and sophisticated instruments. Santiago accessed them only to do his job: "I was going to the compartment where they needed me. Of course, I could see everything, the tubes, the magenta color which signals that you are in the proximity of the nuclear reactor. . . ." Santiago wore a dosimeter aboard the submarine to measure his radiation exposure. Upon disembarking, he would deliver the dosimeter to medical personnel, but never inquired into or was informed of his doses: "They would not tell you. Sometimes I wondered but never reached the point when they had to put me on a break. You understood that you got enough [radiation] when they moved you to another job for a while so that you did not reach your limit."

Habituation to nuclear risk was central to Santiago's accounts of his quotidian engagements with submarine repair on board navy tenders. "You get used to it [risk], it is part of your work. If you don't have trust, you get phobias. This is how I feel about it. After all, they send you to school and give you all you need." I objected that I would want to know my level of exposure and that I felt it would be my right to know, but he reassured me that the navy had qualified personnel—called the R5 division—who monitored radiation exposures. R5 is the acronym for "Reparation Five," the navy division also known as RADCON, which stands for Radiological Control. Back in La Maddalena, until 1983, Santiago worked in the Reparation Two (R2) department (the machine shop) on board the navy tender *Orion*: "Sometimes we wore protective gear and gloves because the parts from the submarines that we repaired could be contaminated. R5 people were making sure that we followed strict safety rules. Sometimes they built tents inside the shop so that the area where we were working was closed off. They gave us all the necessary equipment. When you wear all that stuff you feel uneasy because you have to work and operate your machines with it." All the measures that R5 personnel took on submarines and on tenders followed precise safety protocols, including the storage of contaminated material from the submarines. Santiago remembers that when he repaired sensitive parts, R5 personnel oversaw him and, after the work was completed, collected discarded material and protective gear, depositing all inside secured containers for provisory storage to be shipped to the United States for definitive disposal.

Santiago's story is certainly not unique. It reveals how work routines, safety protocols, and sociotechnical hierarchies structured US sailors' practices and understandings of nuclear risk. More specifically it offers insights on how US Navy personnel's experiences of radiological work shaped their understandings of risk, and how the latter differed from other groups' (experts, politicians, and residents in general) views of radiocontamination risks in the archipelago. Even if residents could not directly witness operations on the base, they formed their opinions indirectly through observations of and interactions with US personnel. On the other hand, US sailors often generated stylized explanations that were, in part, the outcome of their indoctrination and technical training. When local antibase movements questioned the safety of the navy operations, Santiago and his colleagues often dismissed those concerns as a reflection of ignorance concerning nuclear technology or as political prejudices.

When repairs on radiocontaminated parts could not be performed on submarines the material was sealed into containers, called "drums," and moved directly onboard the navy tenders. According to the US Navy report on the decommissioning of the Santo Stefano site, issued in January 2008, "All nuclear-powered vessels were moored directly to the tenders in lieu of the pier. This allowed sealed radioactive material to be moved directly from the submarines to the tender (both of which are categorized as sovereign US property) without being transported across Italian land."[21] Thus, the US Navy could claim that "no radiological work was performed on Italian land areas ashore," preventing antibase movements from scrutinizing these operations with more conventional legal and health-based frames.[22] The US Navy in La Maddalena selectively chose not to mention that in addition to serving as radiological work platforms, submarine tenders were also used to receive "low-level radioactive material associated with maintenance on naval nuclear-powered plants from tended units."[23] The encoded vocabulary that US Navy personnel used for describing hazardous procedures (like radiological work performed on the Navy tender) was difficult for the local population to decipher. To avoid the proliferation of "unreasonable fears" about the fact that radioactive stuff was being repaired a few hundred yards from the urban center of La Maddalena, the US command systematically elided "radioactivity" and "nuclear" from official explanations of its activities on the base.

During our conversation, Santiago repeated narratives and arguments I heard from several colleagues of his in other interviews. "The problem with the nuclear stuff is that when people hear the word they start worrying. So, especially here sometimes we heard people saying that we were the cause of illness or cancer, and so forth. It is not true. Rather, it is well known that these islands are full of granite rocks that are radioactive."[24] Anyone who has lived in La Maddalena and Palau for several years has heard contrasting narratives about the risks of radiological exposure. They became tropes that random interlocutors still pick up in informal conversations at the bar, in the central square, or on Facebook about "the time when the Americans were here." Those Maddalenini who confess to be nostalgic about the US Navy base elide the problem of radioactive contamination, and often accuse antibase activists of creating alarms and mobilizing public fears to kick the Americans out of the archipelago. Likewise, retired US Navy personnel I interviewed believe that radioactivity in La Maddalena was not a real problem. When I asked them what they made of local rumors about radioactivity, I received explanations like Jack's: "Sometimes you hear stories about nuclear subs leaving a fluorescent wave behind them at their passage or stories about contaminated fish or seaweeds changing color because of radiation, but of course these are all fantasies that people generate because they do not know what they are talking about or have wrong information. The granite rocks here are the only radioactive stuff."[25]

Jack retired from the US Navy a few years before the closure of the base and decided to stay in the archipelago to raise his family. He was young when he enrolled in the navy in the mid-1980s: "I lived in a suburb near Philadelphia and was still a high school student. At the time I also worked in a mall where a recruiter from the navy was coming often. I guess he has just been persistent enough to convince me." After he attended the regular boot camp, the navy sent him to a technical school for training as a "torpedo man." His job was maintaining weapons, including nuclear ones, repairing missile tubes onboard submarines, and supplying submarines with weapons stored on tenders. At the end of his course at the SUBROC (submarine rocket) missile school he and the four other torpedo men in his course were assigned their first duties. Jack, like other former and retired US Navy personnel I interviewed, often described his job with formulas such as "the world knows" or "it is public knowledge that," to emphasize that he was not revealing confidential or classified information.[26] Jack specified that his job, loading weapons aboard submarines, "was performed in the open, so people knew what I was doing." But, in this case, "people" is limited to US Navy personnel. The days when Jack provisioned submarines with missiles, even US personnel were prohibited from entering the docking area and sometimes the entire base was off limits for non-US employees. On those occasions, Italian civilian employees on the base understood that something was going on that they could not witness for safety and security reasons. These exclusionary practices triggered rumors about nuclear weapons being stored in reinforced concrete bunkers excavated below the thick granite of the island of Santo Stefano. "Yes, people here thought that we were loading nuclear weapons all of the time, but mostly we were just using 'trainers,' you know, those fake missiles that we employ for training. So, most of the time they were not real and not nuclear, which I can neither confirm nor deny that we had there."[27]

Jack's wife, Laura, was not particularly upset when Jack talked about the risks he ran on the job: "I did not even know what he was doing until many years after we married. He did not share information about his service and frankly I did not ask. The only time I felt insecure and scared was during the first Gulf War, when Jack came home with a gun that he kept secured in our home." Indeed, it was during international crises that the familiar presence of the base for nuclear submarines became suddenly a source of unusual alarm. During those moments of tension, the Maddalenini perceived the tangible dangers associated with living around a potential US military target.

Like Santiago, Jack confirmed that when working on subs and maintaining weapons he was required to wear dosimeters that medical personnel checked each month to assess his level of exposure. Medical examinations were thorough, he explained: "I never reached my limit, but if one does, he is not allowed to be on the program any longer." Jack was confident when he recited formulas memorized during his training in the navy: "Time, distance, and shielding are the main factors that determine your exposure. Radioactivity on board ships is not a hazard in normal conditions. Of course, there would be a problem if subs were dumping nuclear waste at sea, but this is never the case."

Over six years of service, Jack closely observed the procedures and the safety checks followed by each member of the US personnel. In his accounts, procedures were rigidly scheduled, monitored by supervisors, and practiced repeatedly during training exercises: "When we load weapons on board subs, we do it in teams of usually four or five members. Two people work [i.e., perform the loading operation]; one reads the steps, another one supervises the operations." At each step the reader announces out loud what must be done next, one of the two working members acknowledges the step and then performs it with his colleague, while the supervisor compares the performance with the protocol. "This is what we call 'reader working routine,' which assures that everybody is on the same page and knows exactly what each member of the team has to do and what the others are doing and are expected to do at any moment." After the conclusion of loading operations, the entire team met around the table to discuss the report compiled by the supervisor. Rules are important to the oral histories of sailors who worked on nuclear submarines based in La Maddalena. Like other "total institutions," the navy shapes the conduct of individuals in ways that also affect their relations with the external world.[28] The navy becomes a way of life that extends into the private sphere and family networks of personnel. While checks, supervision, hierarchies, and surveillance on the job minimize the risks connected to individual initiatives or deviance, the navy also emphasizes personal responsibility, which necessarily includes the private conduct of its personnel. According to Jack, this ethic enforces individual compliance with safety protocols and allows military personnel to focus on their work without the interference of personal problems. "We had PRP, personal reliability program, which assured that all of us were personally and financially stable all the time. For example, if the commanding officer received notes or reports that somebody working in the nuclear program had problems at home, with the family, or was not paying the bills, or had addiction problems that could potentially put safety at risk, he was immediately put out, at least until a committee in charge of reevaluating the case concluded that the problem was solved."

The navy's emphasis on rules, routines, evaluation mechanisms, and personal responsibility intensifies during submarine missions, as described by Greg, a former US Navy submariner based in La Maddalena in the 1980s. "When you are underwater, you literally depend on each other. Mistakes are not allowed because details make a huge difference. So, you trust each member of your crew unquestionably, especially the 'nukes.'"[29] Greg was a radioman in FBM submarines. He came to La Maddalena in 1983 toward the end of his career, when he was not working on submarines anymore: "The alternative I was given was Diego Garcia—can you imagine living there?" Until 1986, when he retired from the navy, he worked aboard the tender stationed in the Santo Stefano bay as an assistant communication officer. Greg became a radioman back in the 1960s. "At that time there were only two ways for a radioman to avoid Vietnam: going to Canada or volunteering on nuclear subs. The navy needed us especially on those places. In Vietnam, radiomen were the favorite target of the enemy, so I thought I'd rather go underwater."[30]

Each nuclear submarine, especially FBM ones, spent the entire year underwater, except when they needed repairs and refueling on dry docks back in the United States. Each submarine had two crews alternating on board every three months. When the other crew was at sea, Greg and his fellow submariners were constantly training. Greg did not attend the navy nuclear program, but serving on nuclear subs got him familiar with the basic knowledge of how a reactor works and what to do in case of accidents. As Greg explained to me, life underwater is physically and mentally exhausting: three months without sunlight, no alcohol, and little privacy, as even beds are shared between sailors who rotate on and off watch. Psychological and physical examinations are severe for "dolphins," the colloquial name for submariners. On-board rituals are equally important: sometimes they contribute to reducing the tension created by the thought of a steel hull full of people and nukes racing along the ocean floor, chasing and being chased by other submarines. As Greg recalls: "When we were halfway through our patrols, usually there was a 'party' during which we transgressed a little bit. We wore strange clothes and wigs and were allowed . . . *shhhh* . . . only one can of beer, which officially should not have been on board."

Greg discussed a few risks encountered on the job, but radioactivity was mentioned only when I solicited it through my questions. He told me that the nuclear program in the navy was "very serious stuff," that Admiral Hyman Rickover personally interviewed each "nuke" serving on his subs, and that he trusted them completely because of their expertise: "I mean, they go to school much longer than the normal navy guys like me, right? They know what they are doing, oh yes." It is for this reason that when he heard antibase activists in La Maddalena denouncing the risks of contamination, he preferred to not even engage with them: "It is not even worth discussing. There's just so much ignorance and so many misconceptions that it is practically impossible to make certain people reason around these things. Silence is a better strategy."

When Greg and Angela, like Jack and Laura, reflected on the risks associated with the US Navy presence in the archipelago they did not think primarily about radiocontamination, but about terrorist attacks. Greg had a delicate role as communication officer, and every two months he drove to Olbia to receive top secret orders in sealed envelopes delivered from the headquarters of Naples: "You know, in the middle of the 1980s there was still some terrorism going on here in Italy. I had to get into my car, with no gun, because we were not allowed to carry guns on Italian territory. And I remember being so scared that something might happen all the time." I asked Greg if he thought the atmosphere of secrecy around the base operations might have been also at the origin of public misunderstandings and misconceptions that he discussed earlier. Angela was faster to respond than Greg: "This idea that everything was secret in La Maddalena is so unrealistic. I worked on the base and with my colleagues we knew everything that was going on. People talk at the bar, rumors spread. Do you think we did not know all these things? We knew exactly when the tender was leaving Santo Stefano, or when they were training. We understood by just looking at certain movements, or hearing people talking about leaving the next day. It is that simple."[31]

For Italian employees on the US base, Angela's argument makes sense. After all, with several exceptions due to security checks and inaccessible areas, they worked in close contact with the US Navy personnel and, after some years, they understood many of the operations on Santo Stefano even if they were not explicitly informed. The abductive practice of inferring what was going on and why from the observation of the surrounding environment and the behavior of other groups of people was frequent in the archipelago, as it is in our daily lives, whether we pay explicit attention to it or not. If they could not fully grasp the meaning of risk as provided in formal expert representations or acquire classified information about the US nuclear submarines, the Maddalenini could still try to understand the meaning of risk by abstracting the implications of potential radiological hazards through their daily observations of the environment and from the occurrence of anomalous events.

ABDUCTIVE REASONING AND THE MATERIALITY OF RISK

Experts define radiological risk as the product of the probability that an odd event takes place and the calculable consequences of that event. The amount of information required for risk assessment practices is largely based on probabilistic models concerning the behavior of reactor materials, systems of containment, and processes of dispersion of radioactive elements in the environment surrounding a nuclear installation, including the possible effects of radiological exposure to which local communities may be subjected. Knowledge of the physical properties of radioactive materials is obviously crucial to experts who perform radiological risk assessments. Thus, general knowledge of the physical laws concerning radioactivity provides the code through which experts establish whether environmental and bodily signs can be interpreted as effects of radiocontamination.

Because "risk" refers to possible adverse outcomes of any activity, thinking of risk requires and entails a cognitive orientation toward the future, which can be expressed in a variety of registers: speculative, probabilistic, affective, and so on. Experts and nonexperts form their conceptions of risk by drawing information from various sources: they can be past experiences with radiation control or exposure; images that circulate through mass media, movies, and novels; scientific theories; and analogies with other events. Lacking general knowledge and reliable information about the risks they were running, residents of the archipelago had to rely on a repertoire of guessing practices to understand what the possible effects of living nearby a nuclear installation could be.

Nonexpert inferences were not randomly undertaken, or simply the reflection of conspiracy theories. Rather, residents formulated hypotheses about the presence or absence or radiocontamination based on the evidence provided by daily observations of other groups' behavior—US Navy personnel and expert authorities, for example—the surrounding environment, and exceptional events, like unprecedented cases of birth defects.

Even in a small and relatively insular community of ten thousand, ideas about risk were often contradictory. Rumors and gossip about accidents inside the base, and partial data provided by Italian expert agencies fed fears and uncertainties about the probabilities of contamination. Even today, ten years after the closure of the base, some members of the community express doubts about the validity of expert assessments. During interviews and informal discussions I had with them, some Maddalenini, including former employees at the base, were ready to dismiss any concern about radiation exposure on the job and in the archipelago more generally, but toward the end of our conversations often asked me: "You have collected data on this, right? So, what did you discover? Did we have nuclear contamination or not?"

Contradictions and uncertainties emerged as long-term residents interpreted material changes/continuities in the archipelago: lacking direct or indirect knowledge of the object (radiation), they had to infer its presence or absence by observing their surroundings to see whether the effects of radiation—as imagined through abstractions of its conceivable consequences—were evident in the archipelago or not.

For example, Giulio worked on the base for thirty years. He managed forty employees who maintained a complex system of diesel electric generators for the US Navy. He and his wife Roberta, also a former US Navy employee, remember that around the base "everything was clean and organized. Safety rules were always respected, and the sanitary conditions were regarded as an extremely important matter."[32] Like many former Italian employees of the US Navy, the couple shared with me their opinions about "the Americans" as diligent and respectful of the rules, in contrast to Italians, "who do things always approximately." In La Maddalena, they told me, the US Navy introduced exemplary norms of environmental preservation that were simply unimaginable for the Italian standards: "They taught us how to recycle, for example. We were simply dumping trash in a hole and burning it!"[33] When I asked them what they thought of the risks of radiocontamination in the archipelago they argued that it was mostly a political issue mounted by the antibase movements, and provided some evidence to convince me of that: "Look, sometimes during our lunch break at the base we could take a walk. They knew us very well and trusted us. We usually watched down the water at the pier . . . and you can't imagine the spectacle! There were always huge *orate* going around undisturbed," emphasized Giulio, an expert fisher. "Do you think that fish would be there if the seawater was polluted or contaminated?"[34]

Yet other experienced fishermen, like Carlo, who did not have access to the base, scrupulously avoided fishing in waters close to the US installation: "In general I do not think that the submarines were contaminating the islands. But honestly, we did not fish in places where the currents were passing through Santo Stefano. You know what I mean? When in doubt, better to avoid problems."[35]

Repeated observations of the environment around the base and cautionary tales about avoiding fishing where the currents from the bay of Santo Stefano could transport radioactive substances reflect contrasting views of the US Navy installation. On the one hand, Giulio and Roberta, two insiders, saw it as a clean place. On the other hand, Carlo did not exclude the possibility that the base could leak radioactive contaminants, although he was ready to reassure me that it did not pollute the environment. These stories and the everyday evidentiary practices to establish the porousness and the environmental impact of the base resemble what Joshua Reno has called "emplacement," referring to a set of epistemological strategies that "involves knowing and narrating the boundaries, leakages, and ongoing transformation" of places designed to contain the "out of place."[36] These evidentiary practices were based on interpretations of environmental signs mediated through broader assumptions about the behavior of the US Navy and of the fish.

Other frequent observations by La Maddalena's residents focused on the fact that every two years some US Navy personnel moved away from the base, assigned to posts back in the US or somewhere else overseas. "Were they sent away because they could not stay near the nuclear subs anymore?" asked someone speculating on the possible radiation exposure to which sailors and workers on the base could be subjected. Those familiar with the mechanisms of duty rotation on military outposts overseas know that US Navy personnel are only exceptionally allowed to serve on the same post for more than two years. Sometimes they can be deployed in the same location after an interval of time, but that is not the norm.

What I am suggesting here is not that the Maddalenini should have asked before making hypotheses about the reasons behind the "sudden" departure of US sailors. The point is that living near a strategic base for nuclear submarines and surrounded by obscure technological and organizational codes can induce anyone to guess about "unusual," "suspect," and "incomprehensible" behavior of military personnel and interpret it considering the pieces of information and assumptions they have about others' intentions.

In all of the above examples there is a common element. Residents gathered evidence about the presence or absence of radioactive contamination through observations of the environment and the behavior of other groups. They relied also on past experiences of the local environment and on their knowledge of the base. For example, Giulio and Roberta took their recurrent observations of the healthy-looking fish near the submarine base as an index of cleanness. Their experiences as fishers and dwellers in the archipelago allowed them to formulate the following syllogism: it's common knowledge that a particular species of fish swims only in clean seawater; if we see that species swim around the base, then we can conclude that the base is clean. The iconic value of the sign (fish that looks healthy) contributes to shaping the meaning of lack of pollution in the archipelago, and its proximity to the base where the nuclear submarines were stationed indexes absence of pollution. But Giulio and Roberta took these episodic observations, although repeatedly made, as evidence of regularity based on assumptions and background information about the behavior of the US Navy personnel, their environmental preservation standards, and practices, and ultimately the US Navy ethics. Thus, Giulio and Roberta interpreted the evidence gathered around the base as an instance of a general regularity: wherever healthy-looking fish swims there is no pollution.

Other Maddalenini interpreted the presence of US Navy servicemen as an index of the safety of the base: "Otherwise," some asked me rhetorically, "why would the US Navy put its personnel in danger? If they felt safe, we were safe too!" Observations of the US Navy personnel's behavior could also induce residents to draw different conclusions, based on opposite views of the US Navy: sudden departures of US personnel from La Maddalena could be attributed to the fact that after a certain amount of time they had to leave to avoid prolonged exposure to radiation—from submarines, and other unknown materials. These contrasting interpretations emerge from mutually exclusive assumptions about the ethical standards of the US Navy—specifically, whether the United States protects its servicemen on military duty by providing all the necessary safeguards on the job or not. Assumptions about the intentions of US Navy personnel provided the background information that made certain hypotheses about their behavior plausible in the eyes of those who advanced them. This seems to confirm that "specific ecological judgments are not separable from moral statements about the way the world 'should' be."[37]

Despite the perfectly logical inferences that residents used to interpret their daily observations as signs of cleanness or potential pollution, their experiences and knowledge of the local environment were not sufficient to produce sound evidence that the archipelago was not in fact contaminated. For example, an expert radioecologist could easily challenge Giulio and Roberta's reassuring conclusions. Based on their specialized knowledge of radioactivity, experts could disregard the same signs (healthy-looking fish swimming near the base) as insignificant because they were not indicative of the levels of radioactivity in the archipelago. Indeed, in radiocontaminated environments what looks healthy and thriving is not necessarily safe or clean.[38] Acquaintance, at least in theory, with the potential effects of radiation and radioecological methods put scientists on a different footing as to which signs are relevant (significant) and how to interpret them. Their semiotic ideology differed substantially from those of residents, not in terms of logic but in terms of what signs should be taken as relevant in the first place, and in terms of the degree of reliable evidence they could draw from inferences based on single observations, for example. As Umberto Eco reminds us, "Detectives are rewarded by society for their impudence in betting meta-abductionally, whereas scientists are socially rewarded for their patience in testing their abductions."[39]

But in circumstances of uncertainty, when the stakes are high, experts do not always have the time to test hypotheses or have the possibility to answer pressing questions right away, like those that concerned the citizens of La Maddalena in the spring of 1976, when three babies were born with cranioschisis (deformation or absence of cranial bones).[40] The parents of the newborns advanced the hypothesis of a causal relationship between the fatal malformations and the presence of the US nuclear submarines—due to the proximity of the nuclear submarines to the urban center and the lack of similar cases before the US Navy arrival. Italian experts instead excluded that possibility but did not offer any alternative explanation.[41] From a semiotic and very practical point of view, a concrete and dramatic experience such as the birth of babies with a fatal malformation was a sign whose object remained unknown, except for the exclusion of radioactivity as a possible cause. The semiotic ideology of Italian experts acted negatively, to exclude a potential object of the sign, but could not really indicate positively an alternative interpretation of the exceptional event.

In the case of radiological hazards, especially if we do not have direct knowledge or conception of them, environmental signs can provide inferential evidence of the presence or absence of radiocontamination. Given the political stakes of risk assessment in La Maddalena in the

1970s, unprecedented and disquieting events like spikes in the rates of birth defects became obvious candidates for evidence of radiocontamination. One feature of semiotic ideologies consists in the ability to use signs and to perform metasemiotic operations to establish and fix their correct function by indicating explicitly or implicitly the ground of their relationship with potential objects. This is what Richard Parmentier—using a concept developed by Michael Silverstein—calls "metasemiotic regimentation, [that is] the semiotic process of stipulating, controlling, or defining the contextual, indexical, or pragmatic dimension of sign function in 'discursive text' by means of the construction of a relatively fixed or coherent 'interactional text.'"[42]

Below I consider the struggles to establish, fix, or challenge interpretations of unprecedented events. Local intellectuals and antibase activists refrained from using instances of birth defects as evidence of nuclear contamination in the archipelago. They policed and controlled the proliferation of hypotheses about the causal nexus between the birth defects and the presence of nuclear submarines. This calculation about the plausibility of the malformation argument operated as a form of social control through semiotic regimentation, delimiting, de facto, possibilities for interpreting the events as consequences of radiocontamination. In doing so, local members of the PCI decided to rely on the expertise of allied scientists to get information about the environmental status of the archipelago to acquire the necessary authority for advancing credible arguments against the permanence of the US Navy in La Maddalena.

CROSS-BOUNDARY ALLIANCES AND THE SOCIAL REGIMENTATION OF RISK

On May 28, 1976, the Sardinian newspaper *L'Unione Sarda* published the first of a series of articles focused on the anomalous deaths of three babies from La Maddalena.[43] The babies, all born between August 1975 and March 1976, lacked a portion of their craniums. According to Sardinian journalist Giacomo Mameli, similar cases of cranioschisis may have happened before but were more scattered across time and some of them were revealed to be due to other causes. Mameli wrote that his report was "a response to many anonymous requests coming from La Maddalena," where rumors of the possible nuclear origins of cranioschisis spread quickly in corner grocery stores, bars, and in the church courtyard. "People, especially women, are scared," admitted Rosanna Abati, director of the municipal register since 1970. "In five years of work in this office, I had never seen something like this. These things leave us with many doubts."[44]

When Mameli interviewed the doctors at the local civilian hospital, they commented that cranioschisis was extremely infrequent, although

L'UNIONE SARDA

LETTURE

Berlinguer secondo Gorresio

INCHIESTE

Nascite anormali a La Maddalena

COLLIRIO ALFA

VIA LA FORFORA

ANTIFORFORA DOPOSHAMPOO Bipantol

6.1. *L'Unione Sarda*, "Nascite anormali a La Maddalena," May 28, 1976. *Source:* Library Archive, University of Sassari.

statistical records were not accurate because, in the past, families tended to keep such dramatic events private. Also, before the construction of the hospital in 1970, many women from La Maddalena and the surrounding towns of northern Sardinia gave birth in the city of Sassari. For this reason, previous episodes, if they occurred, had probably been recorded there. Dr. Romeo Milani, the director of La Maddalena's hospital, and his colleagues did not advance hypotheses about the causes of the malformations. According to them, these events, although infrequent, were not unprecedented. All of them concurred that what happened should not alarm the community, even if, they admitted, "the problem has been dramatized because of the presence of the US nuclear submarines."[45] In sum, local doctors interpreted the episodes as "natural events."

The parents of the babies agreed to speak publicly about their misfortunes. When interviewed by Mameli, two of the three mothers said that their pregnancies went well, while only the third had experienced a near miscarriage during the fourth month. All three women had previously given birth to healthy babies and, in their families, like those of their husbands, there was no record of malformations or diseases that pointed to an obvious genetic origin of the infant deaths. According to the doctors in La Maddalena, the painful experiences of the three families were random tricks of nature that did not constitute a scientific problem, but mothers and fathers argued otherwise and sought scientific explanations for their losses, which even doctors working in the larger hospital of Sassari could not provide. In the words of one father: "We are scared; we think many things; maybe the nuclear submarines have something to do with what happened, but what can we know about it? The scientists should tell [us]."[46]

Two days after the first article was published, the same journalist for *L'Unione Sarda* reported on the opinions of four scientists from the University of Cagliari, two geneticists and two pediatric neurologists.[47] They agreed that the episodes of cranioschisis should not be dismissed as simply random cases; rather, they called for careful investigation. Citing the uncertain etiologies reported in the existing scientific literature, the specialists advanced prudent hypotheses correlating cranioschisis with poor socioeconomic and sanitary conditions of the archipelago, in addition to pharmacological and environmental factors. Only one of the four experts suggested that ionizing radiation could be correlated with episodes of genetic mutation. Dr. Pinna, a pediatric neurologist at the University of Cagliari, did point to the higher frequency of malformations in the regions of Hiroshima and Nagasaki as a consequence of the nuclear bombing.

After surveying the explanations offered by doctors, the attention of Mameli returned to the still open question of radioactive contamination of the archipelago:

> The opinion of [the four experts] is that a serious investigation of the causes of the malformations is necessary. But it remains to be established whether the environment of La Maddalena, since the arrival of the US submarines, has really been modified. To better understand this, studies were conducted at the Physics Institute of the University of Cagliari, directed by Professor Mario Ladu. Recently, in a letter sent to the Sardinian assessor of health, the

> expert announced that the results of the radioecological exams conducted on samples from La Maddalena excluded the presence of higher levels of radiocontamination in the environment and that the traces found were so small that they are practically insignificant from a public health perspective.[48]

Notwithstanding the reassuring statements of experts like Ladu, doubts about the reliability of the radioecological campaigns conducted in La Maddalena continued to circulate. In the same article, Mameli returned to the conference on "nuclear pollution" held in La Maddalena in February 1975, during which Ladu asserted that current scientific knowledge could not rule out the possibility of negative health effects from low radiation exposure over an extended period. After reminding the readers of this apparent contradiction in Ladu's argument, Mameli added political flair to the story by citing a polemic flier that the La Maddalena chapters of the PCI and PSI circulated in March 1975: "The idea that, if they want real responses to their questions, the people of La Maddalena have to be transformed into guinea pigs sadly emerges as an objective reality, even if nobody will ever admit it publicly."[49]

The cases of cranioschisis are examples of how unusual and disquieting events offered the possibility to articulate new arguments about the lack of radiosurveillance measures in the archipelago and the potential effects of radiocontamination. National newspapers emphasized the point with sensationalist titles, such as, "Babies born deformed in La Maddalena. Is radioactivity the cause?"[50]

The reticence of military institutions on the malformations, and the brusque dismissal of local concerns about radiocontamination by expert agencies (who did not offer alternative explanations) caused the proliferation of hypotheses, rumors, and blatantly fake news about the status of the radiosurveillance program in the archipelago. In the week following Mameli's controversial article, a few local and national newspapers portrayed the imminent arrival of the *Odalisca* as an initiative that CNEN undertook to assuage the local population and to find a solution to the mystery of the "strange deaths."[51] In reality, the boat-laboratory was already scheduled to arrive in La Maddalena for a series of sampling procedures planned months before the cranioschisis events.[52]

In this atmosphere, CNEN and ISS experts deemed it necessary to intervene in public debates to clarify their position and to respond to newspapers' interpretations of data and events. On June 1, Osvaldo Ilari, head of the Environmental Radioactivity Division of CNEN, sent an official note to the press, in which he affirmed that the expedition of the *Odalisca* had nothing to do with the deaths of the three babies and that radioactive contamination could not be the cause behind the cases of cranioschisis.[53] Of the same tenor was the interview that professor Carlo Polvani, head of the Radioprotection Division of CNEN, gave *La Stampa* on June 2, 1976: "Even if there are traces of radiocontaminants, and this is yet to be demonstrated, they are so small that they could not cause any genetic effect on individuals in such a short time span."[54]

Lack of univocal communication, both from expert agencies and local institutions, encouraged the proliferation of rumors. Journalists reported extensively on the "strange cases of La Maddalena" to illustrate, sometimes with dramatic emphasis, the lack of a serious program of radioprotection around the base, and to comment on the difficult conditions under which nuclear experts had to work. On June 1, *L'Unità* quoted Eugenio Tabet, from ISS, and Ilari, who painstakingly denied any causal link between radioactive contamination and the cases of cranioschisis, and at the same time warned that the radioprotection program in La Maddalena was not yet completed and that data remained necessarily incomplete.[55] To those living in the archipelago, Tabet's and Ilari's statements could sound quite contradictory if they were not closely and patiently examined.

At the local level, political leaders like vice-mayor Franco Tamponi, a longtime PSI member, expressed ambivalent thoughts about the connections between the episodes of malformation and the presence of the US Navy: "There are no elements to establish a direct causal relation between our dead kids and the presence of the US base, but this event cannot do anything else but increase our doubts. Honestly, we cannot stop thinking about it because our community does not have memory of cases like these. . . . Also, I received confirmation from one of the obstetricians here that in the last years there have been suspect miscarriages."[56] In Tamponi's ambivalent interpretation of the episodes of cranioschisis is evident a certain degree of self-restraint: on the one hand he alluded to the possibility that malformations could be caused by radiocontamination; on the other hand he invited the public to reflect on the formal requirements of scientific hypotheses testing for the construction of solid evidence, which would preclude the possibility of asserting a causal nexus between malformations and the US Navy presence in the archipelago. Thus, after voicing the concerns of his fellow Maddalenini, Tamponi switched focus on the critical elements that made the current flaws of risk management in La Maddalena unacceptable. He borrowed both legal and scientific arguments from nuclear experts and educated activists: "We want all the necessary safety measures around this base. We need to know immediately if there is an increase in the levels of radioactivity and to design an adequate emergency plan in case an accident occurs."[57]

While unofficial reports and rumors fed hypotheses about the possible health effects of the nuclear submarines stationed in the Santo Stefano bay, local political leaders issued public statements that portrayed these hypotheses as "unscientific" and unlikely to travel far beyond the limits of the archipelago. Local opponents to the US Navy thought that to acquire credibility they needed the robust evidence of scientific facts provided by reliable and independent experts.

In La Maddalena, the processes of discursive elaboration, of which Tamponi's is just one example, did not occur spontaneously, but corresponded to precise strategies deployed by local political elites to construct their arguments and to win the consensus of public opinion, especially at higher levels of the technopolitical debate.

The spread and incorporation of scientific arguments and expert opinions with the common understanding of citizens was an objective of both pro- and antibase parties in La Maddalena. Both sides looked for expert opinions that could back their arguments (in favor or against) about the safety of the US Navy base. But antibase activists saw the proliferation of uncontrolled rumors and hypotheses as detrimental to their cause. Unchecked statements or exaggerated/distorted versions of facts and episodes were double-edged swords. Sometimes they could be used to provoke public reactions and shake the militarized community of La Maddalena. At other times, they could be political boomerangs because they could be easily dismissed by local and national parties in favor of the US Navy base backed by conniving scientific authorities. After all, local opponents experienced firsthand how in 1972–1974 the Italian government and the municipal administration, both ruled by a strong pro-NATO Democrazia Cristiana majority, tried to black-box safety issues in the archipelago by using ad hoc reports produced by expert and military authorities. For this reason, leftist political leaders, in particular, attempted to police the proliferation of hypotheses surrounding episodes like the cranioschisis ones and to filter sensationalist reports in general.[58]

Local antibase elites used a communicative register to shape and steer the debate over the risks of radiocontamination based on three tactical elements: (1) they exploited the political valence of events and episodes that could relaunch the debate over the fallacies of the radioprotection system; (2) they policed the use of rumors and unchecked statements about nuclear contamination in order to avoid their rebuttal by official authorities and the consequent loss of credibility of their arguments; and (3) they relied on allied experts to verify the technical accuracy of their interpretations and to enroll expert opinions that national, local, and military authorities could not easily dismiss. This set of strategies constituted what I call "politics of coherence."

These tactical moves—sometimes collectively organized and at other times individually undertaken—are evidenced in both the arguments and the omissions of newspaper articles and editorials of the 1970s I found in local archives.[59] For example, it is striking that *La Nuova Sardegna* (the second Sardinian newspaper), whose editorial line was usually critical of the US Navy operations in La Maddalena, did not publish any article about the episodes of cranioschisis. Struck by this surprising finding, I decided to ask Gian Carlo Tusceri, who was the newspaper's local correspondent from the early 1970s to 1991, to explain this silence. After a brief experience with activism among the young Christian Democrats, Tusceri became a sympathizer of the PCI, and his critical position toward the US nuclear installation was well known in La Maddalena. However, due to his professional and intellectual curiosity, he had preserved an independent stance on the US Navy base, to the point that he established a friendly relationship with Captain Edward Burkhalter, commander of the 22nd US Navy submarine squadron.[60] This liminal position allowed him to consider the pros and cons of competing arguments, but his choice for coherence and thorough documentation did not attract the sympathies of either opponents or supporters of the US Navy in La Maddalena.[61]

Consider the following excerpt from our interview:

> Orsini: Why didn't you write anything about the babies born with cranioschisis?
>
> Tusceri: Look, I have been among the strongest opponents of the US Navy in La Maddalena, but I refused to fall into the trap of the malformations. Everybody was using the story for political reasons. Many colleagues of mine wrote about the malformations because that was "the news" of the moment! It was making it into the first pages of local and national newspapers, which sold more copies. But I knew that emphasizing those stories was scientifically incorrect and consequently detrimental to the antibase political strategy.
>
> Orsini: But couldn't you try to explain and contextualize better those episodes?
>
> Tusceri: In that situation it was impossible to get into the debate and present more accurate accounts. I decided to not participate in what I thought was a weak political move and a conformist attitude of many colleagues. The argument that the radioactivity supposedly released by the submarines could cause the malformations, in general, was exposed to easy counterarguments and was ultimately indemonstrable. It was the wrong way to conduct the right political battle against the permanence of the US Navy.[62]

Tusceri was not alone in resisting the temptation to use the dramatic cranioschisis events as a political argument. Gianfranco Dedóla, a former member of the PCI and a worker at the Italian military arsenal, remembers that inside the party there was reluctance to use sensationalist stories for political propaganda. This was also the case for the cranioschisis episodes. In June 1976, when Italians voted in general elections, Giovanni Berlinguer gave a political speech in La Maddalena. Dedóla recounts the day in detail. "We were in front of the city hall. The square was full of people waiting for Berlinguer's speech. A few minutes before the start, Berlinguer asked me personally if there were themes in particular that the comrades in La Maddalena considered important to address. He asked if he should mention the story of the babies and I immediately said no! He agreed with me that that wasn't the right thing to do."[63]

Dedóla's and Tusceri's explanations are examples of how by severing the indexical link of the episodes of cranioschisis with nuclear contamination local antibase intellectuals and party activists attempted to regiment the meaning of risk in the archipelago. They also accomplished this with the help of allied experts from CNEN.

Both at the local and regional levels, high-ranking activists were extremely accurate when assembling arguments that could stand against the propaganda of the Democrazia Cristiana and the attempts by national authorities and the US Navy to minimize the problems of the radiosurveillance system in La Maddalena. Salvatore Sanna, speaker of the PCI group in the city council, was a prominent political figure of the archipelago. For decades he had been on the frontline against the US Navy installation and his efforts to document the environmental impact of military activities in Sardinia eventually made him an expert on the matter. During the 1980s, in fact, the Region of Sardinia appointed him as a member of the Committee on Military Easements, an organ composed of military and civilian representatives that examines the legal, economic, and environmental aspects of the presence of Italian and NATO military training camps, garrisons, and other strategic assets on Sardinian territory.[64]

During the 1970s Sanna kept an active correspondence with Carlo Papucci, the radioecologist from the CNEN Laboratory of Marine Contamination. The two had met during Papucci's repeated visits to La Maddalena, for CNEN radioecological campaigns and for conferences. Papucci's leadership in the Confederazione Generale Italiana dei Lavoratori (CGIL)-Ricerca (the union of research employees affiliated with the PCI and other leftist parties and labor unions) facilitated his exchange with Sanna and other political leaders in the PCI.[65] Sanna relied on Papucci's expertise and position in CNEN to get technical information about the radiosurveillance plans for La Maddalena.

Their private correspondence reveals that Sanna was sending Papucci his speeches and official notes to make sure that the technical details were accurate.[66] But Papucci's role was not confined to checking the technical accuracy of Sanna's arguments. For example, when the municipal councilor sent letters to CNEN's director, to the president of the Province of Sassari, and to the Italian navy requesting official data and updates on the implementation of the environmental monitoring program, Papucci gave Sanna his opinion and commented on the credibility of the answers he received.[67] Papucci's affiliation with the CGIL-Ricerca and his expert role in the radioecological campaigns were crucial resources that Sanna used to strengthen his communication strategy. For example in a letter dated March 11, 1977, the PCI councilor illustrated the "points of attack" on the existing gaps in the radiosurveillance plan, and asked Papucci to back his arguments in public in case the Italian government or military authorities tried to dismiss them: "If these points are exact in their formulation, it would be necessary that the *tecnici* [experts] confirm them when (if we will be successful) the Ministry of Health and the other authorities involved will say that everything is under control and that [the radiosurveillance plan] is in an advanced state of implementation."[68] In sum, Papucci not only provided technical mentorship but also actively contributed to shaping the public debate about the US Navy installation.[69]

In June 19, 1976, a few weeks after the episodes of cranioschisis broke into national reporting, an official note from CGIL-Ricerca (inspired by Papucci) reframed and reordered the sequence of events, clarifying (as previously done by CNEN personnel) that the sensationalist approach of the newspapers and the polemic around the causal relationship between the death of the three babies and the supposed nuclear contamination of the archipelago was shifting public attention away from the real problem: the preparation of an emergency plan.[70] "The mission of the *Odalisca*," said the official statement, "has not yet started because of bureaucratic delays due to the complex decisional mechanisms within the agency [CNEN]. . . . Thus the campaign programmed for the past spring could not be done, which means that the entire radioecological survey will be delayed for one year."[71] The CGIL-Ricerca, explained the article, stigmatized the fact that the supposed arrival of the *Odalisca* was used [during the elections] to assuage the anxieties of the public. The note pointed out that the activity of CNEN was part of a larger radioecological campaign with the objective of generating basic knowledge of the environmental characteristics of the archipelago for a more complex evaluation of the risks connected with the US Navy presence: "Without this preliminary survey and with only partial data, experts cannot do their job to evaluate scientifically the dangerousness of radioactivity."[72]

When I asked Carlo how he could manage to be in CNEN and, at the same time, organize a sort of internal opposition through CGIL-Ricerca, he answered that he did not have political weight inside the agency: "I was the low man on the totem pole; I could not decide anything really. But I could express my opinion, and the unions were pretty active inside the agency."

Sanna and other communist party members were competing with already powerful structures and with a diffuse sense of identification with the military at the local level, induced also by the economic and political influence that the Italian and the US Navy used as effective tools of social control. To challenge this power structure, Sanna and others had to rely on the careful and accurate construction of technopolitical arguments that could not be dismissed as simply prejudiced ideological positions against the US Navy. That is why the technoscientific aspects of the controversy concerning the deficiencies of the radiosurveillance system in La Maddalena were so crucial from the beginning. In this context, the political organization of the party, use of coherent scientific explanations, and control over the diffusion of popular arguments based on rumors and anecdotal evidence were key weapons in a battle between unequal adversaries.

In La Maddalena, radioecologists and health physicists, US Navy personnel, policymakers, and long-term residents engaged in public debates about the environmental consequences and the possible health effects of the presence of the US base. US Navy personnel doing radiological work relied on direct experiences of radioprotection where personal badges, gloves, protective clothes, written instructions, and training programs made risk visible. During disputes over the meanings of nuclear contamination associated with the presence of the submarines, experts' discourses focused on the technical and decontextualized characteristics of risk (the probability that an accident would occur multiplied by the calculable harm of that event). Abstract expert definitions and lack of previous direct experience with radiation and its effects left residents without a clear idea of what radiocontamination could look like. They adopted other strategies to objectify risk. For radiological risk to become visible, material signs must be available for interpretation. But material signs, such as radiometric results, measuring instruments, environmental changes, and unusual events cannot be interpreted without what Webb Keane calls semiotic ideologies—some shared codes or assumptions about the ontological order of the world, moral values, agency, intentionality, and political orientations that regiment their meanings.

How do different understandings of risk coexist and influence each other during a controversy? How do exchange and circulation of technical information between experts and nonexperts work? And how do more fluid and porous interactions between experts and nonexperts contribute to shape activist arguments vis-à-vis official narratives of expert and public authorities? The archival and ethnographic material discussed here reveals that, when analyzing the formation of meanings of risk, scholars need to consider how communication moves, and is policed, between experts and nonexperts engaged in technopolitical controversies. In the case of La Maddalena in the 1970s, political party organizations, labor unions, and the church were loci of pedagogy, cultural elaboration, and exchange between different levels and forms of knowledge production. More importantly, the cases discussed above show that experts and nonexperts are not necessarily opposed and homogeneous social entities, but that internal differentiations emerge when we take a closer look at the epistemological conflicts, information control, and political processes that enable both groups to reduce complexity and present stylized arguments in sociotechnical controversies. In terms of method, this involves understanding how experts and nonexperts form meanings of risk in interaction with one another, rather than assuming that incompatible sociocultural identities or cognitive dichotomies are at the core of their epistemic conflicts. I did not assume that people of La Maddalena knew the effects of radiological risk better than scientists or that they could deploy alternative forms of knowledge about it based on their close daily interactions with the local environment or—even more vaguely—because of their cultural identity. My goal was to show how nonexperts deal with invisible risks and how they formulate hypotheses concerning phenomena about which they lack direct experience and knowledge.

In a place such as La Maddalena, in which only radioecology experts and US Navy personnel trained in radioprotection had acquired common understandings of radiological risk, residents made sense of risk in an uneven representational economy shaped by what Ulrich Beck calls "relations of definitions." This imbalance did not prevent residents from formulating hypotheses about the presence or absence of radiocontamination based on observations of the environment, others' behavior, and unprecedented disquieting events. These observations were often contradictory and could not be immediately taken as signs of radiological risks (that is, as effects of radiocontamination) because they could not be explained coherently and could not be interpreted as recurring manifestations of some physical regularity.

Moreover, understandings of risk do change over time, even when the representational economy of risk is dominated by expert discourses and semiotic ideologies. New facts, such as the episodes of cranioschisis, became potential signs of risk that could be politically mobilized, especially in the absence of conclusive scientific explanations, but the unstable (because not demonstrated) causal relationships between the malformations and the presence of the US nuclear submarines led local antibase elites to abstain from using them in their technopolitical arguments. The organizational efforts of local antibase political elites to present sound arguments against the presence of the US Navy involved policing rumors about the cranioschisis episodes to exclude unscientific evidence that expert and central political authorities could easily dismiss. Antibase elites regimented the meanings of risk by assimilating scientific discourses into their representational strategies. They did so with the help of allied experts, who shared a democratic view of nuclear technology and communication with the public.

CHAPTER 7

Scientific Controversies and Political Mobilization after the *Hartford* Accident

On October 25, 2003, the nuclear submarine USS *Hartford* hit a rock near the US Navy base in the Archipelago of La Maddalena. US military authorities reported the accident a full eighteen days after the event. Harsh reactions from the Italian government and Italian representatives followed, criticizing the lack of transparency in the US Navy's conduct. The US Navy report excluded any damage to the reactor and when interviewed to comment on the accident, the Italian minister for the environment Altero Matteoli reassured the public that it was "a serious accident" but "first reports did not mention environmental problems."[1]

One month after the event, Sardinian health authorities and the personnel of the laboratory managing the radiosurveillance system in the archipelago reported that following the submarine's grounding they did not detect any increase in radioactivity levels. In that interval of time, Commission de Recherche et d'Information Indépendantes sur la Radioactivité (CRIIRAD), an independent French research institute, and Fabrizio Aumento, a retired Italian marine geologist, had already begun to collect and analyze samples of seawater, rocks, and algae, with the financial and logistical support of local environmentalist groups. Instead of assuaging the anxieties of the Maddalenini, their interventions raised further questions about the possible consequences of the accident.

Controversies among experts, and the uncertainty that they produced, had obvious reverberations in public debates. The mismanagement of the *Hartford* accident by local and national authorities, including the Italian and US navies, unfolded in a moment of renewed political mobilization in the archipelago. For example, in early 2002 a small group of citizens had already formed to oppose a US Navy project for the expansion of the submarine base. For local activists, the accident was an ideal occasion to mobilize sectors of their community traditionally unwilling to confront military institutions. Consisting of schoolteachers, public employees, students, merchants, and craftsmen, the Comitato Civico Spontaneo (CoCiS, Spontaneous Committee of Citizens), was not simply the product of a contingent political battle. Among its members was a new generation of Maddalenini who grew up during the 1990s, when the Italian navy was gradually downsizing its presence in La Maddalena. They had political experiences outside the archipelago but maintained strong ties with their community of origin. In the meantime, the establishment of the National Marine Park in 1996 provided an alternative to the military-industrial economy. While they went largely unfulfilled, the promises of a progressive demilitarization of the local economy and culture legitimized the aspirations of a new generation of Maddalenini who imagined a different future for the archipelago.

SILENT SERVICE, SILENT ACCIDENTS

US military authorities did not report the *Hartford* accident until November 10, when the spokeswoman of the Sixth Fleet headquarters in Naples announced that the commander of the 22nd Squadron and the captain of the submarine had been suspended from duty. The *Hartford* remained in La Maddalena for assistance and left the base a few weeks later, when it could safely cross the Atlantic, headed to Norfolk, Virginia, for major repairs. The Italian press agency Agenzia Nazionale Stampa Associata picked up the news from the *Day* of New London, a small town near Groton, Connecticut, the *Hartford*'s home base.[2] Italian authorities were caught by surprise when national newspapers reported on the accident, relying on details provided by the *Day*.[3] Indeed, it is striking that a local newspaper based in Connecticut was able to report on the accident on October 28, only three days after the event, while Italian authorities did not access this information until three weeks later, and only through an official press release of the US Navy. In his first article on the event, Gianluca Di Feo, a journalist for *Corriere della Sera*, speculated that "the first indiscretions about the accident appeared on the *Day* . . . probably because the families of the *Hartford* crew informed local reporters of the unexpected return."[4]

In La Maddalena the secrecy surrounding the accident undermined the credibility of military and civilian authorities. The mismanagement of communication after the accident had disruptive effects within a community that had proudly formed its identity along two centuries of cohabitation with the Italian navy and where the US Navy was largely perceived as a reliable institution. The reaction of Rosanna Giudice, the first woman to be elected mayor of La Maddalena, exemplifies how the accident catalyzed tensions between military and civilian authorities. On November 14, 2003, Giudice organized a press conference explaining that both the US and the Italian navy acted irresponsibly by keeping the *Hartford*'s accident confidential. The mayor's attempt to dissociate herself from the behavior of the navy command was not successful. Both political opposition and important voices from within her party accused her of being incompetent or, worse, directly involved in the coverup.

Giudice made clear that the regime of secrecy and the paternalistic attitude of the Italian Navy could not be tolerated further, especially when episodes like the *Hartford*'s accident put the safety of the local population at risk. During the press conference on November 14 she openly addressed her vulnerable position as a woman responsible for a community traditionally dependent on male-dominated military institutions. She interpreted the attempt to cover up the accident by the Italian navy not only as an act of disrespect toward her community and its elected representatives but also as a sexist maneuver "that excluded a woman from the manly business of military command."[5]

Rosanna Giudice was an easy target for the complaints of the Maddalenini. The anxiety of the community had ramped up during the three weeks preceding the news of the submarine grounding. La Maddalena's administration was on the defensive, trying to respond to rumors and polemics about an unexplainable "earthquake" along the northeastern coast of Sardinia. As reported by local newspapers, on the night of October 20 "a boom followed by a tremor woke up half of the residents of La Maddalena who ran down the streets in panic."[6] The day after, hypotheses, conjectures, and contradictory explanations followed. Mayor Giudice and the prefect of Sassari announced that the bang was provoked by a military jet breaking the wall of sound during a training operation, but a few days later they revised their explanation: the tremor was provoked by an earthquake detected near the northern coast of Sardinia.[7] The biggest fear of the Maddalenini was that something had happened on the US base. The local radio station received numerous calls from alarmed citizens asking if, in the event of an accident, the local administration had an emergency plan.[8] Rumors spread that an accident on the US base was the source of the bang. Citing anonymous sources (but most likely an Italian employee on the US base), Stefano Filigheddu, a member of the opposition in the city council, "revealed" that on the night of October 20 "an air compressor exploded inside one of the caves of Santo Stefano, where the US Navy stores weapons and other hazardous materials."[9] US and Italian navy officers rebutted Filigheddu's claim, which was never verified.

After reading the news of the *Hartford* accident on November 12, some Maddalenini interpreted the unexplained "earthquake" as the result of a submarine accident.[10] So strong and generalized was the suspicion that the archipelago had lived under a regime of lies and deception that many were ready to believe that the earthquake-bang-explosion of October 20 was connected to the submarine grounding, and that the US Navy was not telling the truth aven about the date of the accident. Years later, during my fieldwork interviews, some Maddalenini still referred to the earthquake-bang-explosion as a probable consequence of the *Hartford* grounding. If residents opposed to the US Navy base took the coverup of the accident as more proof that civilian and military institutions could not be trusted, other Maddalenini were taken by surprise and incredulity. When the details about the dynamic of the submarine accident started to emerge mostly thanks to newspaper reports, it appeared that the cause had to be ascribed to a series of banal individual errors. Even experienced US Navy sailors who served in La Maddalena reacted with astonishment: "I can't imagine what happened—it's an easy channel," declared a navy veteran to the *Day*.[11] Other veterans, interviewed years later, gave me similar opinions: "That was such a strange accident. I have gone through that channel many times. That point of the archipelago is tricky if you do not steer at the right time. Obviously, something went wrong in the communication. Even if the instruments fail, trained submariners should do it by just reading the charts. The fact that the captain was immediately removed from duty and other six members of the crew have been punished tells you that whatever they have done was considered a serious mistake."[12] The accident contributed to eroding the image that many Maddalenini held of the US Navy as a reliable, efficient, and technologically advanced organization. In a place where almost everybody has a boat and is, on average, a skilled sailor, people responded acrimoniously to the grounding of a submarine in a charted shallow point of the archipelago. "Who could be so stupid to crash over a bunch of rocks that are signaled on every map? If this is the level of their personnel in charge of nuclear submarines, we cannot really feel safe!"[13]

POLITICAL MOBILIZATION BEFORE THE ACCIDENT

The mismanagement of the *Hartford* accident by local and national authorities, including the Italian navy and US Navy, also unfolded in a moment of renewed political mobilization in the archipelago. In early 2002 a small group of citizens had already formed to oppose a US Navy plan for the expansion of the submarine base.

Rumors about the "doubling of the base" became a major concern of the local administration. News was not official, but Italian employees on the US base and the personnel of the local administration directly involved in the appraisal of the project leaked bits and pieces of information during conversations with friends and families. The US Navy project became an official topic for local political debate in November 2002, when regional newspapers reported indiscretions about an imminent massive deployment of US Navy personnel to La Maddalena, which was destined to become a strategic outpost for launching attacks in the upcoming war against Iraq.[14] The increasing percentage of unemployed Maddalenini, who were hoping for a job offer from the Americans, reacted positively to the news about the expansion of the base. Many others instead—including several members of the local administration—had serious reservations about La Maddalena becoming a giant military base invaded by thousands of Americans.

Giudice had recently won the elections with the support of a center-right coalition. The US Navy project caught the newly elected mayor by surprise and provoked her immediate reaction. After all, it was not unusual for the local administration to receive notice of ongoing or already stipulated agreements between the US Navy and the Italian Ministry of Defense only after the fact. In a letter she sent to the minister of defense dated November 20, 2002, Giudice solicited an official response regarding the alleged relocation of the US naval forces from Turkey to La Maddalena.[15] The minister of defense denied that such a plan had ever been submitted and that in any case such a dramatic change was practically impossible.[16]

Although the Ministry of Defense affirmed otherwise, the project for a sizable expansion of the US base was indeed on the table and continued to be discussed among experts and delegates of national and local institutions in the Comitato Misto Paritetico della Regione Sardegna sulle Servitù Militari (COMIPA, Regional Mixed Committee for Military Easements).[17] US Navy and the Italian military authorities presented the plan as a simple restoration of already existing buildings and infrastructures dating back to the 1970s and in need of major updates to satisfy new safety requirements. Representatives of the Region of Sardinia and technical personnel of La Maddalena's administration instead interpreted the project as a substantial expansion of the base. During a heated meeting of the COMIPA in July 2003, the technical personnel voted against the approval of the project, asserting that it had not been previously submitted to the attention of civilian authorities and that it implied a substantial modification of the status quo: an expansion of constructed buildings almost three times as big as the current installation.[18] The process for the approval of the project continued despite the opposition of local and regional authorities and, following an established praxis of disregard of formal procedures, the Italian military commands "convinced" the minister of the environment to let the project move forward.

When the controversy over the doubling of the US base became a public issue, a small group of Maddalenini created an informal network of citizens opposed to the US Navy project. I met some former members of the CoCiS in 2012 during a dinner organized by one of the group's leaders, Antonello Tovo. During our conversation CoCiS members recalled, "We did not want to become a colony of the United States! Can you imagine La Maddalena, with a local population of ten thousand, being invaded by five or six thousand US military personnel? What about our freedom to circulate in the archipelago and our lifestyle, our traditional activities? If we accepted the logic of total dependence from the US base, we could even say goodbye to the tourists. We did not want to radically change our identity. That's why we started to organize."

The formation of CoCiS made evident that the doubling of the US Navy base was a matter of public concern, not only a problem discussed within restricted circles of military and ministerial authorities. Most importantly, it represented a significant departure from the traditional patterns of political party mobilization organized by local elites and inspired by national and regional leaders. The antibase protest at the beginning of the 2000s transformed the classic contraposition between pro- and antibase parties in the archipelago into a more fluid and open arena. New actors appeared on the scene: CoCiS members (initially six) were schoolteachers, public employees, students, merchants, and craftsmen. Their political agenda was not inspired by any national movement outside of La Maddalena and had the only goal, at least initially, of opposing the expansion of the US base. During the weekly meetings "usually hosted inside the workshop owned by one member," they gathered information, audited former local administrators, constructed their strategies, made possible alliances, and shared many frustrations about the traditional apathy of their community. They also had to solve internal conflicts concerning the best course of action to be undertaken. CoCiS was a "group of concerned citizens" united around a common objective, but its members had different political experiences and motivations. Tovo, for example, recalls that he joined because he "had always been an anti-imperialist."[19] He knew about CoCiS while still in Pisa, Tuscany, where he got a BA in history with a thesis on the US Navy base of La

Numero unico *The little Paris* *aggratis*

TORTA GIALLA

La rivista per non prendere uno scoglio

FOSFORESCENZE

Questo esempio di giornalismo neorealista nasce alle ore 00.00 o' cloc del 1° Aprile 1997 d.c. per mano di un'accozzaglia di pendolari da forca. Per comunicare con voi abbiamo accuratamente evitato di rileggere gli articoli, autoimponendoci tre regole: la prima: non la ricordiamo; la seconda: 45 secondi a cartella; la terza: inebriarsi con ogni sorta di stupefacenti. Ma questo documento non è l'espressione del pensiero di vecchi avvinazzati, bensì quello di giovani avvinazzati. Se pensate che la nostra goliardia superi le vostre cognizioni intellettuali sull'argomento, sappiate che la prima fase di preparazione dei combustibili a uranio ci (e vi) regala l'URANATO AMMONICO (la torta gialla). Per gli amici UJO8. Non potevamo quindi evitarci questa estenuante maratona pseudoinformativa soprattutto dopo che, circondati da future colate di cemento, prospettive di caporalato militare, offerte di organi è arrivato il contingente che ha fatto traboccare il vaso. La torta di cui vi parliamo non è di quelle da decorare con le candeline, anche perché, una volta spenta la luce, brilla benissimo da sola. Desiderando quindi festeggiare con altre leccornie i nostri compleanni vi auguriamo di gustare ogni singola fetta e guarnirla a piacere.

A pagina 2

IDEE PER SALVARE UN'ISOLA

E' doveroso far iniziare questo breve contributo con un ringraziamento ai redattori di questa testata per averci invitato ad intervenire su uno dei problemi di La Maddalena: i problemi di Maddalena. Essendo noi stessi i redattori, non facciamo che stringerci la mano e offrire un giro di birra a testa.

Uno dei problemi di Maddalena è l'inadeguatezza della rete viaria, specialmente nel periodo estivo e in particolare d'inverno. Che fare? Si potrebbe, come con rara efficacia è stato suggerito, impedire l'accesso alle macchine dei non residenti (milanesi, torinesi, romani, americani, etc.), oppure far sì che i servizi pubblici di trasporto siano realmente servizi. Certo, se le circa 700 autovetture targate AFI non fossero sull'isola il problema sarebbe parzialmente risolto, ma questo è un altro problema ...

Un altro problema di Maddalena è: l'inquinamento dell'acqua, dell'aria, della terra e del fuoco. Di nuovo, che fare? Sarebbe opportuno impedire il transito di petroliere nelle Bocche; sarebbe inoltre

segue a pag. 5

7.1. Cover page of the first issue of *Torta Gialla*. The original was printed in yellow. *Source:* Marco Leoni, private archive.

Maddalena: "Joining this group was a way to reintegrate myself upon my return. It was also an opportunity to give back to my community. With my knowledge of the history of the US Navy presence I was able to contextualize and analyze the political and diplomatic dimensions of the problem. I gained respect and credibility inside the group. Or at least, I saw that people usually paid attention to what I said. I was excited about the fact that finally, after so many years a group of people from La Maddalena, not from outside, took the lead of the protest."

Growing up at the end of the 1980s, Tovo witnessed local teenagers confronting groups of young pacifists and anarchists who came to La Maddalena from all over Italy to protest the US Navy. Like antinuclear movements in the 1970s, they "invaded" the archipelago because it was a symbol, a sort of practice theater where they could stage epic battles against the state, imperialism, war, and nuclear power: "Our community usually has not responded well to this kind of exogenous protests. There was a general rejection of the disorders and the tensions they provoked. This is why CoCiS strategically changed the local dynamics. We were not punk anarchists from, say, Milan. In our group there were 'respectable' members of the community, so to speak."

Among them were other young activists who formed their political experiences far from La Maddalena. Marco, for example, was already an employee of the local administration. Four years earlier he had graduated in economics at the University of Florence with a master's thesis on the touristic and environmental perspectives of the newly established Marine Park in the archipelago.[20] Like others of his generation, he hoped for a drastic break from the traditional dependence of La Maddalena on the military economy. During his years in Florence, he kept in touch with other students and friends from La Maddalena with whom he coauthored short satirical cartoons and pamphlets about the "noxious presence" of the US Navy base. One of these short self-produced publications in the mid-1990s (mostly circulated among friends and in bars of the archipelago) was titled *Torta Gialla* (Yellow cake), an explicit reference to the US nuclear submarines and the risks deriving from their presence (fig. 7.1). *Torta Gialla* emulated similar initiatives that local political activists organized during the 1970s, the most famous of which was *Il Cobalto è Blu* (The cobalt is blue), a mimeographed satirical journal produced by the Federazione dei Giovani Socialisti di Sassari (the Italian Socialist Party Youth Movement).[21]

Torta Gialla issued sarcastic recriminations against state and military institutions that, according to the editors, de facto ruled over the subservient local community, silenced and controlled through the distribution of a few jobs and other economic benefits. The effects of secrecy and radioactivity, symbols of the American military power, were made visible through cartoons, poems, and rhymes, as if their representations on paper could make them real and force residents to sharpen their senses and awake their conscience. The risk of radioactive contamination, for example, was contextualized using common tropes, such as the passivity of the residents, and transformed into active political claims about the negative effects of the US Navy presence. Like a cancer, the Americans subtly invaded the body of the local community, who lost its sense of identity, spoke a hybrid idiom (local dialect mixed with English), and overlooked its health and environmental problems.

The initiative of the young students editing *Torta Gialla* was a symptom of the more profound socioeconomic transformations that were taking place in the archipelago during the 1990s. With the gradual but inexorable retreat of the Italian navy (the arsenal then counted fewer than two hundred workers) the local administration was facing for the first time the perspective of mass unemployment and a crisis of identity. In 1992 Pasqualino Serra, a dissident Christian Democrat historically critical of the US Navy base, was elected mayor of La Maddalena.

Serra's was and still is one of the most respected families in the archipelago. His grandfather, a descendant of the Corsican shepherds who inhabited La Maddalena before the arrival of the Piedmontese army in 1767, bought the island of Santo Stefano at the beginning of the twentieth century and transformed it into his entrepreneurial dream. In a few years, an islet of granitic rocks separated by three hundred yards of seawater from the major island of La Maddalena became an industrial site of granite extraction surrounded by beautiful vineyards, audaciously transplanted from the interior of Sardinia. When the Italian navy in the 1960s expropriated a portion of Santo Stefano to build a naval pier and a system of protected caves carved into the granite, Pasqualino Serra protested and refused any emolument offered as a payback. When I interviewed Serra in the summer of 2009 (he died a few months later in 2010) he explained to me that when the US Navy installed the base, right on the expropriated portion of Santo Stefano, his opposition to the military rule over the archipelago grew even stronger. Coming from a family of successful entrepreneurs, whose business flourished independently of the military contracts in the archipelago, Serra had always seen the presence of the Italian and the US Navy as a disadvantage because it excluded other possibilities of economic development. When elected in 1992, the new mayor did not oppose the US Navy openly: "In fact, since they came here, I had the most cordial relationships with them. Some, like Commodore Burkhalter, became personal friends for life. My opinion about their presence was not dictated by any negative preconception. Rather, I was opposed to the logic of the Italian state and our military institutions who wanted to impose their rule over this silent community."[22] In his political agenda, Serra envisioned the demilitarization of the local economy and the transition toward a more tourist-oriented future. For this reason, his administration worked restlessly at promoting La Maddalena's candidacy for the institution of a national marine park, which was formally established in the mid-1990s.[23] "I thought that this wonderful archipelago could be preserved and promoted as a national reserve that could attract tourists and make possible the conversion of the local economy, especially in a moment of crisis due to the progressive disengagement of the Italian navy. Second, I thought that the establishment of the park would make the presence of nuclear submarines incompatible and would push the Italian government to rethink its agreements with the US."[24]

To convince his community that it was time to turn the page, the mayor commissioned a study of the economic benefits of the US Navy presence, including indirect revenues, taxes, and local commerce.[25] The document, more than a hundred pages long and dense with statistics and charts, argued that the US Navy did not bring major economic advantages to La Maddalena. Instead, it represented a cost for the local community. Serra was ambitious. He wanted to change the mentality of an entire community by dismantling with numerical evidence deep-seated narratives about the innate military vocation of the archipelago. His project did not bring him immediate consensus but insinuated some doubts within his community and opened a discussion of an alternative economic future.

The socioeconomic and political changes that were taking place in the archipelago at the beginning of the 1990s nurtured the aspirations of new generations, like Marco's and his friends'. To understand the reactions to the submarine accident in 2003 one needs to keep in mind that at the local level a reserve of political capital was ready to be activated. The accident amplified already existing concerns among the local community. More importantly, it materialized all the conjectures and the discourses about risk that the Maddalenini had previously heard of through the voices of opposite factions in the city council. Fears became palpable and justified.

UNEXPECTED EVIDENCE AND THE CHALLENGE OF INDEPENDENT EXPERTS

The post-*Hartford* crisis in La Maddalena challenged in unprecedented ways the rationale of radioecological surveillance policies that Italian agencies established around the US base in the mid-1970s. The modalities of intervention and communication that expert institutions adopted for the occasion failed to respond to the expectations of local community in two ways. First, local and national agencies did not provide any evidence of the radioecological situation until a month after the accident. This opened the space for the interventions of independent experts, who challenged established methods of risk assessment and management. Second, state and regional experts tried to reassure the public about the reliability of the radiometric system by dismissing the preoccupations of the local inhabitants and attributing them to their lack of scientific literacy. Even worse, some Sardinian scientists interpreted local citizens' efforts to open a public debate with the help of independent experts as an attempt to use unscrupulous methods and openly bad science to question the legitimacy of the US Navy presence in the archipelago.[26]

When the *Hartford* accident was officially reported, the personnel of the radiation surveillance laboratory of La Maddalena conducted a supplementary environmental survey. In a laconic report dated November 22, 2003, the directors, Floriana Manca and Giuseppe De Luca, wrote, "Our personnel conducted gamma spectrometry analyses on biological samples collected in the area where the accident allegedly occurred. The results are negative. Therefore, we can conclude that the accident did not provoke any radioactive contamination of the environment."[27]

In their report, Manca and De Luca described their sampling procedures in the zone of the accident as they had done for the past twenty years. These protocols followed specific guidelines based on the environmental studies and risk assessment that the Comitato Nazionale per l'Energia Nucleare (CNEN) and the Istituto Superiore di Sanità (ISS) conducted in the mid-1970s. Given the regime of secrecy surrounding the technical details of the US naval reactors and the discharge formula associated with their operations (that is, the types of radiocontaminants dispersed into the environment as a consequence of their routine operations), Italian expert institutions suggested a radiosurveillance protocol similar to inland civilian nuclear plants. For La Maddalena they hypothesized that US Navy submarines were propelled by pressurized water reactors of 70 to 80 megawatts of power and that the radioactive elements most likely dispersed into the environment (especially in case of accident) would be fission products like iodine-131 and cesium-137, and activation products such as manganese-54 and cobalt-60.[28]

Since the late 1970s, the guidelines illustrated above, combined with sampling procedures, instrumentations, and expert knowledge transmission, produced what Michelle Murphy has aptly phrased a "regime of perceptibility" that structured the way the radiosurveillance system operated and the kinds of evidence it produced. As Murphy argues, different measurement devices and epistemic approaches can render visible, or perceptible, widely different risk factors in different environments.[29] The radiosurveillance system established in La Maddalena shaped the way in which radioactivity and nuclear risk were made visible.

Before Manca and her colleagues in La Maddalena's lab published their first report, in November 22, the northern Sardinia chapter of the World Wildlife Foundation and the Corsican environmentalist group ABCDE joined forces to collect samples of seawater and algae in the areas where the accident supposedly took place.[30] They sent these samples to the

French agency CRIIRAD, an independent laboratory founded after the Chernobyl accident in response to the ambiguous conduct of the French government in assessing the risks from radioactive fallout.[31] On January 15, 2004, the director of CRIIRAD, the engineer Bruno Chareyron, communicated the results of the gamma spectrometry performed on the samples.[32] While the analyses revealed that no radioactive elements usually associated with reactors' activities were present, anomalous concentrations of thorium-234 were discovered in the algae.[33]

The findings of CRIIRAD posed a problem that was not contemplated in the radiosurveillance protocol of the Agenzia per la Protezione dell'Ambiente e del Territorio (APAT, National Agency for Environmental Protection), the Agenzia Regionale per la Protezione Ambientale della Sardegna (ARPAS, Regional Agency of Environmental Protection of Sardinia), and consequently by the local laboratory of La Maddalena. As Manca explained during an interview released twenty days after the first CRIIRAD report (February 6, 2004): "Our lab is perfectly equipped to detect thorium in the environment but in the past we never did so because thorium is a natural radioactive element that has nothing to do with the operations of nuclear reactors; therefore it is not included in the list of radiocontaminants that we look for around nuclear installations."[34]

While the personnel of the local lab addressed why they were not looking for thorium, the puzzle of its high concentrations in the red algae remained unresolved: "We will make a comparison with other sites with the help of other research centers to understand what is going on with thorium, to make sure that seaweed is not experiencing the same phenomenon and that it is not affecting the food chain."[35] In La Maddalena Manca's words did not sound reassuring enough.

The new evidence that CRIIRAD provided introduced elements of uncertainty and provoked disquiet among La Maddalena's residents, who were already alarmed by the silence of civilian and military authorities on the *Hartford* accident. Importantly, CRIIRAD was not the only independent institution to intervene after the accident. At the beginning of 2004, Fabrizio Aumento, a retired marine geologist, sponsored by the Italian environmentalist group Legambiente and the Association of Italian Scientists Against the War, started to collect samples of rocks, sediments, and algae from the archipelago and from various other Mediterranean locations (reaching more than 160 samples). Dissatisfied with the results published by CRIIRAD and with the first report of Sardinian experts, Aumento adopted a comparative approach to determine whether the radioactivity of the samples coming from in La Maddalena was due to the abundance of granite rocks or was introduced by anthropogenic sources. As Aumento recounts:

> I wanted to increase the number of samples and to include material coming from different places to make a more reliable comparison between the concentrations of radioactive elements present in La Maddalena and elsewhere. This allowed me to distinguish the natural radioactivity emitted from the granite rocks of La Maddalena from the anthropogenic radioactive elements. I am a marine geologist, and my approach is quite different from that of a classic nuclear expert. Only many high-quality samples can give you reliable results. Therefore I was not happy with the first assessments made by the CRIIRAD and the other Italian research centers.[36]

Aumento's preliminary results were even more worrisome than the ones announced by CRIIRAD a few months earlier.[37] Based on wide-spectrum autoradiography techniques, Aumento's measurements revealed the presence of alpha-emitting particles, which he took as evidence of possible plutonium contamination.[38] These findings renewed the alarm among La Maddalena's residents and pushed the Ministry of the Environment to ask APAT to conduct an extraordinary radioecological survey in the archipelago with two main goals: verifying and explaining the high levels of thorium-234 in the algal samples analyzed by CRIIRAD and assessing whether traces of plutonium were present or not in the area.[39]

APAT published its results in June 2004.[40] In the meantime, CRIIRAD delivered its third and final report, which substantially agreed with APAT's conclusions: 1) the high concentration of thorium in the algae was due to natural bioaccumulation not related to anthropogenic emissions; 2) the traces of plutonium present in some of the samples were not alarming because they were probably the result of nuclear test fallouts from the 1950s and 1960s.[41] CRIIRAD added that the hypothesis of radioactive discharges in La Maddalena could be excluded once and for all only by comparing current data with the levels of radioactivity before the arrival of the US Navy. CRIIRAD thus invited both US and Italian military authorities to break the regime of secrecy surrounding the base's activities.

WHO IS AN EXPERT?

Aumento's intervention introduced new evidence and scientific methods to radioecology in La Maddalena. Expert communities, especially Italian radioprotectionists and radioecologists who had worked in La Maddalena for more than two decades, resisted them. At the local level, even the World Wildlife Foundation activists who promoted the CRIIRAD study publicly contested Aumento's thesis that the traces of plutonium in some of his samples derived from the US submarine base.[42] In La Maddalena, retired radiochemist Giancarlo Fastame dismissed Aumento's results as methodologically flawed.[43]

Aumento's thesis did not get much public attention outside La Maddalena until 2005, when the well-regarded *Journal of Environmental Radioactivity* published his study, provoking negative reactions from American and Italian radioecology experts.[44] Introducing the series of comments sent in response to Aumento's article, the editor of the journal, S. C. Sheppard, openly acknowledged the controversial nature of Aumento's conclusions. All the letters sent to the editor complained that an article "with such problematic methodological flaws and unsupported conclusions could even be published in a highly respected scientific journal."[45] Carlo Papucci, the veteran of CNEN radioecologists working in La Maddalena (now retired), and Roberta Delfanti, director of the Center for the Study of Marine Environment of ENEA, argued that the autoradiography methods used by Aumento's team did not reliably measure plutonium concentrations. Papucci and Delfanti stressed that the "conversion factor" used to calculate concentration levels on the basis of the evidence provided by the "hot spots" in the radiography films was arbitrary, in that it did not account for the possible contribution of natural alpha emitters to the data set. Therefore, they insisted, Aumento's conclusions were to be taken as "wishful thinking" and "personal opinions," and could not be accepted as evidence of plutonium contamination.[46] Arrigo Cigna, also retired, voiced similar critiques. Even if traces of plutonium could be revealed through radiography, Cigna argued, "hot particles were distributed throughout the world from the radioactive fallout [of nuclear atmospheric tests] and may still be available in the environment since the radioactive decay is not affected by weathering, as the authors must know."[47]

Aumento's reply to Papucci, Delfanti, and Cigna reveals the substance of the controversy and helps explain why his approach and that of Italian radioecology experts diverged so significantly to lead to different data interpretations. First, Aumento and his coauthors justified the adoption of radiography not for its indisputable reliability or for the sake of methodological innovation, but rather as a necessary response to an emergency:

> An accident, that was kept secret by the military authorities, . . . could have provoked a nuclear disaster in La Maddalena National Marine Park. A couple of laboratories initiated investigations on possible nuclear contamination of the area: inexperienced local citizens collected a handful of algal samples selected at random from the vast area in question, and sent them to these laboratories for gamma ray spectrometry analyses. The analyses of

> several key radioisotopes were omitted. The gamma ray spectra did not reveal any particular anomalies, except for the thorium-234 excesses in two samples. As marine geoscientists we could not accept the validity of the data of this line of approach.[48]

Indeed, Aumento and his colleagues worried that the granitic composition of the archipelago, notoriously associated with higher levels of background radioactivity, could lead to biased results. For this reason, they adopted a comparative approach involving the selection of larger samples from equally granitic environments that did not present any direct source of anthropogenic radioactivity (such as nuclear plants or industrial activities). This broader ecological design, according to them, was key to discerning the real status of La Maddalena's environment. According to this approach, the fact that higher concentrations of supposedly transuranic elements appeared in the samples coming from near the US Navy base in La Maddalena was a clear indication of their strong correlation with the submarines' activity.

The comparative model was also justified on the grounds that geographic discrimination across possible sources of radioactivity was crucial in the absence of more precise and time-consuming radiochemical procedures: "Our multidisciplinary approach, based on expert fieldwork performed by the very same analysts (both on land and at sea) who subsequently performed the lab work on the extensive sampling repeated over time, combined by rapid laboratory analyses, were and are the only ways to proceed in the event of a possible nuclear catastrophe!"[49]

Aumento's article clearly contested the scientific canon of Italian radioecologists and expanded the debate by shifting the focus of the initial controversy from the concentration of thorium-234 in the red algae (first discovered by the French CRIIRAD) to a broader discussion about the evidence of nuclear pollution provoked by the activities of the US Navy in the area.

Italian nuclear experts instead countered Aumento's arguments by relying upon established epistemic models developed in years of radioecological campaigns in the area and during their disciplinary training. A holistic approach, such as the comparative ecological model based on large spectrum radiography (as the one adopted by Aumento), appeared inappropriate to expert radioecologists like Papucci, Delfanti, and Cigna, habituated to more thorough investigations of radiocontaminants dispersion into given ecosystems. According to former ENEA experts, broader ecological comparisons alone could not discriminate among natural and artificial sources of radioactivity: only specific and time-consuming radiochemical analyses designed to detect radioactive traces associated with the activities of nuclear reactors could give reliable results. Equally important, they interpreted the possible presence of radioactive elements (plutonium and cesium) in the waters of La Maddalena as evidence of the long-term effects of atmospheric experiments (in the 1950s and 1960s) and of the Chernobyl disaster. This interpretation was not contingent on the case of La Maddalena but was a well-established paradigm of Italian radioecology expertise since the 1950s, when the National Committee for Nuclear Energy (CNEN) organized a national network of radiometric stations specifically designed for measuring the levels of environmental radioactivity determined by nuclear tests' fallout. Cigna, one of the experts who contested Aumento's findings, was a pioneer of these studies and brought his expertise within the radioprotection laboratories of CNEN and ENEA, of whom he became director in the early 1970s.

In 1984, the team of ENEA experts (formerly CNEN) who by then had conducted radioecological campaigns in La Maddalena for more than ten years, published a final report providing a comprehensive synopsis of their results. They wrote that since 1972 the radiometric exams in La Maddalena had never evidenced health-relevant accumulations of radionuclides that could be linked to the operations of the US submarines.[50] According to the authors, including Cigna, comparative analyses of samples coming from La Maddalena demonstrated that the concentration of fission products in the archipelago's environment were similar to those coming from other sites. Therefore, they concluded that "the generalized presence of artificial radionuclides must be caused by a ubiquitous phenomenon such as the deposition of radioactive fallout from experimental atmospheric explosions."[51] This interpretation of the data, I argue, must be understood as the outcome of deep-seated protocols in radioecological research during the Cold War, as it was primarily focused on the environmental dispersion of radionuclides coming from atmospheric explosions.

Aumento's intervention, even if problematic in many respects, contributed new elements for public discussion. His approach was justified on the grounds that the US submarine accident generated questions to which Italian public institutions were not able to respond, both politically and scientifically. He deemed the consolidated models of Italian nuclear experts "outdated and corresponding to old conceptions of radioecological monitoring, especially in cases of dramatic events such as an accident."[52]

While Aumento's article, published in a reputable international scientific journal in English, intensified the controversial debate within specialists' circles, in La Maddalena it did not acquire the same relevance as CRIIRAD's findings. Scientists more traditionally identified as nuclear experts—namely nuclear physicists, radioprotectionists, and nuclear engineers—rejected Aumento's approach because it was not in line with their disciplinary standards.[53]

LOCAL POLITICAL MOBILIZATION AND THE QUEST FOR DEMOCRACY

From January to June 2004, uncertainty over the radioecological analyses conducted by both independent and state-sponsored laboratories undermined the credibility of La Maddalena's radiosurveillance system. Exposed to alternative approaches and data interpretations from independent experts, residents of the archipelago started questioning the rationale and the efficacy of the radiation surveillance policies adopted for over thirty years. Editorials and investigative reports in local and national newspapers began to inquire into how the radiometric monitoring system worked. Was it reliable? Who were the experts in charge of it? What was the safety protocol if accidents occurred? Was there an emergency plan? Were any cases of cancer in the archipelago connected to the presence of the submarines? Were epidemiological studies conducted to rule out this possibility? As questions multiplied, so did the number of experts who were involved in the debate.

Giancarlo Fastame, a retired radiochemist from La Maddalena, frequently addressed these issues from the columns of *Il Vento*, a popular local weekly magazine. Having accumulated long managerial and scientific experience in large petrochemical companies throughout Italy and Europe, Fastame had a reputation as an expert within his community: "Because of my education and professional experience I certainly have concrete notions about radiation and radioactive contamination. So, what I did after the accident was to get basic information about its dynamic, the US Navy safety procedures, and the radiosurveillance system of La Maddalena, including the reports of the local laboratory and of the national authorities."[54] Even for somebody like Fastame, gathering official documentation on the radiosurveillance of the archipelago was not easy. When he asked the local administration for access to the monthly reports of the laboratory directed by Manca, the personnel of the city hall refused, insisting that Fastame, a private citizen, could not present any legitimate reason for looking at the data. With the intervention of the municipal ombudsman, a week later the radiochemist obtained the reports anyway. The data he gathered showed that "the radioactivity was at the natural level, so there was no problem at all. Dr. Manca confirmed this also when we met. She explained to me the situation and the routine radiosurveillance protocol. We are both experts, we use the same language, so we understood each other immediately."

Fastame explained to me that in his opinion, first CRIIRAD and later Aumento provoked alarm with their interventions: "In the city hall, full of people, they said that their analyses showed the presence of radiocontamination. You can imagine the reaction of this community." It was in response to these alarms that Fastame began his work of divulgation. Local journalists and members of the community started to call him to ask for his opinion. The director of the local radio station Radio Arcipelago and of *Il Vento* decided to host weekly programs in which Fastame explained basic concepts of radioactivity and nuclear energy and responded to frequent questions about the situation after the accident. "Sometimes I was asked bizarre questions, like if people could eat the fish of the archipelago. Someone told me that he started to buy frozen fish coming from the Atlantic! Surely that fish was contaminated, unlike the one caught here."

During our conversation, Fastame told me that he does not blame his fellow Maddalenini for their "ignorance." Rather, he laments the fact that the local authorities and the national expert agencies did not have any communication strategy: "If I, a radiochemist, could barely access the data, you can imagine how hard it would be for other people to know what is going on and what to do in case of a real emergency." For Fastame the crisis of trust and the fears of radiocontamination in the archipelago had two origins. First, military and expert authorities failed to have a transparent relationship with the Maddalenini. Second, this lack of transparency caused uncertainties and doubts, which traditional opponents of the American presence could use to amplify the alarms and create a stronger antibase front. Like other members of the community involved in the debate over the US base since the beginning of the 1970s, Fastame came to know firsthand the culture of secrecy that, almost fifteen years after the end of the Cold War, still reigned in the archipelago: "Secrecy is counterproductive. There is no need to keep secrets about the levels of radioactivity around a nuclear installation. Unfortunately, especially in Italy, and particularly within the navy and among military circles, there is the idea that keeping secrets is the basis of authority." Fastame recalled that when he worked in France, he experienced another communicative approach that the state adopted and organized to make the population part of the national nuclear culture. The French model, according to the radiochemist, should have been taken and directly transplanted in Italy. Instead of protecting the public by keeping it ignorant—Fastame insisted—public authorities should have organized credible emergency plans:

> The one that the Prefect of Sassari presented here in late 2004, during a small meeting exclusively reserved for the authorities (I was invited only be-

> cause of my reputation as a local expert but could not intervene or ask questions) was simply ridiculous. There were very naïve evacuation strategies, unrealistic contamination maps and dispersion models, and poorly organized interventions. For example, according to the plan, a group of firefighters specialized in radiocontamination emergencies should have come from the north of Italy to monitor the levels of radioactivity after a hypothetical accident. Can you imagine that? What could have they done after one or two days? Nothing.

Fastame proposed, almost provocatively, to install a monitor in the public square to display in real-time the radiometric data collected by the local laboratory. The military and the local administrators never gave an official response to this suggestion. Three elements emerged in Fastame's analysis of the postaccident reactions: the unpreparedness and disorganization of public safety agencies, the ignorance of the public, and the lack of transparency of military and civilian authorities. He communicated this view in a report that he sent to local, regional, and national authorities, concluding that of all the institutions involved in the mismanagement of the radiosurveillance system the US Navy was the most transparent. Considering the attempt to cover up the *Hartford* accident, Fastame's statement sounds hyperbolic, but it is understandable if interpreted from the perspective of an educated Maddalenino who worked in highly regulated industrial environments and with a comparative view of safety protocols. His review of the safety organization in La Maddalena was an explicit critique of Italian bureaucratic mentality and inefficiency, juxtaposed with the efficiency of foreign "nuclear cultures."

Local activists also desired to access information after the *Hartford* accident. CoCiS was initially formed by six Maddalenini opposed to the expansion of the submarine base proposed by the US Navy. When they learned of the accident, their strategy changed: "They made a huge mistake, and we could not do anything else but jump on it. Our choice was not simply strategic. People here were in fear, like I had never seen before. It was real."[55] Fabio, a former CoCiS member, who later became assessor of public works in the local administration, told me: "We did not even think of using the problem of radioactivity when we started our campaign against the expansion of the US base. But when the accident happened, clearly that became our major topic, together with the emergency plan, the efficiency of the radiosurveillance system, and the epidemiological data."[56]

CoCiS strategy involved promoting a democratic approach to information access, intended as a collective effort to understand what was going on. The public dimension of the problem also needed a stage for a pedagogical representation of a collective drama: a community that had always been surrounded by secrecy and paternalistic calls for obedience finally woke up and decided to stand up for its right to know. CoCiS members did not want to embrace, prejudicially, any particular argument or scientist opinion. Rather, they wanted to involve the public and make their fellow citizens feel engaged with the problem of risk after the accident, but also with the other dimensions of the US Navy presence that had been neglected over time.

While this group of citizens had no specific expertise on radiation detection or environmental monitoring, it mobilized traditionally apathetic sectors of the local community to confront secrecy by opening a balanced public debate on the scientific and safety problems that emerged after the accident. If, during the 1970s and the 1980s, mostly exogenous groups went to La Maddalena to openly protest the US Navy presence, CoCiS changed this pattern. Now the Maddalenini felt confident enough to participate in public. Antonello Tovo, one of group's leaders, recalls this with pride: "During our first rally in the public square, we counted around five hundred persons. I was desperate and started to cry. I thought that it was a complete failure. After all the work we had done, gathering only five hundred protesters seemed like a defeat. Then I received a call from Salvatore [Sanna], who, with an excited tone of voice, congratulated me on the formidable result! He told me that getting those people out to openly protest the mismanagement of the *Hartford* accident was just unthinkable ten years earlier."[57]

Local activists also took the lead by organizing and moderating a series of public events during which experts of different positions discussed the first results of the radioecological surveys performed after the accident: "Our attempt to open a public debate on the risks related to the US Navy presence in the archipelago was based on three methodological principles. We had to be credible, transparent, and objective, unlike the mayor, the Italian navy, and the scientists who came here to tell us that everything was fine and that we did not have to worry."[58]

CoCiS exploited the window of opportunity offered by the intervention of independent experts to open the black box of the radiosurveillance system put in place by Italian experts and authorities. In so doing, CoCiS members acquired credibility among the local population in a moment of disorientation and lack of trust toward military and civilian institutions.

SARDINIAN EXPERTS FACE THE PUBLIC

In February 2004, CoCiS organized two public assemblies in the city hall. Vincenzo Migaleddu moderated the first event (on February 8), titled "La Maddalena and Nuclear Power: a Dialogue for More Thoughtful Policies." At the time, Migaleddu was a radiologist in Sassari and active member of World Wildlife Foundation Gallura, one of the nongovernmental organizations that sponsored CRIIRAD's independent analyses of the *Hartford* accident. During the second debate held on February 14, CoCiS invited Giorgio Cortellessa and Mario Ladu, retired physicists with opposite views of the environmental effects of the US Navy base. Cortellessa made his career in the ISS. In the past he had vocally opposed the US Navy installation, denouncing presumed evidence of radioactive discharges in the archipelago with interviews and op-eds on national and Sardinian newspapers.[59] Ladu was a preeminent Sardinian physicist and radioprotectionist involved since the 1970s in La Maddalena's radiosurveillance program as a consulting expert for the Region of Sardinia. His position, like that of most Italian nuclear physicists, was markedly antimilitary. But after the *Hartford* accident, he took a strong position against the alarm created by CRIIRAD's results. According to him, the controversy over the "radioactive algae" was another political tactic of antibase activists to kick the US Navy out of the archipelago. As he stated in a newspaper interview during the controversy, "I have monitored the levels of radioactivity in La Maddalena repeatedly for the last twenty-five years and the only radioactivity I found there is due to the abundance of granite, as various researchers had already noticed since the 1930s. . . . [Politicians] have brought up the problem of radioactivity instrumentally, invoking scientific arguments for mere political purposes. This is not honest."[60]

Ladu voiced the frustration of his Sardinian colleagues, who felt accused of being reticent about the real risks of contamination after the accident, and even in collusion with the US Navy and Italian military authorities. By contrast, the initiatives of CoCiS offered an outlet to the more active sectors of the local community, who felt that they could finally express their anxieties and pose questions they never asked before.[61] Passages from an epistolary correspondence between a member of CoCiS and a Sardinian nuclear physicist offer a glimpse of the contrasting points of view about nuclear risk that existed in La Maddalena after the *Hartford* accident.

A few days before the second public assembly organized in La Maddalena, Ladu announced that he could not participate in the event. In his place, CoCiS invited Paolo Randaccio another nuclear physicist and radioprotectionist (a friend of Ladu) from the National Nuclear Physics Institute of Cagliari. On the evening of February 14, the large room where the city council usually assembled was so full of people that hundreds of citizens had to wait outside of the city hall.[62] For the first time, the personnel of the radiosurveillance laboratory faced a crowd of anxious and suspicious Maddalenini, who interrogated, interrupted, accused, and even booed whoever tried to dismiss their preoccupations. As Manca recalls: "With that atmosphere it was impossible to make people understand that from a scientific viewpoint their anxieties were unjustified because there was no evidence whatsoever of any sort of contamination after the accident. But they preferred to support—yes, support, like soccer fans—those who were creating alarm!"[63]

A few days later, Randaccio, who participated in the event, started a correspondence with professor Giovanna Sotgiu, a member of CoCiS, a teacher of French at the local classical lyceum, and also a highly active public intellectual. In his letter Randaccio addressed his disappointment and preoccupation for the atmosphere of suspicion breathed during the assembly:

> The people approved en masse all the interventions that emphasized the risks of nuclear contamination . . . while they disapproved and put under accusation all those interventions that tried to prove that there is no risk. . . . What struck me even more is the fact that nobody seemed to know that in La Maddalena there is an efficient and well-equipped laboratory of radiosurveillance. I do not know whether your request of further analyses to the French lab is due to simple lack of information or distrust in the local institutions. Distrust toward the laboratory directed by Dr. Manca has been expressed several times, with requests for new measurements, new methodologies, different instruments, without saying a word about the reasons why [the personnel of the local lab] should modify the way in which they perform their analyses. Dr. Manca has explained very clearly that her lab has never looked for thorium because it would be like trying to detect the presence of salt in the [sea] water: we know that thorium exists and does not provoke any harm; therefore we need to look at the real polluting factors. The cause of the higher or lower concentrations of thorium is only a collateral scientific problem. I am more and more convinced that the people are simply afraid of nuclear things and ionizing radiation because they do not know anything about them. Trying to make them reason in a meeting like the one we attended on 02/16/2004 is simply impossible.[64]

Randaccio not only repeated the mantra of local Sardinian experts since their first radiometric report in November 2003; in trying to prove the good faith and competence of the personnel of the local lab, he justified overlooking the presence of thorium by arguing that it did not constitute a new phenomenon and was irrelevant for the radiosurveillance system. By comparing the presence of thorium in the archipelago to that of salt in seawater, Randaccio represented thorium as a ubiquitous, natural, and therefore innocuous element because totally extraneous to anthropogenic radioactive contamination. One could argue that the equation natural = innocuous when speaking of radioactive elements is quite simplistic and misconceived, but Randaccio's explanation (and brutal simplification) concerning the mundane presence of thorium in the environment was probably meant to convey the idea that, contrary to nonexperts, experts know what signs are worth paying attention to and which are not.[65]

At the end of his letter, Randaccio asked Sotgiu to collaborate with an experiment he wanted to conduct in La Maddalena. The initiative consisted in distributing among the student population dosimeters for detecting natural radioactivity and a series of lectures about laboratory techniques for understanding its real presence and danger: "The youngest part of the population, more sensitive to scientific problems and less affected by political distortions, shall develop a correct culture of nuclear things and natural radioactivity, so that these themes can be addressed with rationality instead of relying on assumptions of total rejection of technology."

Sotgiu responded to Randaccio's pedagogical proposal and tried to address some of his criticisms by explaining the point of view of the local community. While conceding that some of the extreme positions observed during the public assembly of February 14 derived from ignorance, misunderstandings, and emotional reactions to uncertainty and to the mismanagement of the *Hartford* accident, Sotgiu maintained that "knowing the technical aspects [of radioprotection] is important to understand rationally what the problems are and how to solve them; but from another point of view, I am asking myself why, after phasing out nuclear plants in our country [after the referendum of 1987] we in La Maddalena have to live with a military nuclear installation potentially subject to incidents, whose consequences are unimaginable."[66]

This contradiction posed a problem of credibility for Italian public institutions, which, in the context of the post-Chernobyl denuclearization of Italy, had to make the presence of a nuclear military installation in La Maddalena appear safe and thus politically acceptable. "Believe me," Sotgiu added in her earnest letter to Randaccio, "the behavior of our politicians and the atmosphere of secrecy, partial information, and promised emergency plans that we have never seen, do not help us approach the correct scientific information that experts like you are trying to divulge."

Sotgiu's explanation for why residents preferred to give credit to alarmist theses rather than accepting the reassuring explanations of Randaccio and colleagues was that "when facing a presumed increase of the cases of cancer in La Maddalena, the idea that the cause is related to the US Navy presence and that it is possible to eliminate it [by closing the base], makes people more or less consciously espouse this possibility." Sotgiu's analysis suggested that her fellow Maddalenini almost wished that submarines were nuclear things affecting their bodies because in that case they could identify the precise cause of their illness and solve the problem by getting rid of the US base. The mobility of nuclear submarines and the fact that they could be easily removed from the archipelago made them perfect nuclear scapegoats.

By contextualizing local perceptions of risk within a political and a cultural point of view, Sotgiu's reply summarizes several different modes of analysis deployed by Sardinian experts and local inhabitants to make sense of the post-*Hartford* crisis. While experts like Randaccio tried to reassure the local population by dismissing their fears and critiques of the political and technical mismanagement of the accident, local activists used the intervention of independent experts to put the system of radiosurveillance under unprecedented scrutiny. When, after the disclosure of the *Hartford*'s accident, CRIIRAD and Aumento produced data that the local laboratory of La Maddalena could not interpret according to established protocols, the controversies about thorium accumulation processes, particularly in red algae, and the possible presence of plutonium, created unprecedented alarm among the archipelago's residents. Understanding the complex mechanisms at play in nuclear risk perceptions demands an analysis of radiosurveillance systems as assemblages of heterogeneous elements: laboratory protocols, instruments, sampling procedures, data interpretations, epistemic cultures, legal and political infrastructures. Collectively, these elements regulate and make possible particular regimes of perceptibility about risk.

The failure of local radioecologists to produce stalwart responses to the concerns underscored by CRIIRAD and Aumento was not a contingent, technical matter, but the result of a complex communicative and sociotechnical short circuit. Local scientists defended established practices relying on disciplinary assumptions that excluded, de facto, the contribution of nuclear submarines to the archipelago's radiation levels. These approaches identified the source of radioactivity in La Maddalena as tied to environmental factors—such as granite rocks and bioaccumulation process of the algae—and residue from the radioactive fallout of nuclear testing in the 1950s and 1960s and from the Chernobyl accident.

By naturalizing the presence of radioactivity in La Maddalena, Italian expert institutions attempted to remove any causality from the US Navy presence, which, according to professor Sotgiu, many local inhabitants were ready to espouse. This strategy, however, was ineffective. After months of uncertainty and distrust in public and military institutions, the credibility of the system of radiosurveillance could not be effectively restored.

CONCLUSION

Natural Histories

UNEARTHING THE MILITARY-INDUSTRIAL LEGACY OF LA MADDALENA

On January 25, 2008, five hundred people attended the "disestablishment ceremony" of the US Naval Support Activity base of La Maddalena.[1] On the stage assembled in front of the US Navy headquarters, the Italian and the American flags waved beside one another for the last time. US and Italian officers, along with political and religious authorities, celebrated thirty-five years of American presence in the archipelago. The US Navy departure was the end of an era—of close friendships, work routines, and shared stories. Italian workers on the US base, like many other Maddalenini, were also worried about their future. A few days after the departure of the last US Navy contingent, a group of activists of Legambiente—one of the environmentalist movements that had opposed the US base since the mid-1980s—traveled to La Maddalena to celebrate the "liberation" of the islands from the nuclear submarines. As they approached the docks of Cala Gavetta on their boat, a group of upset merchants and former employees on the US base welcomed them: "Go back! You come to celebrate but here there are people without jobs!" they yelled, while launching rotten eggs and tomatoes at the environmentalists.[2]

This episode epitomizes the conflicts around the ongoing attempts of the Maddalenini to transform what was once considered the "natural"

C.1. The decommissioning ceremony of the US Navy in La Maddalena in 2008. *Photo credit:* Andrea Nieddu.

C.2. The departure of the US Navy from La Maddalena. *Photo credit:* Andrea Nieddu.

military vocation of the island—a military-industrial economy—into one oriented toward ecotourism and the "natural" beauty of this Mediterranean outpost.

For some regional and national politicians, the closure of the US nuclear base and the decommissioning of the Italian navy arsenal wedged open new possibilities. Once attractive for its strategic military position—sandwiched between Corsica and Sardinia, with deep coves that could hide navy ships and submarines—La Maddalena could now attract tourists, rather than soldiers, to its crystalline waters. As this book shows, representations of La Maddalena as a "natural" military outpost are deep-seated and alive, but in recent years they have become ambivalent. Despite the nostalgia for the "golden age" of the archipelago and the concrete economic benefits of the military presence, the closure of the Italian navy arsenal in the early 2000s, and of the US submarine base in 2008, have brought to the surface old debates and tensions over the military role in the local economy and possible alternatives to it, and unveiled the toxic legacy produced by the archipelago's military-industrial activities.

For almost ten months, between 2008 and 2009, 1,600 construction workers, organized in three daily shifts, labored without pause to restore the vast area (155,000 square meters on land and 120,000 square meters docks) of the decommissioned Italian navy arsenal, built at the end of the nineteenth century on the southeastern tip of the major island of La Maddalena. The massive restoration project was part of a larger reconversion plan whereby the Region of Sardinia and the Italian government intended to revitalize the fading economy of the archipelago. The Italian government proposed to host in La Maddalena the thirty-fifth meeting of the G8 group scheduled for July 2009 to inaugurate the new era. Despite the acceleration of the restoration works and the careful media campaign to represent the archipelago as the next tourist mecca of the Mediterranean, a few weeks before his inauguration prime minister Silvio Berlusconi made the decision to relocate the G8 meeting in the city of L'Aquila, in the Abruzzi region, recently hit by a catastrophic earthquake. The image of US president Barack Obama and the other world leaders walking through the ruins, shepherded by Berlusconi, appeared on television with all its dramatic impact. The Maddalenini felt disappointed by the missed opportunity to become the center of international relations for a week, but mostly feared that the decision of Berlusconi could jeopardize the promised economic relaunch of the archipelago. Unfortunately for them, the ruins of L'Aquila buried La Maddalena's reconversion plan as well.

The relocation of the international meeting exacerbated the decline of the new arsenal project and inaugurated a series of scandals that defrauded residents of the future they were promised after the dismantlement of most of the military bases and their economic influx.[3] After massive investments for transforming the arsenal into a resort area, the site soon became an object of litigation concerning episodes of corruption, mismanagement, and the discovery that the companies who should have

C.3. "La Maddalena will not vote" poster to protest the transfer of the G8 event. Photograph by the author.

cleaned up the seabed area did not remove the contaminants.[4] Sardinian judges ordered the sequestration of the resort complex—declaring it unfit, until proven otherwise, for public use. Thirteen years later, the area remains inaccessible. The company that was granted a forty-year exclusive use of the tourist reception and naval dock facilities abandoned all the structures, leaving them deserted and exposed to the corruption of the winds. In a few months the promise of new jobs for many Maddalenini became a chimera.

On June 13, 2011, the Associazione Italiana Esposti Amianto (Italian Association of Workers Exposed to Asbestos) organized a conference in La Maddalena. The choice of location was deliberate. In 2007, Sardinian health authorities had already classified the area of the former Italian navy arsenal as one of the most environmentally exposed sites of the region, due to the quantity of chemicals stored inside the military facility and leached into the sea and soils over more than one hundred years. Decades after their retirement, former arsenal workers reflected on the hazards they faced on the job and described how safety regulations were often absent, transgressed, or overlooked. "I worked inside the arsenal for almost twenty years and can testify that asbestos was stored everywhere, and workers used it for repair operations on the ships without any precaution or protection," declared former arsenal employee and city councilor Gaetano Pedroni.[5] In the face of a growing number of colleagues dying from mesothelioma, retired workers have reinterpreted the Italian navy arsenal as a toxic workplace.[6] This has further destabilized notions of local identity tied to military culture and its economic benefits. Some archipelago residents now mobilize politically to challenge the island's military-industrial history and its long-term impacts on physical health and the local economy.

Two dramatic ironies have emerged from the decommissioning of the archipelago's military-industrial sites. First, long-term inhabitants often remember and celebrate with nostalgic tones La Maddalena's military vocation, arguing that the "natural preservation" of the archipelago was only possible because of its centennial military occupation, which impeded the development of other industrial activities that would have degraded its environment on a larger scale. These selective memories, though, omit the fact that the Italian navy intervened quite actively in shaping the land and seascape of the archipelago, sometimes with massive works of embankment that transformed or deleted entire beaches, and other times by simply denying access to zones of strategic military importance and making them unavailable to civilian uses.[7]

The closure of the Italian navy arsenal first, and later of the US submarine base, has revealed that those areas of the archipelago were potentially toxic and polluted. The second irony is that while some residents have expressed concerns about the risks of nuclear contamination around the US Navy installation for over thirty years, it is now clear that the mundane operations of the Italian navy arsenal were also major sources of environmental pollutants and health hazards—even perhaps the only source. "When facing a presumed increase of the cases of cancer in La Maddalena, the idea that the cause is related to the US Navy presence and that it is possible to eliminate it [by closing the base], makes people more or less consciously espouse this possibility," admitted Giovanna Sotgiu in 2004, after the *Hartford* accident. Perhaps, scapegoating the US Navy was a way to preserve intact the image of the Italian navy arsenal as the beating heart of the community. It is difficult to imagine or to admit that the place and the institutions that sustained so many lives in La Maddalena were also poisoning them.

In sum, the environmental restoration of the former arsenal and the US Navy base should have brought the military-industrial legacy of the archipelago to an end, but instead it worked to unearth toxic remains and to make visible the damaging health and environmental effects of the military-industrial economy around which the local community constructed its life and its sense of identity for well over two centuries. The "natural" history of the archipelago, tied to its strategic role in the Italian and the US military-industrial complex seems finally decoupled for the sake of its (potential) value as a tourist destination, incompatible with the presence of polluting industries. This now visible bifurcation puts the Maddalenini in front of new choices. The costs of the military economy become available for different political interpretations as they emerge through the physical signs of environmental pollution and health effects.

Even now, years after decommissioning, the legacy of the nuclear base is still haunting the future of the community. Local activists who opposed the US Navy presence for decades did not express interest in monitoring the decommissioning process of the US submarine installation, as if, after the departure of the American ships in 2007, the problem of radioactive contamination ceased to be relevant. Some long-term residents argued that this confirms their suspicion that "the problem of radioactivity was raised only instrumentally, for political reasons."[8] Yet local administrators and businessmen express ambivalence about the future uses of the site. A local administrator told me: "Officially the site is clean, but the Italian navy decided to maintain its military prerogatives on it, with the consequence that it cannot be used for other purposes. But even if it were available to civilian uses, how many tourists do you think would want to go there knowing that until five years ago there were nuclear submarines?"[9]

Any time national media present stories featuring the storage of hazardous material (from chemical weapons from Syria to nuclear waste coming from decommissioning Italian nuclear plants) the Maddalenini think that the granite bunkers of Santo Stefano are still there, available to the military, protected by secrecy, and inaccessible to the public, but still incorporated in the life of the community. Once used for the storage of weapons and other hazardous material, they can be reopened as a storage facility, probably for other projects through which the Italian state,

C.4. Fortifications of the artillery batteries of Punta Rossa, Island of Caprera. Photograph by the author.

C.5. Fortifications of Punta Rossa, Island of Caprera, detail. Photograph by the author.

once again, will reclaim the possession of the archipelago in the name of its centrality to the future of the nation. La Maddalena rejoins Sardinia and its composite identity: the island of ancient civilizations, archaeology, nature, and uncontaminated sea, and the very contaminated nature of its colonial role within Italy's military-industrial complex.

In this book I argue that certain risks are not immediately identifiable. Objectifying and representing risk requires work, infrastructures, expertise, technical equipment, and organization. Without material evidence and knowledge of their potential effects, certain risks remain undefined, undistinguishable from the everyday background. The collective work of making them identifiable is both technical and political, relational, and connected to the material and cultural circumstances in which they spread unevenly. For this reason, objectifications and representations of risk change over time, even within the same community. The question of (in)visibility is key to interpreting La Maddalena's past and the long-term effects of the military presence on its future. According to architect Stefano Boeri, who coauthored the reconversion project of the former Italian navy arsenal, the archipelago's unique environmental features have forced military architects "to create unusual and sophisticated forms of camouflage and positioning (through the use of local material, in their choice of outlooks and viewpoints, and in adapting access points to the land) while keeping faith with their original minimalist concepts of construction."[10] La Maddalena has been marked by the military presence, and this is immediately (im)perceptible if one pays attention to the various military forts, bases, ammunition depots, watch towers, and bunkers that have been erected throughout the past three centuries. But in the archipelago defensive architecture had to adapt in unique ways to rock formations levigated for millennia by the winds and the washing of the seawater, resulting in unique shapes—what Boeri calls the "excesses of nature."[11] All the military structures of the archipelago present two main characteristics. First, they blend into the natural landscape of granitic rocks—they were built with a mimetic intention, to render them imperceptible from the exterior. The second characteristic of these forts, towers, and artillery batteries is that they offer a privileged point of observation toward the same landscape in which they blend. In sum, the military architecture of La Maddalena is a plastic manifestation of the idea—and strategic need—of "nonreciprocal visibility."[12]

Mimesis and (in)visibility are recurring themes throughout the history of La Maddalena. From the Corsican shepherds who inhabited the most inland and elevated areas of these islands to the pirates who absconded inside their granitic coves, from the erection of towers and forts

C.6. Main conference building "House of the Sea." La Maddalena. *Photo credit:* Fabio Presutti.

to the secrecy of military operations, and the seclusion of entire portions of territory for exclusive military use, the occupants of the archipelago have lived through various forms of concealment and revelation. Memory and forgetting work in the same way. Historical narratives about the birth of the community—as a direct emanation of the Sabaudian occupation of 1767 and subsequent military installations—make the military presence coincide with the social and natural life of the archipelago. But the military presence did not just blend into the natural landscape: it has assembled, constructed, excavated, expropriated, and disciplined the environment with the intention of controlling it and subjecting it to its strategic needs. The factitious and the natural are purposefully mixed.

The effect and the purpose of "origin" narratives produced by military historians like Aristide Garelli—and reproduced by large sectors of the local residents—have been to represent the strategic military function of the archipelago as the only possible past, present, and future of the community. But what if instead of this deterministic reading of La Maddalena's past, we unlocked it to imagine a different future? The power of history making reveals itself during controversies, in times of uncertainty, like the one that the inhabitants of the archipelago are experiencing as the military monoculture is progressively receding from the center of their lives. The failed attempt to reconvert the function of this center—the former military arsenal—into a resort area oriented toward an alternative touristic future has opened a profound scar that has rendered the toxic legacy of the military presence visible, in unexpected ways.

In the intentions of Boeri and the architects of the economic reconversion of La Maddalena, the military arsenal—once a secluded place—should have become open to the public: a vanguard touristic port, with annexed reception structures, shipyard, and storage facilities. The House of the Sea, the most representative building of the new resort area, is a glass box, protected only by a thin hive-shaped glass-metal structure (fig. C.6). Designed to be a privileged point of observation, thanks to its physical characteristics (partially suspended over the sea), the house would offer unique conditions to gaze upon the natural elements of the archipelago (winds, sea, islands, sun) and at the same time would be open to the scrutiny of external observers. This is in sharp contrast to the military function of forts and bunkers. In fact, Boeri's design was meant to reverse the strategic logic of defense architecture (seeing without being seen).

The compressed (ten months) and secret construction process that reshaped the entire area of the former navy arsenal, though, contradicted from the beginning Boeri's logic of transparency. During the work of reconversion, the site remained off-limits, with military personnel surveilling the area and allowing only workers and authorized personnel to access it. The mandate under which the head of the Italian Protezione Civile, Guido Bertolaso, acted as plenipotentiary manager of the area declared "of national interest," testifies to the continuity of the Italian state's patronizing presence in the archipelago.[13] La Maddalena's case is only the umpteenth instance of the dangerous practice of the Italian government to remove budgetary, organizational, and executive functions of its agencies from public scrutiny and normal procedures of accountability, with the excuse that the implementation of grandiose and ludicrous public works for big events corresponds to a situation of emergency, like a natural calamity.[14] The use of the archipelago as a strategic platform in the interest of the state is a problem that the community of La Maddalena continues to deal with even after its progressive demilitarization. The militarily controlled life of the archipelago has excluded the proliferation of alternative economic opportunities, has deprived La Maddalena of the right to know, and has infused a false sense of protection at the expense of the community's capacity to define and choose its own destiny. My perception is that for the Italian state La Maddalena has to remain open, undefined, a sort of perennial stem cell potentially ready to receive and to adapt to whatever function someone, somewhere else decides to assign to the archipelago regardless of what the Maddalenini think. The future cannot be mass tourism and indiscriminate exploitation of the local environment either. The preservation of La Maddalena's environmental equilibrium is a crucial condition for the economic and cultural survival of the local community. The struggle ahead is about rediscovering the multiple meanings of identity and purpose after the military have gone away.

I offer these concluding considerations with the awareness of how difficult for all the Maddalenini this moment is, and with only the hope that somehow, sometime, this book, too, can contribute to open a debate about the future of these beloved islands.

Notes

INTRODUCTION: THE ATOMIC ARCHIPELAGO

1. France's request that all Italian military bases in the proximity of its borders be removed was withdrawn after Italy entered the NATO alliance in 1949.

2. These submarines were of the Sturgeon and Los Angeles class. Smaller and faster than Polaris submarines equipped with Inter-Continental Ballistic Missiles, fast attack submarines were called "hunter killers" because their mission was to surveil the activity of the enemy, to chase down enemy vessels, to patrol specific sectors of the open sea and of the coasts, and to escort their fleet. Polmar and Whitman, *Hunters and Killers*. Sanna, *La Maddalena, Sardegna*; Sanna, *La Base Atomica di La Maddalena-Santo Stefano*; Tovo, "L'America a La Maddalena."

3. Sanna, "Quadro storico sull'origine e l'evoluzione," 69–92.

4. Hecht, *Radiance of France*, 15.

5. Johnstone and Cramer, "Burdens of the Glory," 199–223; Heller and Lammerant, "U.S. Nuclear Weapons Bases in Europe," 96–130.

6. Dobson and Whitman, "Project Lamachus," 169–92.

7. Engel and Carté Engel, "On Writing the Local, within Diplomatic History," 20.

8. Krige, "Hybrid Knowledge," 337–57; Mateos and Suárez-Díaz, "Technical Assistance in Movement," 345–67; Cassata, "Building the Casaccia Gamma Field," 67–101; Oreskes and Krige, *Science and Technology in the Global Cold War*.

9. Gabrielle Hecht defines *technopolitics* as "the strategic practice of designing or using technology to constitute, embody, or enact political goals." *Radiance of France*, 15.

10. Curli, *Il Progetto nucleare italiano (1952–1964)*; Silvestri, *Il Costo della Menzogna*; Bini and Lordero, *Nuclear Italy*.

11. See, for example, the excellent volume *Le due scienze*, by Francesco Cassata, on the Italian debate concerning Lysenkoism.

12. Berlinguer, *Politica della scienza*; Zorzoli, *La ricerca scientifica in Italia*; Società Italiana di Fisica, *La scienza nella società capitalistica*; Ciccotti, Cini, de Maria, and Jona-Lasinio, *L'ape e l'architetto*.

13. Moore, *Disrupting Science*.

14. An illustration of the motives behind the protest can be found in a document redacted by the occupants: "Esperienze e prospettive di lotta nei centri di ricerca," 173–91. Other examples of the activity of such groups in Italy can be found in Cassata, *L'Italia intelligente*, especially part III. The episode of the occupation of the International Laboratory of Biophysics and Genetics of Naples, directed by Adriano Buzzati-Traverso, was described by the protesters in "Una corporazione della scienza," 121–46. Ciafaloni, "Le corporazioni della scienza," 79–90; Berlinguer, *Politica della scienza*, esp. ch. 7, "I Lavoratori della scienza."

15. De Castro Pietrangeli, Marsili, and Modigliani, *Storia e identità di un ente di ricerca*.

16. Similar debates were also taking place within other political parties, even before 1968. For example, in 1961 the Central Office for Cultural Activities of the Democrazia Cristiana organized a conference titled Una politica per la ricerca scientifica. The contributions converged on the idea that stronger investments for scientific and technological research, in coordination with a substantial reform of the school system, should be organized in view of the economic growth of the nation. See Democrazia Cristiana, *Una politica per la ricerca scientifica*.

17. Partito Comunista Italiano, *La ricerca scientifica e la società in Italia*.

18. Carlo M. Santoro, "La ricerca scientifica e la società italiana," *L'Unità*, March 11, 1970, 10.

19. Partito Comunista Italiano, *Uomo Natura Ambiente*.

20. Barca, "Work, Bodies, Militancy," 115–33. See also Hardenberg and Pelizzari, "Environmental Question, Employment, and Development," 77–105.

21. A short history of the ecological movement in Italy can be found in Della Seta, *La difesa dell'ambiente in Italia*; Corona, *Breve storia dell'ambiente in Italia*, esp. chs. 3 and 4; Candela, *Storia ambientale dell'energia nucleare*.

22. Krige, *American Hegemony and the Postwar Reconstruction*.

23. See, for example, Mary Nolan's discussion of the use and the analytical currency of Americanism as concept for the historiographical interpretation of the American presence in Europe, and especially in Germany, in "Anti-

Americanism and Americanization in Germany," 88–122. Nolan, *Visions of Modernity*; De Grazia, *Irresistible Empire*; Stephan, *Americanization of Europe*.

24. Johnstone and Cramer, "Burdens of the Glory," 199–223; Heller and Lammerant, "U.S. Nuclear Weapons Bases in Europe," 96–130; Simich, "Corruption of a Community's Economic and Political Life," 77–94.

25. Lutz, "Empire Is in the Details," 593–611; Coronil, "After Empire," 241–71.

26. Oldenziel, "Islands," 13–41. See also an interesting RAND Corporation analysis of the use of overseas bases as strategic military platforms for launching aerial attacks toward the USSR: A. J. Wohlstetter, F. S. Hoffman, R. J. Lutz, and H. S. Rowen, *Selection and Use of Strategic Air Bases*, Report R-266 (RAND Corporation, April 1954).

27. Kelly, "U.S. Power, after 9/11 and before It," 347–69; Steinmetz, "Return to Empire," 339–367; Vine, *Island of Shame*, esp. chs. 2 and 3. Sandars, *America's Overseas Garrisons*, in particular chs. 1–5; Harkavy, *Great Power Competition for Overseas Bases*; Lutz *Homefront*.

28. Kaplan, "Where is Guantanamo?" 831–58.

29. For the history of the early phase of US military presence in Italy after World War II, see Nuti, "U.S. Forces in Italy, 1945–1963," 249–72.

30. Nuti, *La Sfida nucleare*.

31. For an overview of US overseas basing agreements see Cooley, *Base Politics*, esp. ch. 2.

32. William Burr, "How Many and Where Were the Nukes? What the U.S. Government No Longer Wants You to Know about Nuclear Weapons during the Cold War," *National Security Archive*, 2006, www.nsarchive.gwu.edu/NSAEBB/NSAEBB197/index.htm#18. See in particular Document 13, Letter from Secretary of State John Foster Dulles to Secretary of Defense Charles E. Wilson (Top Secret), April 12, 1956, and Document 14, US Embassy Rome Dispatch 525 to Department of State, Transmitting Documents Constituting Military Atomic Stockpile and "Consent" Agreements (Secret), January 17, 1962.

33. See Nuti, *La Sfida nucleare*, esp. chs. 5 and 8, and Simich, "Comiso."

34. The original article reporting on the ceremony is Joe W. McFadden, "Italian, American Airmen Honor 50th Anniversary at Ghedi Air Base," *Spangdahlem Air Base*, February 5, 2014, https://www.spangdahlem.af.mil/News/Article-Display/Article/730327/italian-american-airmen-honor-50th-anniversary-at-ghedi-air-base/. Hans Kristensen, "Italy's Nuclear Anniversary: Fake Reassurance for a King's Ransom," *Strategic Security* (blog), June 30, 2014, Federation of American Scientists, http://fas.org/blogs/security/2014/06/ghedi/. For a detailed discussion of the presence, origins, goals, and legal status of US and NATO military bases in Italy see Desiderio, "Viaggio nelle basi USA in Italia," 59–72, and "La fatal Vicenza," 397–406 and "Paghiamo con le basi la nostra sicurezza," 27–42; Nuti, "U.S. Forces in Italy, 1945–63," 249–72.

35. Duke, *United States Military Forces and Installations in Europe*, 200.

36. Duke, *United States Military Forces and Installations in Europe*; Arkin and Fieldhouse, *Nuclear Battlefields*.

37. See Cooley, *Base Politics*, esp. ch. 6, "Italy and Japan: The Politics of Clientelism and One-Party Democratic Rule," 175–216.

38. The honorable Massimo De Carolis was a fervent supporter of NATO. Editor of the weekly conservative magazine *Il Settimanale*, De Carolis went to La Maddalena to write a reassuring article on the innocuousness of US nuclear submarines and on that occasion was offered a short cruise underwater aboard a nuclear submarine by the US Navy command. De Carolis, "A bordo del 'mostro nucleare,'" *Il Settimanale* 5, no. 5, February 1, 1977, 29.

39. Kuchinskaya, "Twice Invisible," 78–96.

40. Kuchinskaya, *Politics of (In)Visibility*.

41. Murphy, *Sick Building Syndrome and the Problem of Uncertainty*.

42. This happened particularly after events, like the accident at Three Mile Island, that made locals worry about the presence of the nuclear base. "Ma può davvero accadere anche a La Maddalena?" *L'Unione Sarda*, April 30, 1979.

43. Hecht, *Being Nuclear*.

44. My argument here is influenced by the reading of Mol, *Body Multiple*, and by Webb Keane's concept of *bundling*; that is, the copresence of various qualities as they are simultaneously embodied in material objects. Keane uses Peircean semiotics to analyze how the potential significance of general qualities gets instantiated through their embodiment in material objects. The important point here is that "bundling is one of the conditions of possibility for . . . the 'biography' of things, as qualisigns bundled together in any object will shift their value, utility, and relevance across contexts." See Keane, "Semiotics and the Social Analysis of Material Things," 414.

45. The literature on compartmentalized knowledge in cold war science, especially in the nuclear sector, is vast. Masco, *Nuclear Borderlands*; Hecht, *Being Nuclear*; Brown, *Plutopia*; Masco, *Theater of Operations*.

46. Gusterson, "Nuclear Weapons and the Other in the Western Imagination," 111–43.

47. These are common tropes that I collected during many informal conversations and interviews I had with friends and local residents over the past four years of fieldwork in La Maddalena. Some of the same tropes are also present in newspaper accounts of life in La Maddalena after the arrival of the US Navy. Among many examples, see "L'Ombra dell'Atomica su La Maddalena," *Rivista Italiana di Ecologia*, February 1973, 37–61; "Ecco La Maddalena: I Nostri Inviati dall'Isola Radioattiva," *L'Europeo*, May 1974, 30–37; "A Bordo del 'Mostro' Nucleare," *Il Settimanale*, February 1978, 26–29.

48. Hecht, *Being Nuclear*, 6.

49. Hecht, *Being Nuclear*, 6.

50. Hilgartner, Bell, and O'Connor, *Nukespeak*.

51. See Brown, *Plutopia*; Hecht, "Invisible Production and the Production of Invisibility," 353–68; Kuletz, *Tainted Desert*.

52. To be clear, I am not trying to deny the importance of the physical characteristics of radiation and radioactive materials, which are doubtlessly relevant. What I want to do is to demystify "the nuclear" as a special field of studies and put it in fruitful conversations with others that are equally concerned with the problems of environmental risk and advocacy, popular epidemiology and non-expert knowledge, and the relationship between science, technology, capitalism, and democracy. For example, see Kirsch, *Mining Capitalism*, esp. ch. 4, "Corporate Science."

53. See Oreskes and Conway, *Merchants of Doubt*; Proctor and Schiebinger, *Agnotology*.

54. Masco, *Theater of Operations*, especially ch. 3, "Sensitive but Unclassified," 113–44.

55. Masco, *Theater of Operations*, 124.

56. Peter Galison discusses this aspect of state secrecy in "Removing Knowledge," 37–54.

57. Moynihan, *Secrecy*, 60.

58. Masco, *Theater of Operations*, 123.

59. Masco, *Theater of Operations*, 123.

60. Masco, *Theater of Operations*, 125.

61. Can secrecy, as a pillar of the security state, be compatible with democratic institutions? Countless commentators have tried to answer this vexed question during the Cold War. For example, warning against the extremes of McCarthyism, Edward Shils argued that the survival of American democratic institutions was possible only at the condition that secrecy and publicity coexisted in a fine balance, guaranteeing freedom and security. He was not alone. Other contemporary commentators underlined with alarming tones the exponential growth of secrecy in American society and its danger for democratic institutions. Expressly connected to nuclear weapons and the advent of national security state during the Cold War, David Wise and Thomas B. Ross opened their bestseller *The Invisible Government* with an urgent admonition: "There are two governments in the United States today. One is visible. The other is invisible. The first is the government that citizens read about in their newspapers and children study about in their civics books. The second is the interlocking, hidden machinery that carries out the policies of the United States in the Cold War." Shils, *Torment of Secrecy*.

62. The criterion of introduction as stated in the Nuclear Non-Proliferation Treaty of 1970 (that is, the evidence of nuclear capabilities provided by nuclear tests), allowed Israeli's elites to maintain an official nonnuclear status.

63. Taussig, *Defacement*, 5.

64. Simmel, "Sociology of Secrecy and of Secret Societies," 441–98.

65. Masco, "Lie Detectors," 441–67.

66. Kaur, "Sovereignty without Hegemony," 3–28; Wynne, *Rationality and Ritual*.

67. Canetti, *Crowds and Power*.

68. De Felice, "Doppia lealtà e doppio stato," 493–563.

69. Bobbio, *Future of Democracy*, esp. ch. 4, "Democracy and Invisible Power," 79–97.

70. Bobbio, *Future of Democracy*, 95.

71. Frickel, "Absences," 86–95. See also Frickel, "Not Here and Everywhere," 263–276.

72. Hess, *Alternative Pathways in Science and Industry*.

73. Frickel and Moore, *New Political Sociology of Science*.

74. See Hacking, *Historical Ontology*.

75. Proposers of the "risk perception" paradigm have recently reconsidered the importance and value of sociocultural variables in public and expert understanding of risk, but their models remain largely schematic when addressing the complex mechanisms of knowledge formation "in context." Moreover, they continue to consider "culture" as an overarching and yet unspecified and decontextualized factor in shaping public understandings of risk. See Kasperson, "Social Amplification of Risk," 153–78. Douglas, *Risk Acceptability according to the Social Sciences*, esp. ch. 3, "Risk Perception"; Krimsky and Golding, eds., *Social Theories of Risk*; Lupton, *Risk*, esp. ch. 2, "Theorizing Risk."

76. Michael, "Ignoring Science," 105–25; Wynne, "Public Understanding of Science," 361–89.

77. See Wynne, *Rationality and Ritual*, and "Misunderstood Misunderstandings," 281–304. See also Jasanoff, *Designs on Nature*, and her recent discussion "Mirror for Science," 21–26.

78. Beck, *Risk Society*.

79. Wynne's alerts about the perils of overstretching his original argument are quite illuminating. See, for example, Wynne, "May the Sheep Safely Graze," and Wynne's reply to Darrin Durant: "Elephants in the Rooms Where Publics Encounter 'Science,'" 21–33.

80. Irwin, Simmons, and Walker, "Faulty Environments and Risk Reasoning," 1311–26.

81. Irwin, Simmons, and Walker, "Faulty Environments and Risk Reasoning," 1312.

82. Michael, "Comprehension, Apprehension, and Prehension," 357–78.

83. Marres, *Material Participation*.

84. Marres, *Material Participation*, 65; Newell, "Affectiveness of Symbols," 1–22. I would like to thank the two anonymous reviewers who pointed out Newell's insightful work to me.

85. Marres, "Issues Spark a Public into Being," 208–217.

86. This point is clearly discussed in Newell's "Affectiveness of Symbols," especially in Paul Manning's comments about affect as a dyadic relation between two entities, which is not yet articulated and nonrepresentational. As such it is not mediated by reflexive thought (thirdness). See Newell, 14–16.

87. Parr, "Working Knowledge of the Insensible," 820–51; Kuchinskaya, "Articulating the Signs of Danger," 405–21; Reno, "Beyond Risk," 516–30; Shapiro, "Attuning to the Chemosphere," 368–93.

88. Kuchinskaya, "Articulating the Signs of Danger."

89. This point has been largely debated and problematized within the field of political ecology. See Horowitz, "Local Environmental Knowledge," 235–248; Agrawal, "Dismantling the Divide between Indigenous and Scientific Knowledge," 413–39; Ottinger, "Changing Knowledge, Local Knowledge, and Knowledge Gaps," 250–70.

90. Manetti, *Theories of the Sign in Classical Antiquity*, esp. 36–53.

91. Charles Sanders Peirce, "How to Make Our Ideas Clear," in Peirce, *Philosophical Writings of Peirce*, 31.

92. Keane, "Semiotics and the Social Analysis of Material Things," 419.

93. Keane, "On Semiotic Ideology," 64–87.

94. Keane, "On Semiotic Ideology," 66–67.

95. Keane, "On Semiotic Ideology," esp. 74–80.

PART I: A STRATEGIC NAVAL OUTPOST

1. For example, see Johnstone and Cramer, "Burdens of the Glory," 199–223; Moon, *Sex Among the Allies*. Johnson, ed., *Okinawa*; Lutz, ed., *Bases of Empire*. McCaffrey, *Military Power and Popular Protest*; and Vine, *Island of Shame*. For a study of the impact of US military bases on Italian local communities, see Simich, "Corruption of a Community's Economic and Political Life," 77–94.

2. Garelli, *L'Isola della Maddalena*; Marmonier, "La Question de La Maddalena," 1–41; Baldacci, *Ricerche sull'Arcipelago de La Maddalena*; Sotgiu and Sega, *Un'isola e il suo ammiraglio*.

3. Ciasca, "Corsi Colonizzatori della Sardegna nel Sec. XVIII," 314–90; Sole, "Contributo alla Storia di La Maddalena (1720–1767)," 35–62; Castangia, *Sovranità, Contiguità Territoriale e Isole*. Sanna, *Il popolamento dell'Arcipelago Maddalenino prima dei Savoia*. Sotgiu and Frau, *Le Bocche di Bonifacio*.

4. Sotgiu, *A Mimoria d'a petra*; Sotgiu, *Arcipelago di La Maddalena (1839–1843)*; Sotgiu and Del Monaco, *La Maddalena e i suoi traffici marittimi*.

5. Sotgiu, *Moneta*; Abate and Nardini, *Il Pane del governo*.

6. Baldacci, *Ricerche sull'arcipelago de La Maddalena*.

7. Garelli, *L'Isola della Maddalena*; Marmonier, "La Question de La Maddalena," 1–41; Michel, "L'Occupazione Sarda delle Isole della Maddalena," 28–64; Ciasca, "Corsi Colonizzatori della Sardegna," 314–90; Castangia, *Sovranità, Contiguità Territoriale e Isole*; Prasca, "Alcune Lettere dell'Ammiraglio Nelson," 374–407;

Prasca, *L'ammiraglio Giorgio des Geneys e i suoi tempi*; Riccio, "La Vittoria de La Maddalena nel 1793," 885–900.

8. This gap has been addressed by a recent study of Sanna, *Il popolamento dell'Arcipelago Maddalenino*.

9. Salvatore Sanna also edited and wrote two volumes about the US Navy base: *La Maddalena, Sardegna* and *La base atomica di La Maddalena-Santo Stefano*.

10. Salvatore Sanna, interview by the author, June 2009.

11. This is the thesis of local historian and linguist Antonio (Tonino) Conti, who published a book on this subject: *Sbirizzendi pe l'Isula*. Conti's thesis about the prevalence of Italian in the linguistic structure of the local dialect contrasts with a previous interpretation of La Maddalena's linguistic identity as derived from the Gallurese, the language of northern Sardinia, by local linguist Renzo De Martino: *Il dialetto maddalenino*.

CHAPTER 1: NOT AN AMERICAN FLEET TOWN

1. For interesting comparisons with US military overseas basing strategies and practices see Gillem, *America Town*.

2. See Vine, *Base Nation*.

3. "U.S. Sailors on Italian Isle Find Amity but No Social Life," *New York Times*, February 12, 1977, 2.

4. "U.S. Fleet on Island Sparks Controversy," *Arizona Republic*, October 28, 1972.

5. "Marinai USA all'assalto della macchina del peccato," *La Nuova Sardegna*, April 4, 1973.

6. "9 Yank Cars Firebombed," *Pacific Stars and Stripes*, August 17, 1976, 8; "March Protests Navy Presence," *Stars and Stripes*, August 20, 1976; "Un unico commando alla Maddalena ha incendiato le auto americane," *Corriere della Sera*, August 18, 1976; "Incidenti alla Maddalena alla Marcia antimilitarista," *Tutto Quotidiano*, August 20, 1976; "Scontri alla Maddalena tra polizia e dimostranti," *La Nuova Sardegna*, August 20, 1976; "Gli antimilitaristi caricati dalla polizia," *La Nuova Sardegna*, August 21, 1976; "La Maddalena: il muretto che ha fatto tremare il mondo," *Avanti!*, August 28, 1976.

7. "A La Maddalena i marinai USA fuori dell'isola le loro famiglie?" *La Nuova Sardegna*, February 1, 1973.

8. "Piovono le ingiunzioni di sfratto. Alle stelle i canoni di affitto," *La Nuova Sardegna*, March 28, 1973.

9. "Nessun americano è in grado di pagare gli altissimi fitti," *La Nuova Sardegna*, April 7, 1973.

10. "Island Life: An Isle off Sardinia to Become Modern Home for Navy Families," *Stars and Stripes*, July 4, 1973, 10.

11. "La Maddalena: si sblocca il problema degli alloggi per le famiglie americane," *La Nuova Sardegna*, January 16, 1973; "In febbraio gara d'appalto per i primi 43 'alloggi USA'," *La Nuova Sardegna*, January 27, 1973.

12. *Benvenuti a La Maddalena*.

13. *Benvenuti a La Maddalena*, 7.

14. Frances K. Porter, "Just Possibly It Could Be Worse," *Stars and Stripes*, letter to the editor, September 10, 1977.

15. Frances K. Porter, "Invitation to Visit La Maddalena," *Stars and Stripes*, letter to the editor, November 27, 1978.

16. R. M. Hughes, "More on Life in La Maddalena," *Stars and Stripes*, letter to the editor, December 16, 1978.

17. Every retired US Navy sailor I interviewed remembered those parties fondly. Sometimes the parties, which US Navy Support Office started to organize in 1975, hosted more than five thousand Italians, as reported by one article: "Navy's Friendship Day on Sardinia: 'It went great,'" *Stars and Stripes*, November 2, 1980, 8.

18. Former US Navy employee, interview with the author, July 2010. Anonymity is maintained to protect the informant.

19. Former US Navy employee, interview with the author, August 2009. Anonymity is maintained to protect the informant.

20. Carlo is not the real name of the former US Navy employee.

21. Former US Navy employee, interview with the author, October 2012. Anonymity is maintained to protect the informant.

22. Former US Navy employee, interview with the author, May 2012. Anonymity is maintained to protect the informant.

23. Previous contestations of US and NATO military bases in Italy took place more explicitly as part of Cold War geopolitical struggles between the United States and the Soviet Union. In particular, the deployment of the Jupiter missiles between 1961–1963 in Gioia del Colle, Puglia, triggered a wave of protests including parliamentary opposition from communists and socialists and the organization of mass protest marches. Two factors probably explain the differences between the case of Gioia del Colle and La Maddalena. First, the installation of the Jupiter missiles happened in a moment of high international tension (with the missile crisis in Cuba). Second, the local population of Gioia del Colle had the immediate perception of the missiles (with nuclear warheads) as nuclear objects and threatening presences. The ramps and the missiles erected in the middle of the flat rural landscape were visible from afar. The meaning of their presence and their dangerousness could be hardly hidden. On this episode see Nuti, *La Sfida nucleare*; Nash, *Other Missiles of October*.

CHAPTER 2: BECOMING NUCLEAR

1. A selection of scenes from the original movie are available at http://www.veoh.com/watch/v18220342h9qsPzX9?h1=benvenutti.

2. Amme, "Soviet Navy in the Mediterranean Sea," 154–59; Roucek, "US and USSR Confrontation," 546–75.

3. "Soviet Thrust in the Mediterranean," *Time*, June 28, 1971, 27–28.

4. See Turchetti, "Sword, Shield and Buoys," 205–31; Camprubí and Robinson, "Gateway to Ocean Circulation," 429–59; Camprubí, "'No Longer an American Lake,'" 428–46. "Choke points" are bottlenecks, constrained points of access into a naval area for both surface and underwater navigation. It is possible to access the Mediterranean through the Strait of Gibraltar and through the Dardanelles. For submarine navigation other choke points are the sea south of Sardinia and Sicily, between Crete and Greece, between Crete and Africa, and between Crete and Turkey.

5. "Cinquemila visitatori sul gigante del mare," *L'Unione Sarda*, September 2, 1972.

6. "Base alla Maddalena per sommergibili nucleari," *Il Messaggero*, September 16, 1972.

7. Documents of the Pentagon insisted that the increased activity of the Soviet navy in the Mediterranean shifted the balance of forces deployed in the area. A new strategy was needed in order to respond to this challenge.

8. For example, the article "Kossighin polemico sulla base USA alla Maddalena," *Il Messaggero*, October 26, 1972, reported that Soviet prime minister Alexei Kosygin expressed frank disappointment in the concession of La Maddalena to the US Navy. *L'Unità* also emphasized Soviets' discontent with the cession of La Maddalena to the US Navy: "Kossighin deplora la cessione agli USA della Maddalena," *L'Unità*, October 26, 1972.

9. "Italians debate U.S. port plan," *New York Times*, October 2, 1972, 14.

10. "C'è chi teme le radiazioni dei sommergibili nucleari," *Corriere della Sera*, September 22, 1972. Also, *Il Messaggero* reported the spread of concerns in other parts of Sardinia about the possible radioactive contamination coming from the submarines stationed in La Maddalena: "Sorpresa e timori in Sardegna per base nucleare alla Maddalena," *Il Messaggero*, September 19, 1972.

11. Senato della Repubblica, *VI legislatura, 38ª seduta pubblica. Resoconto Stenografico, Venerdì 6 Ottobre* 1972, 1848–1903.

12. Senato della Repubblica, *VI legislatura*, *38ª seduta pubblica*, 1855. Here Medici referred to the visit of the merchant ship *Savannah* in 1964, also documented in a film produced by CNEN for an instructional project. The documentary is available in the historical section of ENEA's webTV page: http://webtv.sede.enea.it/index.php?page=listafilmcat2&idfilm=173&idcat=30. On that occasion on November 23, 1964, the contracting parties signed an indemnity agreement concerning any liability that an Italian court may have found in relation to any nuclear accident deriving from or caused by the operations and repair of the NS *Savannah* in Italian territorial waters: United Nations, *Treaty Series*, vol. 532, 1965, 133. A new agreement between Italy and the United States was registered by the United Nations in 1965 according to which "the United States will provide compensation by way of indemnity for any legal liability which an Italian court

may find for any damage to people or goods deriving from a nuclear accident in connection with, arising out of or resulting from the operation, repair maintenance or use of the N.S. *Savannah*, in which the N.S. *Savannah* may be involved within Italian territorial waters, or outside them on a voyage to or from Italian ports if damage is caused in Italy or on ships of Italian registry. Within the $500 million limitation in such public laws, the operator of the ship shall be subject to the jurisdiction of the Italian court and shall not invoke the provisions of Italian law or any other law relating to the limitation of shipowner's liability." "Exchange of Notes Constituting an Agreement Concerning Liability during Private Operation of N.S. *Savannah*. Rome, 16 December 1965," United Nations, *Treaty Series*, vol. 534, 1966, 140–42.

13. Senato della Repubblica, *VI legislatura, 38ª seduta pubblica*, 1855–56.

14. Senato della Repubblica, *VI legislatura, 38ª seduta pubblica*, 1873–75.

15. Senato della Repubblica, *VI legislatura, 38ª seduta pubblica*, 1876.

16. "La base della Maddalena: protestano fisici e biologi," *Paese Sera*, October 4, 1972; "Protestano scienziati ed ecologi per la base alla Maddalena," *Il Messaggero*, October 5, 1972.

17. "Severa critica del fisico Amaldi," *L'Unità*, October 8, 1972.

18. In his biographical essay, Amaldi explains very clearly how the developments of nuclear physics in the years preceding World War II and immediately after it created a firm opposition within the Italian scientific community toward the military applications of nuclear science and technology. This position prevailed within the Italian school of nuclear physics especially after Hiroshima and Nagasaki. See Amaldi, *Da via Panisperna all'America*. See also Leopoldo Nuti, *La Sfida nucleare*, esp. ch. 1, 15–44.

19. "Protestano scienziati ed ecologi per la base della Maddalena," *Il Messaggero*. Italia Nostra (Our Italy) is one of the most important cultural associations of Italy. It was founded in 1955 by a group of scholars, philanthropists, and politicians belonging to elite circles of Roman society. Its initial goal was to promote and protect the cultural, archaeological, and environmental patrimony of Italy, challenged by massive speculations connected to the booming of the construction industry.

20. Consiglio Nazionale del SANN-CGIL, "Risoluzione contro l'installazione di una base USA per sommergibili nucleari nell'isola della Maddalena," SANN-CGIL, Sindacato Assembleare Nazionale Nucleari, Rome, October 23, 1972, 2, held in the personal archive of Carlo Papucci.

21. Consiglio Nazionale del SANN-CGIL, "Risoluzione contro l'installazione di una base USA."

22. These are some of the titles that appeared in national newspapers during the five days following the debate at the Senate: "I pericoli della radioattività dai sommergibili nucleari," *Il Messaggero*, October 6, 1972; "Medici: 'Alla Maddalena nessuna base nucleare,'" *Il Giornale D'Italia*, October 6, 1972; "Protesta fuori del

Senato di radicali e pacifisti," *Il Messaggero*, October 8, 1972; "I sommergibili del silenzio," and "Vivaci contrasti al Senato sulla base militare alla Maddalena," *Corriere della Sera*, October 8, 1972; "L'inquinamento nucleare," *L'Unità*, October 8, 1972; "Nuove pesanti accuse al governo per la cessione della Maddalena," *L'Unità*, October 8, 1972; "Sommergibili atomici in Sardegna: nuova conferma del ministro Medici," *L'Unità*, October 9, 1972; "Medici continua a 'imbrogliare' sulla Maddalena," *Paese Sera*, October 9, 1972; "Confermate le rassicurazioni per La Maddalena," *Il Mattino*, October 9, 1972; "Colloqui di Medici in Sardegna sulla base navale Usa," *Il Messaggero*, October 9, 1972; "La Sardegna una base avanzata per la strategia USA nel Mediterraneo," *L'Unità*, October 10, 1972; "Tre consigli comunali sardi votano contro la base USA alla Maddalena," *L'Unità*, October 12, 1972; "Senato: la Commissione ecologica indaga sulla base della Maddalena," *L'Unità*, October 13, 1972.

23. My sources on the central political role of the priest are local historiographies and formal and informal conversations with many Maddalenini who knew him personally. A first account of the political influence of Monsignor Capula can be found in *Il governatore*, the detailed biography published by local historian and writer Gian Carlo Tusceri. A more recent book by local historians and journalists Francesco Nardini and Salvatore Abate, *Il pane del governo*, describes the role of the priest during the anticommunist hunt inside the military arsenal in the early 1950s.

24. Tusceri describes the episode in *Il governatore*, 171–89. Giuseppe Deligia confirmed Tusceri's version in various interviews. See, for example, Ronchi, "Quando vennero gli americani," 37–43. I heard the same story directly from Deligia during a personal interview in July 2010. He confirmed that he and Capula received a confidential phone call about the arrival of the US Navy before May 16, 1972. The US Navy sent secret scouting missions to La Maddalena well before Deligia and Capula were informed, as Antonio Cocco, the commander of the naval base of La Maddalena in the early 1970s, wrote in his memoir *Per la Patria e per il Re*, 172–74.

25. Comune di La Maddalena,"Verbale di Deliberazione del Consiglio Comunale di La Maddalena," 37, October 19, 1972, Municipal Archive of La Maddalena.

26. "Verbale di Deliberazione del Consiglio Comunale di La Maddalena."

27. This poster and other political documents I use in this chapter come from the Municipal Archive of La Maddalena. I want to thank the administrative personnel of the municipal archive for their assistance and generosity. Many thanks to Angelo Comiti, former mayor of La Maddalena, for allowing me to consult all the documentary sources I requested access to. Many thanks also to Antonello Tovo for sharing this document with me and for allowing me to use some of the archival material he collected during his master's thesis research on the US Navy base of La Maddalena. His detailed analysis of the impact of the US

Navy on "his archipelago," and the impressive number of documents he consulted, allowed me to start my research with great advantages. For this, and for his insights and friendship I will be thankful forever.

28. Comune di La Maddalena, "Proposta di Piano special di sviluppo di La Maddalena predisposto dalla Giunta Municipale," October 20, 1972, Municipal Archive of La Maddalena. It is worth reading the opening of the proposal to understand the strategy of Deligia's administration: "For a Special Development Plan of La Maddalena prepared by the Municipal Administration delegated by the City Council during the assembly of October 19, 1972. In reference to the attached deliberation unanimously approved by the City Council during the extraordinary assembly of 19.10.1972, the municipal administration, preoccupied for the possible consequences that the permanence of the American tender ship can have on the security and integrity of the entire territory of the Archipelago of La Maddalena, and above all for the possible negative effects on the local economy and on tourism; interpreting the anxiety of its citizens and their legitimate aspirations for a peaceful progress, after careful examination of the social and environmental situation of the Island, submits to the attention of the Superior Authorities directly interested a proposal for a special plan, articulated in the following points, which have been previously illustrated but for the most part not adequately considered."

29. Salvatore Sanna, private archive.

30. "Proposta di Piano special di sviluppo di La Maddalena predisposto dalla Giunta Municipale," 1.

31. Here Medici clearly accuses the opposition of using the problem of nuclear contamination instrumentally. The letter from Medici to the mayor of La Maddalena, October 17, 1972, is included in the "Verbale di deliberazione del Consiglio Comunale di La Maddalena," 38, November 21, 1972, 3, Municipal Archives of La Maddalena.

32. CAMEN's report, illustrated by mayor Deligia during the meeting of the city council; Comune di La Maddalena, "Verbale di deliberazione del Consiglio Comunale di La Maddalena," 38, November 21, 1972, 6–8.

33. It's possible to compare CAMEN's documents with the US Navy report NT-72-12 of November 1972, which I found in Carlo Papucci's private archive. Similar reports titled "U.S. Navy Report of Monitoring Near La Maddalena, Italy for Environmental Radioactivity" for the years 1989, 1990, 1994, 1995, 1996, and 2003 were available at the Municipal Archive of La Maddalena. In 1976 CNEN published, in Italian, the US Navy radiosurveillance report for La Maddalena in its official monthly magazine with the title "Rapporto della Marina americana relativo ai controlli sulla radioattività ambiente 1975 nelle vicinanze de La Maddalena," *Notiziario CNEN* 22, no. 8–9 (1976): 87–90.

34. "Rapporto della Marina americana," 24.

35. "In tremila a La Maddalena per contestare la base U.S.A.: Massiccio

intervento delle forze dell'ordine—La battaglia dei manifesti—Indifferenza tra gli isolani," *La Nuova Sardegna*, November 21, 1972.

36. "Polemiche sulla cessione della base alla Maddalena," *L'Unità*, November 28, 1972.

37. The so-called *battaglia dei manifesti* (battle of posters) was the way in which Deligia's majority and the Socialist and Communist opposition tried to mobilize their respective members and constituencies in La Maddalena. Given the control that the Italian navy exercised in the archipelago, the community had never experienced (apart from the elections of 1948 and the protests after the anticommunist hunts inside the arsenal in 1952) the staging of intense political contrapositions on public squares and through the streets. The Democrazia Cristiana and their allies asked the "peaceful and quiet" population of La Maddalena to stay home and the protesters coming from outside to go back to their towns. On the other side, the local opposition knew that only an "invasion" of protesters from the rest of Sardinia could guarantee a critical mass, which symbolically would demonstrate the existence of consistent support for their antibase mobilization.

38. Società Italiana di Fisica, "Estratto del verbale della 184° riunione del Consiglio di presidenza della S.I.F.," *Bollettino della Società Italiana di Fisica* 96, April 2, 1973, 4–5.

39. "Il congresso nazionale dei fisici contro la base alla Maddalena," *L'Unità*, November 2, 1972; "I fisici italiani condannano l'aggressione USA nel Vietnam," *Avanti!*, November 3, 1972. The document approved by the plenary assembly of SIF on November 1, 1972, and the transcripts of the debate can be found on the official bulletin of the society: *Bollettino della Societa' Italiana di Fisica* 97, no. 11 (June 1973): 29–52.

40. *Bollettino della Societa' Italiana di Fisica*, 44.

41. *Bollettino della Societa' Italiana di Fisica*, 46. Eugenio Tabet was a health physicist working at the Radiation Laboratory of the Italian Higher Health Institute. Tabet would become a member of the team selected by CNEN and ISS to conduct the preliminary radioecological studies they were instructed by the minister of health to carry out in the summer of 1974.

42. Salvatore Sanna, personal archive.

43. *Bollettino della Societa' Italiana di Fisica*, 44.

44. "'Nessun pericolo di radiazioni dai sommergibili' dicono gli esperti," *Corriere della Sera*, November 15, 1972; "Nessun pericolo di radiazioni dai sottomarini atomici a La Maddalena," *La Nuova Sardegna*, November 23, 1972.

45. "Per La Maddalena manifestazione al centro nucleare," *L'Unità*, November 23, 1972. Giovanni Berlinguer has been an important political figure in the PCI. The younger brother of Enrico Berlinguer, secretary of the PCI from 1972 to 1984, Giovanni was a doctor and university professor of occupational health. His political activity as a member of the central committee became of crucial im-

portance when he was made responsible for the organization of the public health and scientific research policies of the PCI.

46. "CNEN: non c'è 'parere favorevole' per la base USA a La Maddalena," *L'Unità*, November 24, 1972; "Irresponsabile si del CNEN per la base a La Maddalena," *Avanti!*, November 24, 1972; "Il governo prepara un altro falso sulla Maddalena," *Paese Sera*, November 24, 1972; "Scandalo sul 'falso parere favorevole' del CNEN per la base alla Maddalena," *L'Unità*, November 25, 1972; "Il ruolo del CNEN deve essere chiarito," *Avanti!*, November 30, 1972. The Commissione Tecnica was an expert committee composed of members of CNEN radioprotection division, ISS representatives, and delegates from ministries who issued safety assessments for nuclear installations.

47. The assembly of the personnel of the nuclear safety and the sanitary protection divisions of CNEN unanimously signed a letter to CNEN president Ezio Clementel and the executive committee of the agency asking for clarifications about the "reassuring assessment that various newspapers attributed to CNEN." "The personnel of the two divisions, and in particular technicians and researchers . . . declare to be completely extraneous to this episode [the production of the contested document] and exclude their responsibility about any assessment allegedly issued by this agency." *Mozione al Presidente del CNEN ed alla Giunta Esecutiva*, Rome, November 20 1972, held in the personal archive of Carlo Papucci.

48. Partito Comunista Italiano, Cellula CNEN della Casaccia, "La base nucleare U.S.A. alla Maddalena: problemi politici e di sicurezza," Rome, November 21, 1972, held in the personal archive of Carlo Papucci.

49. Here I use the adjective *technopolitical* to point out the use and deployment of technical arguments to justify and reach political goals. In her study on the French nuclear program Gabrielle Hecht defines *technopolitics* as "the strategic practice of designing or using technology to constitute, embody, or enact political goals." See Hecht, *Radiance of France*, 15. For more examples on the politics of technological designs see Winner, *Whale and the Reactor*, esp. ch. 2, 19–39. For similar uses of the concept of technopolitics see also Slayton, *Arguments that Count* and Schmid, *Producing Power*.

50. After the press conference conservative newspapers guaranteed an abundant media coverage in favor of the government's positions: "Si sgonfia la speculazione sulla base USA alla Maddalena," *Il Tempo*, December 9, 1972; "Nessun pericolo di radiazioni per la base Usa della Maddalena," *Il Giornale D'Italia*, December 10, 1972; "Per La Maddalena (assicurano gli esperti) nessun pericolo," *Il Giorno*, December 10, 1972; "La Maddalena: nessun pericolo, ma grossi problemi di sviluppo," *Il Globo*, December 10, 1972; "È escluso che provochi Danni la base navale alla Maddalena," *Gazzetta del Popolo*, December 10, 1972; "La Maddalena: nessuna contaminazione," *Avvenire*, December 10, 1972.

51. "Un rapporto tecnico del CNEN ammette: 'Non sappiamo nulla sulle navi nucleari USA,'" *L'Unità*, December 10, 1972; "Maddalena: nuovi tentativi del

governo di servirsi di un falso 'si' del CNEN," *L'Unità*, December 10, 1972; "Forte imbarazzo del governo per la base della Maddalena," *Avanti!*, December 10, 1972; "La Maddalena: reali i rischi di contaminazione," *L'Unità*, December 17, 1972.

52. "Maddalena: il Sindacato CNEN smentisce il Presidente," *Paese Sera*, December 24, 1972; "Il parere del CNEN fu atto personale del presidente," *L'Unità*, December 24, 1972.

53. Carlo Papucci, personal interview with the author.

54. "I fisici italiani condannano l'aggressione USA in Vietnam," *Avanti!*, November 3, 1972. On the condemnation of the "Jason Group" by other professional associations and scientist activists in the United States, see Moore, "Organizing Integrity," 1592–627; Burhop, "Scientists and Soldiers," 4–8.

55. Edmondo Bruti Liberati, "La magistratura: la 'radicale svolta' della metà degli anni sessanta del Novecento," *Il Politico* 251, 2 (2019): 77–104; Antonella Meniconi, *Storia della magistratura italiana* (Bologna: Il Mulino, 2013).

56. Amendola held the office of *pretore*, a magistrate devoted to the ordinary administration of justice on various matters in both civilian and penal causes, in Rome. During the 1980s he also started a political career within the Italian Green Party and was elected to the European Parliament for one legislature (1989–1994).

57. "Basi infette," *Il Messaggero*, March 22, 1974.

58. "Waste Data on N-subs Found False," *Mainichi Daily News*, January 30, 1974; "Gov't Admits False N-sub Waste Data," *Mainichi Daily News*, February 6, 1974; "On Science and Scientists," *Mainichi Daily News*, February 7, 1974; "Who'll Check on N-subs Waste," *Mainichi Daily News*, February 7, 1974.

59. The article cited by Amendola is Weinberg, "Science and Trans-science," 209–22. This is the quote through which Amendola "enrolls" Weinberg to make his argument on La Maddalena more compelling: "Let's take as reliable the data collected so far in La Maddalena, which say that the levels of radioactivity are well below the limits for human beings. One of the most important nuclear experts of the world, the American Alvin M. Weinberg, former director of the grandiose Oak Ridge nuclear laboratories where the first atomic bomb has been produced, in an article published in 1972 with the suggestive title 'Science and Trans-science,' affirms that it is impossible to determine with certainty the long-term genetic effects of radioactive exposure even of the lowest levels." Amendola, "Basi infette," 3.

60. "Allarme! L'onda radioattiva," *Il Messaggero*, April 11, 1974.

61. For example, in a confidential telegram sent on April 11, the day in which *Il Messaggero* published Amendola's article "L'Onda radioattiva," Volpe informed Kissinger that the president of CNEN was disappointed about the leaking of information about the radiometric report in preparation and that the agency decided to not comment on national media. Cable, U.S. Embassy in Rome to Secretary of State, ID: 1974ROME05156_b, Declassified/Released US Department

of State EO Systematic Review June 30, 2005, accessed November 20, 2021, https://wikileaks.org/plusd/cables/1974ROME05156_b.html.

62. "Interrogazione socialista sulla Maddalena inquinata," *Il Messaggero*, April 12, 1974; "Radioattività alla Maddalena," *Paese Sera*, April 12, 1974; "Escluso alla Maddalena l'inquinamento atomico," *Il Tempo*, April 12, 1974; "Maddalena: sale la radioattività provocata dai sommergibili USA," *L'Unità*, April 12, 1974; "Alla Maddalena aumenta la radioattività," *Avanti!*, April 19, 1974; "Maddalena: radioattività in aumento," *Rinascita*, April 19, 1974; Giorgio Tecce, "Una farsa sotto il mare," *L'Espresso*, April 28, 1974.

63. "Il Comune chiede a Roma accertamenti," *La Nuova Sardegna*, April 9, 1974; "L'inquinamento nucleare sotto i limiti di sicurezza," *La Nuova Sardegna*, April 13, 1974.

64. Comune di La Maddalena, "Verbale di Deliberazione del Consiglio Comunale di La Maddalena n.4 del 20.4.1974," 3–4, Municipal Archive of La Maddalena.

65. "Verbale di Deliberazione del Consiglio Comunale di La Maddalena n.4 del 20.4.1974," 4.

66. "Ciclostile proprio—Movimento Giovanile D.C.—Via R. Margherita—La Maddalena," May 3, 1974, held in the private archive of Salvatore Sanna.

67. "Mozione conclusiva—Studenti Istituto Tecnico Nautico 'D. Millelire,'" May 6, 1974, held in the private archive of Salvatore Sanna.

68. Some members of the local Democrazia Cristiana commissioned the anonymous flier. The document is not signed and does not have any political symbol on it. Salvatore Sanna, personal archive.

69. "Ecco La Maddalena. I nostri inviati nell'isola radioattiva," *L'Europeo*, May 2, 1974.

70. "Ecco La Maddalena," 1.

71. U.S. Embassy in Rome, May 30, 1974—Telegram ID: 1974ROME07431_b, Declassified/Released US Department of State EO Systematic Review June 30, 2005, accessed November 20, 2021, https://wikileaks.org/plusd/cables/1974ROME07431_b.html.

72. "La Maddalena radioattiva: Qualcosa si muove," *Il Messaggero*, April 14, 1974; "L'esempio giapponese," *Il Messaggero*, May 3, 1974.

73. "Arriva la commissione e parte la 'Gilmore,'" *La Nuova Sardegna*, April 26, 1974.

74. "Tutela sanitaria della popolazione contro i pericoli derivanti dalle radiazioni ionizzanti e controllo sulla radioattivita' ambientale dell'isola de 'La Maddalena,'" Ministry of Health, Doc. 400.5 / A 5/1/75, June 10, 1974, Municipal Archive of La Maddalena. Various newspapers illustrated the program of studies: "Istruttoria della Sanità alla Maddalena," *Avanti!*, June 13, 1974; "I fisici chiedono indagini senza veli per La Maddalena," *L'Unità*, June 15, 1974; "Maddalena: indagine sulla base nucleare," *Il Messaggero*, June 16, 1974.

75. *Indagine ambientale nell'intorno della base nucleare navale situate nell'isola de La Maddalena*, report of the ISS and CNEN expert committee for the implementation of a radioecological study of La Maddalena, n.d., held in the Municipal Archive of La Maddalena.

76. *Indagine ambientale nell'intorno.*

77. "Un punto di partenza," *Il Messaggero*, June 23, 1974.

PART II: TECHNOPOLITICS OF RISK

1. Buffoni, Gasparini, and Zattera, "Uno studio dei contaminanti in mare nell'area de 'La Maddalena.'"

2. Storia del rapporto tecnico "Analisi di un ipotetico incidente nucleare nell'Arcipelago de La Maddalena" di Buffoni, Gasparini, Zattera, ENEA, internal communication, November 23, 1984, held in the personal archive of Giuseppe Buffoni.

3. Storia del rapporto tecnico, 2.

4. Nota per la direzione. ARCIPELAGO DE LA MADDALENA: Rete di Monitoraggio in continuo della radioattività ambientale, March 13, 1991 [signed Carlo Papucci], Archive Center for the Study of Marine Environment, ENEA, Santa Teresa.

CHAPTER 3: RADIOECOLOGY AND THE ITALIAN NUCLEAR REGULATORY REGIME

1. Cigna, Domenici, Malvicini, and Vido, "Radioattivita' dei prodotti di fissione," 73–78.

2. For example, see Cigna and Malvicini, "Determinazione della data dell'esplosione di una bomba nucleare," 162–69; Malvicini, "Le ricadute radioattive nella zona di Ispra," 267–76.

3. See Cigna, "Ripercussioni anche in Italia degli scoppi atomici di Las Vegas?" 84–87.

4. Santomauro and Cigna, "Prime misure sulla radioattivita' delle precipitazioni atmosferiche."

5. Higuchi, *Political Fallout*, 2020.

6. Hamblin, *Poison in the Well*.

7. See Edwards, *Vast Machine*; Creager, *Life Atomic*, 1–23; Rainger, "Wonderful Oceanographic Tool," 93–131.

8. See Edwards, *Vast Machine*, 208. See also Creager, *Life Atomic*, esp. ch. 1, "Tracers," 1–23.

9. Odum, "Ecology and the Atomic Age," 27–29.

10. Hagen, *Entangled Bank*. The birth of radioecology in the context of the development of the biological and oceanographic studies regarding the effects of nuclear tests and nuclear waste disposal at sea has received considerable attention by historians of science. In addition to Hamblin's volume, see Rainger, "Won-

derful Oceanographic Tool"; Rainger, "Science at the Crossroads," 349–71; Rothschild, "Environmental Awareness in the Atomic Age," 492–530.

11. During the 1960s one of the emergent analytical parameters in marine radiobiology became the "concentration factor." It was developed by Ukrainian scientist G. G. Polikarpov in his doctoral dissertation "Radioekologiya Mosrkikh Organizmov by Atomizdat," translated into English and published in 1966 as *Radioecology of Aquatic Organisms*. Polikarpov defined the concentration factor as "The capacity of an organism to accumulate radioactive substances . . . expressed by the ratio of its radioactivity to that of the aqueous medium or the preceding food link to which the radionuclide was concentrated" (27). Polikarpov's scholarship attained international recognition beginning in the early 1960s, when bioaccumulation factors in marine and terrestrial ecologies became closely studied. In September 1961 the first National Symposium on Radioecology took place at the University of Colorado, Fort Collins, under the auspices of the American Institute of Biological Sciences and sponsored by the divisions of Biology and Medicine of the US Atomic Energy Commission. See Schultz and Klement, *Radioecology*.

12. See Hamblin, *Poison in the Well*, esp. chs. 1 and 2.

13. Walker, *Permissible Dose*.

14. Dr. Cigna graciously clarified this aspect during an interview in February of 2012.

15. For a more detailed analysis of Buzzati-Traverso's scientific and organizational role see Capocci and Corbellini, "Adriano Buzzati-Traverso and the Foundation of the International Laboratory of Genetics and Biophysics in Naples (1962–1969)," 489–513; Cassata, *L'Italia intelligente*, esp. part I.

16. A preliminary program of the laboratory was outlined by Buzzati-Traverso and Michael Bernhard (the director of the Fiascherino center from 1959 to 1974) during the first IAEA conference on the disposal of radioactive wastes, in Monaco, at the end of 1959. See Buzzati-Traverso and Bernhard, "CNRN Research Programme for the Study of Radioisotope Accumulation"; Carlo Papucci, "Il laboratorio di Fiascherino: la nascita," Presentation at "Below the Surface" symposium, Mentone, October 21, 2018.

17. One of the first reports on the activity of the Italian marine radiosurveillance network is Argiero, Del Corso, Manfredini, and Palmas, "Controllo sistematico della radioattivita' lungo le coste italiane," 261–67.

18. Bernhard et al., *Rapporto sull'attività scientifica e tecnica*.

19. P. Cagnetti, M. Bernhard, and A. Zattera, "Radioecological investigations in the Gulf of Taranto: general circulation in the Gulf of Taranto and diffusion processes offcoast the Trisaia Center," paper presented at the Simposio internazionale di radioecologia applicata alla protezione dell'uomo e del suo ambiente, Rome, September 7–10, 1971.

20. By the beginning of the 1960s Italy had already installed three nucle-

ar power plants with different reactor technologies, financed by three different industrial groups: one in Garigliano (General Electric water boiling reactor), one in Latina (near Rome, with a British graphite reactor), and one in Trino Vercellese (Westinghouse pressurized water reactor). A fourth power plant was built in the Po Valley, in the town of Caorso, at the beginning of the 1970s. For an overview of the Italian nuclear program and more detailed historical accounts of specific nuclear power plants, I suggest Paoloni, *Energia, ambiente, innovazione*; Elli, *Atomi per l'Italia*; Curli, "Energia nucleare per il Mezzogiorno," 317–51; Curli, *Il Progetto Nucleare Italiano (1952–1964)*. For more recent accounts of the Italian nuclear program see Gerlini, "Rise and Fall of Nuclear Italy," 170–237, and Bini and Lodero, *Nuclear Italy*.

21. CNEN documented the monitoring activities performed by its personnel at the arrival of the merchant nuclear-powered ship *Savannah* (USA) in the port of Naples in December 1964. The documentary *Nave senza fumo* is available at the ENEA WebTV page in the section "Film Historical Archive": https://webtv.sede.enea.it/index.php?page=listafilmcat2&idfilm=173&idcat=30.

22. Amaldi and Campos Venuti, *Aspetti sanitari della protezione delle popolazioni dalle radiazioni*.

23. In CNEN's and ISS's documentation and reports the two terms are often used interchangeably.

24. The first complete legislation regulating the pacific employment of nuclear energy in Italy was the Decree of the President of the Republic no. 185, 1964 (DPR 185, 1964), followed by subsequent modifications and updates in 1971 and 1977. Despite the fact that IAEA regulations prescribed specific exposure limits and safety protocols, at the beginning of the 1970s regulatory regimes concerning nuclear plants siting varied nationally. For a comparative analysis from an Italian perspective see Amaldi, Campos Venuti, Frullani, Maiani, and Tabet, "Criteri di scelta dell'ubicazione delle installazioni nucleari," 626–46.

25. The process of multidisciplinary integration into the field of marine and continental radioecology became the benchmark of this (now established) scientific sector also in Italy. This is evident from the reports of the First National Symposium on Radioecology, held at the University of Parma in 1970. The proceedings were published as *Atti del primo convegno sullo stato di avanzamento della radioecologia in Italia, organizzato dall'Istituto di Zoologia dell'Università degli Studi di Parma e dall'Associazione Italiana di Fisica Sanitaria e Protezione contro le Radiazioni, 5–6 Novembre, 1970*.

26. Sara Pritchard defines *envirotechnical systems* as "the historically and culturally specific configurations of intertwined 'ecological' and 'technological' systems, which may be composed of artifacts, practices, peoples, institutions, and ecologies." Pritchard, *Confluence*, 19. Other volumes discussing the usefulness of the envirotechnical approach are: Reuss and Cutcliffe, *Illusory Boundary*; Jørgensen, Jørgensen, and Pritchard, *New Natures*.

27. Here it is worth illustrating an important aspect of the composition of

the Commissione Tecnica. The committee was established by DPR 185, 1964 on radioprotection with the clear intent of creating a collegial body of experts made up of CNEN, ISS, and other agencies, that could evaluate radiological protection and nuclear safety matters independently from CNEN. A detailed description of the siting regulations in Italy compared with other countries in the 1970s can be found in Amaldi, Campos Venuti, Frullani, Maiani, and Tabet, "Criteri di scelta dell'ubicazione delle installazioni nucleari."

28. For a critical discussion of the state of the art of radioecology and radioprotection in Italy during the 1970s, see Cigna and Ilari, "Role of the Radioecological Protection."

29. As most nuclear plants are installed near sources of water (usually lakes, rivers, and along coastlines), knowing the geological and hydrological variables affecting the processes of radionuclides' dispersion has clear implications for siting policies, and sampling and surveillance strategies.

30. Aldo Brondi, personal interview with the author, May 2010.

31. For a synthetic illustration of the "global approach" promoted by CNEN/ENEA geologists see Brondi and Ferretti, "Efficacia ed applicazione di concetti mesotipologici," 451–59. For a more technical explanation of concepts and methods applied in the "global approach" see Anselmi et al., *Studi sui parametri geologici rilevanti*.

32. Law no. 1860, December 31, 1962.

33. See Turchetti, "Most Active Customer,'" 495–98. For a different interpretation of US-Italy relations in the civilian nuclear field see Bini, "Atoms for Peace (and War)," 23–40.

34. At the end of July 1955, the governments of Italy and of the United States had already signed a bilateral agreement according to which CNRN was allowed to lease from the US Atomic Energy Commission (USAEC) a certain amount of enriched uranium and to purchase a research reactor of the type CP-5 similar to the one installed by the USAEC at the Argonne National Laboratory. The reactor was then installed in Italy at the research center of ISPRA, which the Italian government lent to EURATOM in 1960.

35. See Curli, "Energia nucleare per il Mezzogiorno," 317–51.

36. See Curli, "Energia nucleare per il Mezzogiorno," 317–51; Paoloni, *Energia, ambiente, innovazione*. The nuclear plant of Trino Vercellese was designed and built by Edison, the biggest electric utility in Italy, and was an expression of the private interests in nuclear development. Together with other private companies, such as FIAT, Montecatini, and Pirelli, Edison contributed to the establishment of the first Italian research center on nuclear technology. Centro Informazioni Studi Esperienze (CISE) was founded in Milan in 1946 with the goal of starting a series of experimental studies for the industrial applications of nuclear energy. With the establishment of CNRN under the supervision of a group of physicists and scientists, two conflicting visions—public and private—of the Italian nuclear

development emerged. For a detailed historical account of the institutional and political conflicts that shaped the beginnings of the Italian nuclear program see Paoloni, *Energia, ambiente, innovazione*, and "Gli esordi del nucleare," 383–408. For a more concise version of Paoloni's study in English, see *Il Nucleare in Italia/Nuclear Power in Italy* (Archivio Storico ENEL, 2009), also available at https://www.enel.com/en-GB/doc/sustainability/nucleare_in_italia.pdf. For an interesting analysis of the conflicts that limited the Italian nuclear program, see the polemical account by Silvestri, *Il costo della menzogna*.

37. For the history of the Latina nuclear power plant see Elli, *Atomi per l'Italia*. A synthetic version of Elli's book in English is "British First Nuclear Export," 27–42.

38. Also, with a boiling water reactor produced by General Electric.

39. For example, Mario Silvestri wrote: "From what I exposed so far it is easy to conclude that the commonplace according to which the first three nuclear plants built in Italy (and still operational) were the outcome of a considered choice to investigate each one of the three types for testing their behavior and acquiring experience with their technology is far from the truth. The people who have directly witnessed those early developments know how false this story is: there wasn't any program or purpose, there was only the intent to sabotage each other. The logic that guided the construction of the three nuclear plants was only that of a struggle between opposing interests, rather than of a constructive competition." Silvestri, *Il Costo della Menzogna*, 190, my translation. See also Zorzoli, "Did the Italian Decision Makers Understand," 91–101.

40. Amaldi and Campos Venuti, "Aspetti sanitari della protezione delle popolazioni dalle radiazioni," 199–217, see esp. 201–4. Commissione Tecnica per la Sicurezza Nucleare e la Protezione Sanitaria, *Relazione sulla Trentennale attività della Commissione*.

41. Also, in the United States the Atomic Energy Commission was both promoter and regulator before the Nuclear Regulatory Commission was founded in 1975 and the two functions were split.

42. See Amaldi and Campos Venuti, "Aspetti sanitari della protezione delle popolazioni dalle radiazioni," 199–217, and Giancarlo Boeri's intervention during the workshop organized by the Partito Comunista Italiano titled *The Energy of the Future*, in *L'Energia del Futuro*, 227–31.

43. The only exception was the nuclear plant of Caorso, for which the Region of Lombardy and the Provinces of Parma and Piacenza established a network of radiosurveillance stations separate from Ente Nazionale Energia Elettrica's (ENEL, National Electric Energy Utility)—the operator of the nuclear plant—and directly controlled by the local administrations with the support of research centers affiliated with the University of Parma and Bologna. On this, and more generally on the problematic aspects of the Italian legislation on radiation pro-

tection see the useful analysis of Spaziante, *Questione Nucleare e Politica Legislativa*, esp. ch. 2, "L'Energia nucleare in parlamento," 96–143.

44. The debate between ISS and the Ministry of Industry over the proposal to give CNEN exclusive control on Italian radioprotection is well explained by Spaziante, *Questione nucleare e politica legislativa*. See especially "I controlli nucleari tra CNEN e Istituto Superiore di Sanità, 114–21.

45. DPR 185, 1964, section 89. For a synthetic view of the norms and regulations of the Italian nuclear program and activities together with those of other European countries, Canada, the United States, and Japan, see NEA-OECD, *Nuclear Legislation Analytical Study*.

46. *Legge 833/78, Istituzione del Servizio Sanitario Nazionale*, December 23, 1978. This process of decentralization started with the institutionalization of the regions in 1970 and with following regulations, in particular: *Decreto Presidente della Repubblica n. 4, 14 Gennaio 1972* and *Decreto Presidente della Repubblica n. 616, 24 Luglio 1977*.

47. The film is available at the ENEA Web TV page in the section "Film Historical Archive": http://webtv.sede.enea.it/index.php?page=listafilmcat2&idfilm=169&idcat=30.

48. About this particular operation, see Hacker, *Elements of Controversy*, 131–58.

49. On March 1, 1954, after the Castle Bravo test, the entire crew of the fishing boat was contaminated by the radioactive wave following the nuclear detonation at the Bikini Atoll. See Hacker, *Elements of Controversy*, 131–58, and the documentary *Atomic Cafe*.

50. Here I use Bruno Latour's analytical insights on "circulating reference." He suggests that the conservation of specimens is not simply the way in which scientists guarantee evidence for the truth of a statement but "rather it is our way of keeping something constant through a series of transformations." Specimens and inscriptions stored in a laboratory are "traces that establish a reversible route that makes it possible to retrace one's footsteps as needed"; they are "references" of the reversible processes of transformations that scientists perform in order to reduce and amplify on manageable scales the characteristics of the phenomena they study. See Latour, *Pandora's Hope*, 70–71.

51. Comitato Nazionale per l'Energia Nucleare, *Verbale della XVII Riunione per l'Organizzazione del Rilevamento della Radioattività Ambientale in Italia*, Rome, October 13, 1972, PROT. SAN. 13/72, held in personal archive of Carlo Papucci. See also D'Amato, "Rapporto Annuale sulla radioattivita' ambientale in Italia," 313–22; Boeri and Giorcelli, "Le reti nazionali per il rilevamento della radioattivita' ambientale in Italia"; Argiero, Del Rosso, Manfredini, and Palmas, "Controllo sistematico della radioattività del mare lungo le coste Italiane," 261–67; Cigna, "Forty Years of Anthropogenic Radionuclides in Surface Seawater," 261–90.

52. For a clear illustration of this point, I defer to the classic Pickering, *Mangle of Practice*.

53. Comitato Nazionale per l'Energia Nucleare, "Misure di radioattività ambientale presso l'isola della Maddalena," 87–90; D'Amato, "Rapporto annuale sulla radioattività ambientale in Italia," 313–22; Boeri and Giorcelli, "Le reti nazionali per il rilevamento della radioattività ambientale in Italia," 28–35.

54. I was able to reconstruct the process through which Italian radioecologists performed environmental studies in La Maddalena thanks to the documentation made available by the ENEA Center for the Study of Marine environment in Lerici, La Spezia (previously named CNEN Center for the Study of Marine Environmental Contamination). I want to thank Roberta Delfanti, director of the center, and especially Carlo Papucci for granting me access to the archival material and for their kind assistance and explanations during several research visits. I also would like to remember Aldo Brondi, who graciously helped me at the beginning of my research in the summer 2010.

55. Brondi, Ferretti, Anselmi, and Rabottino, "Arcipelago della Maddalena," 1293–318.

56. Latour, *Pasteurization of France*; Murphy, *Sick Building Syndrome*.

57. In the archipelago locals used to catch *Pinna nobilis* both as an edible mollusk and for other purposes; their shells sometimes were used as decorative elements—for example, to build table lamps. The use of *Pinna nobilis* as bioindicator in radioecological campaigns from the 1950s to the 1970s is well documented in Italy. See for example, Ghiretti, Salvato, Carlucci, et al., "Manganese in *Pinna nobilis*," 232–33; De Franceschi, Gentili, Guidi, et al., "60Co in the Marine Mollusc *Pinna nobilis*," 376–77.

58. Carlo Papucci, personal communication with the author.

59. Carlo Papucci, "Intervento di Carlo Papucci / C.G.I.L.-Ricerca," held in the personal archive of Carlo Papucci.

60. CNR Oceanography Center of Saint Terenzo (Lerici) to Ezio Clementel, letter, May 10, 1976, Doc. CNEN Prot. n. 3887 M-C-1, 10/5/76 [uncatalogued], Archive ENEA Center of the Study for Marine Environment (formerly CNEN Center for the Study of Marine Contamination), Lerici, Italy.

61. A letter handwritten by Papucci in May 1976, probably addressed to other union members or to a newspaper, makes evident the stalled-out situation in which CNEN personnel found themselves: "The laboratory is ready to do the campaign planned for this spring, but supposedly bureaucratic obstacles impede its actual implementation. The boat of CNEN is adequately equipped and ready to go [to La Maddalena] but at this time it is not operative because there is no crew governing it. The decision to recruit an external crew needs to be taken by the executive offices. The most urgent risk is that the planned spring campaign is delayed until next year with the consequent postponement of the entire scientific program. The problem must be solved now, in a very short time." Handwritten document, held in the personal archive of Carlo Papucci.

62. See, for example, Comitato Nazionale per l'Energia Nucleare, *Relazione sulla campagna radioecologica attorno* (1977); Sacripanti and Fameli, *Centrale nucleare del Garigliano*.

63. Comitato Nazionale per l'Energia Nucleare, *Relazione sulla campagna radioecologica attorno* (1981), 72.

64. Here I refer to Scott Frickel's work on the production of ignorance and his (and others') contribution to a new political sociology of science that asks "how, where, and why ignorance, once produced, becomes institutionalized beyond science." Frickel, "Not Here and Everywhere," 263.

65. Polvani, "L'ubicazione degli impianti nucleari e la protezione della popolazione," 33.

66. For a detailed analysis of this problem see Amaldi and Campos Venuti, "Aspetti sanitari della protezione delle popolazioni dalle radiazioni," 199–217. Two specific studies on nuclear siting and the political relevance of the different industrial projects of ENI and CNEN are Curli, "Energia Nucleare per il Mezzogiorno," 317–51; and Elli, *Atomi per l'Italia*. For a description of siting procedures, construction authorization, and safety assessment before 1964 in Italy, see Comitato Nazionale per le Ricerche Nucleari, *V Congresso Nucleare, 1960*, esp. 17–30.

67. Arrigo Cigna, personal interview with the author, Cocconato d'Asti, February 2012. Cigna rightly cites the case of Montalto di Castro, but also the initial design of the nuclear power plant of Caorso (built between 1971 and 1981) had to be altered substantially during construction because CNEN, on the basis of its technical assessment, required a reexamination of the project. In addition to DPR 185, 1964, the legislation on nuclear siting of 1975 (Law no. 393, art. 23) established that within three years (by 1978) CNEN should elaborate a map of possible nuclear sites based on preliminary studies of the geomorphology, hydrography, and seismic activity of the national territory. In 1979 the accident at Three Mile Island pushed Italian experts to reconsider for the second time certain features of the boiling water reactor of Caorso, similar to the one at Three Mile Island, and this led to costly and time-consuming modifications of the initial design.

68. Eugenio Tabet, personal interview with the author, January 2012.

69. Italian regulation for the peaceful use of nuclear energy was organically introduced in DPR 185, 1964. The legislation about nuclear ports was drafted only in 1979, after La Maddalena. Before then, the presence and transit of nuclear naval units, both civilian and military, in Italian harbors was evaluated on a case-by-case basis with protocols tailored to the characteristics of the naval units, and an emergency plan was crafted on each occasion.

CHAPTER 4: A TECHNOPOLITICAL COMPROMISE

1. "Non sono stati resi noti i dati dei rilevamenti effettuati dall'Odalisca," *La Nuova Sardegna*, July 19, 1975.

2. For example, on July 19, 1975, the local newspaper *Tutto Quotidiano* reported on the state of uncertainty and frustration surrounding the radioecological campaigns in the archipelago: "Limitata in partenza l'indagine del Cnen sull'inquinamento atomico."

3. "Limitata in partenza l'indagine del Cnen sull'inquinamento atomico."

4. US Embassy in Rome, Subject: NPW OPERATIONS AT LA MADDALENA, July 15, 1974, 15:50, Telegram ID: 1974ROME09746_b, https://wikileaks.org/plusd/cables/1974ROME09746_b.html.

5. US Department of State, Subject: NUCLEAR POWERED SUBMARINES AT LA MADDALENA, Telegram ID: 1974STATE076272_b, accessed November 20, 2021, https://wikileaks.org/plusd/cables/1974STATE076272_b.html.

6. Annual reports titled "Environmental Monitoring and Disposal of Radioactive Wastes from U.S. Naval Nuclear-Powered Ships and Their Support Facilities" were publicly available in *Environmental Data and Reports*, a monthly bulletin of the US Environmental Protection Agency, from 1970. In addition to these compendia, the US Navy produced more specific and detailed environmental analyses about each operational facility, at home and abroad.

7. *REPORT NT-90-4, January 1990: U.S. Navy Report of Monitoring near La Maddalena, Italy for Environmental Radioactivity 1989*, 1. The Command of Marisardegna in La Maddalena sent the report to the mayor of La Maddalena on March 2, 1990.

8. US Department of State, Subject: NPW OPERATIONS AT LA MADDALENA, May 10, 1974, 22:21, Telegram ID: 1974STATE097721_b, accessed November 20, 2021, https://wikileaks.org/plusd/cables/1974STATE097721_b.html.

9. On this point, see Nuti, *La Sfida nucleare*, and Ciglioni, *Culture atomiche*, esp. 273–94.

10. Arkin and Fieldhouse, "Focus on the Nuclear Infrastructure," 11–15. For an extended survey of the US global nuclear infrastructure during the Cold War see, by the same authors, *Nuclear Battlefields*.

11. "Ministero della Sanità. Oggetto: Approdo messi navali a propulsione nucleare—compilazione piano emergenza a norma degli artt. 112 e segg. del D.P.R del 13/2/64," n. 185, June 25, 1974, held in the personal archive of Carlo Papucci. MINISTERO DELL'INTERNO, Oggetto: "Isola di La Maddalena—Base nucleare U.S.A.—Piano di emergenza esterna," April 15, 1975, held in the personal archive of Carlo Papucci.

12. "Ministero della Sanità, Doc. 400.5 / A 5/1/75: Tutela sanitaria della popolazione contro i pericoli derivanti dalle radiazioni ionizzanti e controllo sulla radioattività ambientale dell'isola de 'La Maddalena,'" June 10, 1974, Municipal Archive, La Maddalena,.

13. US Embassy Office in Rome, Subject: NPW OPERATIONS AT LA MADDALENA, July 15, 1974, 15:50, Telegram ID: 1974ROME09746_b, accessed

November 20, 2021, https://wikileaks.org/plusd/cables/1974ROME09746_b.html. The diplomatic official mentioned by Volpe in the confidential cable was Eric Da Rin, at the time a member of the executive cabinet of the Italian Ministry of Foreign Affairs, led by Christian Democrat Aldo Moro. Da Rin was the ministerial coordinator for NATO and disarmament affairs.

14. US Embassy Office in Rome, Subject: NPW OPERATIONS AT LA MADDALENA, September 5, 1974, 09:44, Telegram ID: 1974ROME12147_b, accessed November 20, 2021, https://wikileaks.org/plusd/cables/1974ROME12147_b.html.

15. US Embassy Office in Rome, NPW OPERATIONS AT LA MADDALENA, September 5, 1974.

16. US Department of State, Subject: RADIOLOGICAL MONITORING AT LA MADDALENA: CNEN REPORT ON FEB 24 ECOLOGY CONFERENCE, March 31, 1975, 14:45, Telegram ID: 1975STATE071792_b, accessed November 20, 2021, https://wikileaks.org/plusd/cables/1975STATE071792_b.html.

17. US Embassy Office in Rome, Subject: NUCLEAR POWER WARSHIP RELATED ISSUES, August 8, 1975, 16:15, Telegram ID: 1975ROME11474_b, accessed November 20, 2021, https://wikileaks.org/plusd/cables/1975ROME11474_b.html.

18. US Embassy Office in Rome, Subject: NUCLEAR POWER WARSHIP RELATED ISSUES: MONITORING AT LA MADDALENA, August 18, 1975, 09:02, Telegram ID: 1975ROME11768_b, , accessed November 20, 2021, https://wikileaks.org/plusd/cables/1975ROME11768_b.html. I report the entire text of the cable here:

> The following information is in response to STATE 193933. 2. In our message 4948 of April 4, 1975, Embassy reported Albonetti's view of his options for reporting on "radiological monitoring at La Maddalena." Albonetti indicated he would prefer to simply run data plus brief introductory statement to the effect that there was no scientific basis for publicly expressed worries. In our messages 4906 and 4619 we transmitted that data which were to appear in the "article." 3. As reported in ROME's 11474, "ITEM" appearing in June issue of CNEN's Notiziario closely followed preference outlined in para two above. Data appeared exactly as transmitted in refs E and F. 4. Informal translation of introductory statement follows: Title: "Measurement of Environmental Radioactivity at La Maddalena." Begin text: In the months of April and July 1974, the CNEN environmental radioactivity laboratory carried out the latest in a series of surveys aimed at verifying the level of environmental radioactivity of the general area of the base at La Maddalena (earlier results were published in the May 1974 issue of Notiziario). The survey results, which appear in the following table, point

out that the levels found in the various samples which were examined do not justify apprehensions from the health point of view not of radioprotection. The data will also appear in CNEN's annual publication "Data on environmental radioactivity collected in Italy." End text. 5. Text followed by small map of La Maddalena area showing "collection points" and by full page and one half of data. Volpe.

19. Centro Applicazioni Militari Energia Nucleare, "ARGOMENTO: Approdo di mezzi navali militari a propulsione nucleare in porti italiani. Compilazione piano emergenza esterna," Prot. N. 01/241/544, May 31, 1975. The letter to CNEN was categorized as "sensitive" and included the cited document: "Sintesi delle ipotesi e dei criteri adottati per la stesura dei 'Piani di emergenza per le unità militari a propulsione nucleare che sostano nelle basi navali,'" box "Porti Nucleari, La Maddalena" folder "La Maddalena," Sezione ENEA-DISP, Archivio Storico ENEA, Casaccia.

20. The limits of ISS and CNEN campaigns were clear also to the local population, constantly (at least partially) informed by local newspapers: "Limitata in partenza l'indagine del CNEN sull'inquinamento atomico," 12.

21. Eugenio Tabet, personal interview with the author, January 12, 2012.

22. Of the first type (fission products) are elements created by the reactions inside the core of the reactor, while the second (activation products) are formed by the irradiation of the materials containing the reactor and those of the primary cooling circuit.

23. "Indagine Ambientale nell'intorno della base nucleare navale situate nell'isola de La Maddalena," APPENDIX 1, internal document, ISS, n.d., [Summer/Fall 1974], held in the personal archive of Salvatore Sanna.

24. Boeri and Giorcelli, "Le reti nazionali per il rilevamento della radioattività ambientale," 29–31. The role of CAMEN in the radiosurveillance system of La Maddalena was somehow independent from the coordinated strategy implemented by ISS and CNEN, although it participated with the other two institutions in the systematic intercomparisons of data starting in the late 1970s. It is not possible to establish with certainty to what extent CAMEN had direct supervision of the annual radiometric reports that the US Navy was performing in La Maddalena, like in any other facility both in the US and overseas (for example in Japan, among other countries). The US Navy published its data in official bulletins and shared them with other US institutions as part of a larger network of radiometric and radioecological monitoring stations across the country.

25. The conference took place in Rome on September 24–25, 1975.

26. Eugenio Tabet, "Il ruolo degli Organi Sanitari Centrali e Periferici nella protezione della popolazione in relazione a installazioni nuclear navali," Relazione presentata al Convegno "Basi nucleari U.S.A., popolazione, e ambiente: il caso de La Maddalena," held in the personal archive of Carlo Papucci.

27. The monthly radiometric measurements began only in 1978, while in installation of a local laboratory for the continuous monitoring of radioactivity levels happened in 1979.

28. Mario Mittempergher, "L'intervento del CNEN a La Maddalena," report presented at the conference "Basi nucleari U.S.A., popolazione, e ambiente: il caso de La Maddalena," held in the personal archive of Carlo Papucci.

29. The *presupposti tecnici* (technical data) include the characteristics of the nuclear reactor considered to elaborate the hypotheses of radionuclides release in case of accident. This technical information is extremely important because each reactor type has specific characteristics based on its construction and operation, which have a clear influence on the kinds of accidents that can happen and their possible consequences. The technical report constituted one of the fundamental elements that CNEN-DISP and the Commissione Tecnica evaluated before issuing the authorization to operate the plant and for the approval of external emergency plans.

30. Stralcio verbale 258 18/2/76: 2) Parere sui presupposti tecnici per il piano di emergenza esterna relative alla base navale militare de La Maddalena, box Porti Nucleari, La Maddalena, folder "La Maddalena," Sezione ENEA-DISP, Archivio Storico ENEA, Casaccia.

31. Stralcio verbale 258 18/2/76.

32. Ministero dell'Interno, Direzione Generale della protezione Civile. OGGETTO: Arcipelago de La Maddalena: Piano di Emergenza Esterna. N.ro 146/030/3 di prot., Roma, lì 26/4/1977, box Porti Nucleari, La Maddalena, folder "La Maddalena," Sezione ENEA-DISP, Archivio Storico ENEA, Casaccia.

33. Stralcio verbale 290, 14/4/78; and Doc. DISP/MADDALENA (78) 1. Impianto: Base Navale militare. Sito: Isola de La Maddalena. INFORMAZIONI AGGIUNTIVE IN MERITO ALL'ESAME DEL PIANO DI EMERGENZA ESTERNA, Rome, February 1978, box Porti Nucleari, La Maddalena, folder "La Maddalena," Sezione ENEA-DISP, Archivio Storico ENEA, Casaccia.

34. This event in Italian diplomatic history was kept secret until the early 2000s. A first account of Italy's participation in this project can be found in Paolo Cacace's book *L'atomica europea*.

35. The lack of regulation concerning the navigation, approach, and anchoring of nuclear vessels at Italian ports was one of the problems that the Commissione Tecnica had to deal with when examining the safety procedures concerning nuclear ships. See Amaldi and Campos Venuti, "Aspetti sanitari della protezione delle popolazioni dalle radiazioni," 199–217.

36. This was done in accordance with the Decree of the President of the Republic no. 185, 1964 (DPR 185, 1964), which still remains, with subsequent modifications, the main Italian nuclear regulatory frame on radioprotection in the civilian nuclear field. External emergency plans elaborated during the 1960s and 1970s for the protection of the population living in the proximity of nuclear

ports can be found in the archive of the Istituto Superiore per la Protezione e la Ricerca Ambientale (ISPRA)—formerly the Dipartimento Sicurezza e Protezione del CNEN (CNEN-DISP)—in Rome and in the (recently rescued) archives of ENEA Casaccia. I thank Lidia Pucciarelli and Silvia Scarpato, respectively former and current curator of the ENEA-DISP and ISPRA archives, for their assistance and clarifications.

37. The case of the US submarine *Scorpion*, which approached the port of Taranto (Apulia) in March 1968, is mentioned in a document of the Ministry of the Interior of 1974 relative to the preparation of the emergency plan for La Maddalena. Ministero dell'Interno, Direzione Protezione Civile, Oggetto: La Maddalena—Approdo mezzi navali a propulsione nucleare—compilazione piano emergenza a norma degli artt. 112 e segg. del D.P.R del 13/2/64, n. 185, June 25, 1974, held in the personal archive of Carlo Papucci.

38. From the external emergency plans prepared by Italian authorities for the visits of the US commercial nuclear ship *Savannah* we know that one of the first safety procedures that radioprotection experts performed before allowing anchorage at port was to ascertain the status of the reactor on board and to assess the regularity of the safety measures followed by the vessel's crew. According to a 1988 memo prepared by Lidia Pucciarelli, secretary of CNEN's Commissione Tecnica, in response to a parliamentary interrogation by MP Publio Fiori about the safety procedures followed at Italian nuclear ports in regard to foreign military nuclear ships, Italian authorities predisposed nine emergency plans, including La Maddalena: "The agreements between the Italian government and foreign governments establish that in case of accident the commander of the damaged vessel should immediately inform the Italian navy command . . . and within one hour the naval unit should be towed away from the port . . . *It is acknowledged that Italian authorities cannot perform any safety assessment on military nuclear vessels. (These procedures, though, were followed for commercial nuclear ships, like the* Savannah, *even by ENEA-DISP)*" (italics added). Appunto per l'ingegner Eletti, OGGETTO: Interrogazione parlamentare On.le Fiori n. 3-00288 relativa ai mezzi navali nucleari, April 1, 1988, Archive ISPRA, Rome. The emergency plans predisposed by the provincial offices for radioprotection of Livorno (1965–67), Genoa (1965), and Trieste (1966) are available at the former ENEA-DISP Archive, now Archive ISPRA.

39. Department of State, Subject: NUCLEAR POWERED SUBMARINES AT LA MADDALENA, April 15, 1974, 21:29, Telegram ID: 1974STATE076272_b.

40. In the same memo she wrote in 1988 about safety measures followed by Italian authorities at nuclear ports, Pucciarelli specified that "radiometric measurements of the air and of sea water are conducted by the Italian navy during the stationing of foreign naval units along Italian coasts." Appunto per l'ingegner Eletti. A detailed description of CAMEN's radiosurveillance protocols along

Italian coastal lines can be found in various reports. See, for example, Argiero et al., "Controllo sistematico della radioattività del mare lungo le coste italiane," 261–67.

41. See "Presupposti tecnici per il piano di emergenza esterna relativo alla sosta di unità navali militari a propulsione nucleare nei porti italiani," Documento DISP | Navi nucleari (1979) 1 del C.N.E.N., Rome, May 1979, Archive ISPRA. The document was updated in 2000 and in 2005. See Lidia Pucciarelli, "Presupposti tecnici per il piano di emergenza esterna relativo alla sosta di unità navali militari a propulsione nucleare nei porti italiani—Revisione 2000," DOC. NUC/NAVI NUCLEARI (2000) 1, May 25, 2000. The document has a very useful historical introduction with a contextual explanation for the rationale and the origins of the current regulations. To my knowledge, the most updated version of the document on the elaboration and approval of external emergency plans in Italian nuclear ports is by Ing. Paolo Zeppa, Doc. EMEPIA/2004/02—Revisione 1, Rome, February 2005, Archive ISPRA.

42. The minister of defense, Attilio Ruffini, responded to several parliamentary interrogations after the accident with reassuring words: "State authorities intervene through monitoring activities to check the levels of radioactivity in the air and in the surrounding environment. The monitoring of radioactivity in the air is currently performed by two radiosurveillance networks: one controlled by the Italian navy, consisting of five monitoring stations, and one controlled by the US Navy. A third monitoring system, moreover, will be installed shortly under the responsibility of the Provincial Laboratory of Hygiene and Prophylaxis of the Region [Sardinia], which will operate in addition to the other two." Camera dei Deputati, *Bollettino delle Commissioni—VII Commissione Permanente Difesa*, Thursday, November 3, 1977, 254. Several newpapers reported Ruffini's words: "Il ministro della difesa rassicura. Misteri e perplessità rimangono," *Tutto Quotidiano*, September 25, 1977, 1, 9; "Affidato ai soli militari il controllo della radioattività nella base atomica," *L'Unione Sarda*, November 4, 1977;

"Nessun pericolo per La Maddalena," *L'Unione Sarda*, September 25, 1977, 2; "Non c'è radioattività alla base della Maddalena," *L'Unità*, October 1, 1977; "La Maddalena non rischia pericoli da radiazioni," *Il Tempo*, October 1, 1977; "Dal ministro della Difesa il sindaco della Maddalena per le navi nucleari," *La Stampa*, October 1, 1977.

43. Gian Nicola Cabizza, "Compendio dei controlli sulla radioattività ambientale alla Maddalena effettuati c/o il Laboratorio Provinciale di Fisica delle radiazioni, dalla sua entrata in funzione ad oggi. Sintesi delle precedenti comunicazioni mensili," June 20, 1980, Archive of the Province of Sassari.

44. Marina Militare, Comando M.M. Autonomo in Sardegna, La Maddalena, "ARGOMENTO: Rete continua di Monitoraggio," September 28, 1989, 1, held in the personal archive of Carlo Papucci.

45. Senato della Repubblica, "Seguito dell'indagine conoscitiva sulla situ-

azione ambientale dell'Arcipelago di La Maddalena: audizione del Direttore del Centro interforze studi ed applicazioni militari (CISAM)." Legislatura 14°–13ª Commissione permanente "Territorio, Ambiente, Beni Ambientali"—Resoconto stenografico n. 384, December 16, 2004, pp. 11–12.

46. Private communication with the author, June 21, 2014.

47. Provincia di Sassari—Ufficio Ecologia, "OGGETTO: Controllo sulla radioattività nello arcipelago di La Maddalena—Relazione informativa," Sassari, November 11, 1985, p. 6, Archive of the Province of Sassari.

48. "Verbale di riunione del giorno 25 Novembre 1986—ARGOMENTO: Rete Monitoraggio Civile Arcipelago di La Maddalena," November 25, 1986 [no folder], Municipal Archive of La Maddalena. A list of participants and their signatures appear at the end of the document. Twenty-six people attended the meeting, including the mayor of La Maddalena, Antonio Fonnesu; Floriana Manca, new director of the local laboratory; a representative of the central government; various representatives of the Region of Sardinia and of the Province of Sassari; technicians of the company who fabricated and installed the radiometric stations; military personnel; and experts from CNEN, ISS, and the Ministries of Health and Defense. These data give a sense of the complexity of the decision-making process, involving the coordination of all the administrative levels and often contrasting interests represented around the table.

49. I found the document inside a random box abandoned in the attic of the Municipal Palace of La Maddalena.

50. "Perché dobbiamo tenerci una base pericolosa? La Maddalena: Convegno sul controllo della radioattività ambientale nell'arcipelago," *Tutto Quotidiano*, November 9, 1977.

51. Carlo Papucci, "'Caso Maddalena' e intervento del sindacato: situazione al Novembre 1977," November 8, 1977, held in the personal archive of Carlo Papucci.

52. Papucci, "'Caso Maddalena' e intervento del sindacato," 4.

53. In part, this was due to the fact that the United States concluded a "free pass" agreement with Italy (and with other European countries) according to which nuclear-powered submarines did not need permission from Italian authorities each time they approached Italian ports. A retired US Navy sailor stationed in La Maddalena during the 1980s has confirmed this during a personal interview. He mentioned that in La Maddalena the transit of nuclear submarines was facilitated by the fact that the US Navy did not need permission to enter the bay of Santo Stefano. This clearly made the refitting and refurnishing operations more agile. Probably, Italian military authorities knew at least in part the frequency of nuclear-powered ships transits at nuclear ports, but certainly this was information that both the US Navy and the Italian navy did not want to share with civilian authorities. In 1988, for example, Sardinian newspapers cited unofficial sources made available by pacifist movements, reporting that "only

in 1985 La Maddalena hosted a total of 832 units." "La Maddalena un 'record nucleare,'" *La Nuova Sardegna*, June 4, 1988.

54. Until 1979 Italy did not have specific regulations on radioprotection and emergency plans for nuclear ports. When civilian nuclear ships (like the US *Savannah* and the German *Otto Hahn*) were expected to approach Italian nuclear ports (Naples, Livorno, Genoa, and Trieste, for example) ad hoc emergency plans were designed and implemented by expert agencies (CNEN and ISS), the prefectures, and military authorities. Only after 1979, when the Commissione Tecnica redacted the external emergency for La Maddalena, did the Italian government decide to establish safety procedures at all nuclear ports, based on the model provided by La Maddalena. A detailed description of the history of the legislation on nuclear safety at Italian ports can be found in "Presupposti tecnici per il piano di emergenza esterna relativa alla sosta di unità navali militari a propulsione nucleare nei porti italiani," Agenzia Nazionale Protezione Ambiente, DOC. NUC/NAVI NUCLEARI, (2000), 1, Archive Istituto Superiore per la Ricerca e la Protezione Ambientale, Rome. For a concrete example of how radioprotection and environmental monitoring worked at Italian nuclear ports during the 1960s, the documentary *Nave senza Fumo* produced by CNEN provides a great historical example. The documentary is available at the online catalogue of ENEA: http://webtv.sede.enea.it/index.php?page=listafilmcat2&idfilm=173&idcat=30.

55. Not until 1978 was CAMEN officially assigned this role in analogy with other nationally accredited institutions, like CNEN and ISS. See *Legge 23 Dicembre 1978, n. 833: Istituzione del Servizio Sanitario Nazionale*, in particular art. 6, section 1, points v) and z), concerning the activity of state agencies, regarding the organization of the health services attributed to military forces.

56. This in fact was the situation lamented by Italian members of parliament on various occasions. For example, during a 1977 audit of the undersecretary of defense Giuseppe Caroli (Democrazia Cristiana) in the Committee on National Defense of the Italian Senate, Communist Senators Carlo Bernardini, Araldo Tomelli, and Protogene Veronesi underlined the state of disorganization and scarce resources under which CAMEN was operating. During his reply, in particular, Bernardini complained that "the question of secrecy cannot restrict our right to be informed, for the simple reason that secrecy cannot exist when we are concerned with the problem of radioprotection of civilians. . . . Before concluding I also want to remember that to day there is no [formal, by law] constitutive act for C.A.M.E.N." Senato della Repubblica, VII Legislatura, Commissione Difesa, "14th Resoconto Stenografico Seduta di Mercoledì 4 Maggio 1977," 84–86.

57. See Nuti, *La Sfida nucleare*, 64–66.

58. A concise history of the center can be found in Vaglini, *Il nucleare a Pisa*. Other "traces" of CAMEN's history, developments, and shortcomings can be found in parliamentary acts. See, for example, the declarations of Italian deputy

Giuseppe Nicolai during a question time addressed to the minister of defense Gui on January 23, 1969: *Atti Parlamentari della Camera dei Deputati, V Legislatura, Discussioni*, p. 4092, cited in Nuti, *La Sfida nucleare*, 65.

59. In 1985, a major reorganization of CAMEN (which then changed its name to Centro Ricerche e Studi Applicazioni Militari, CRESAM) happened through another ministerial decree (Decreto del Ministro della Difesa 13 Luglio, 1985). The same thing happened again in 1994, when CRESAM became CISAM (Centro Interforze Studi per le Applicazioni Militari), through another decree of the minister of defense (Decreto del Ministro della Difesa 28 Aprile, 1994).

60. Legge 29 Settembre 1962, n. 1483: "Autorizzazione ad assumere personale laureato per ricerche e studi nel campo dell'energia nucleare e istituzione presso il Ministero della difesa, di un ruolo di personale tecnico di concetto per l'energia nucleare," https://www.gazzettaufficiale.it/atto/serie_generale/caricaDettaglioAtto/originario?atto.dataPubblicazioneGazzetta=1962-10-30&atto.codiceRedazionale=062U1483&elenco30giorni=false.

61. Decreto del Presidente della Repubblica n. 90 del 15 Marzo 2010.

62. Laura Bottici et al., "Atto di sindacato Ispettivo n. 4-01202, Seduta n. 141 del 26 Novembre 2013," Senato della Repubblica, Atti Parlamentari, http://www.senato.it/japp/bgt/showdoc/showText?tipodoc=Sindisp&leg=17&id=728578.

63. See Decreto del Presidente del Consiglio dei Ministri, n. 183, 24 Giugno 2005, http://www.difesa.it/Legislazione/Norme_in_rete/Pagine/urn_nir_parlamento_decreto_ministeriale_2005-06-24_183_07_11_200501_06_2011_15_08_38.aspx. See in particular Article 4.

See also Roberta Pinotti, "Risposta all'interrogazione n. 4-01202, Fascicolo 67, 16 Dicembre 2014," Senato della Repubblica, Atti Parlamentari, http://www.senato.it/japp/bgt/showdoc/frame.jsp?tipodoc=Sindispr&leg=17&id=826234. For a detailed description of CISAM's procedures for the decommissioning of its nuclear reactor, and in particular the problem of substantial lack of external control over the activities of the center, it is worth reading Senator Manuela Granaiola's question to the minister of the environment: "Al Ministro dell'ambiente e della tutela del territorio e del mare, Seduta n. 141 del 26 Novembre 2013," Senato della Repubblica, Atti Parlamentari, Resoconto Stenografico, Allegato B, 280–82, http://www.senato.it/service/PDF/PDFServer/BGT/728587.pdf.

CHAPTER 5: ITALIAN BUREAUCRACY AND THE PRODUCTION OF IGNORANCE

1. The Region of Sardinia is one of the five Regioni autonome (autonomous regions) of Italy (together with Sicily, Valle D'Aosta, Trentino Alto Adige, and Friuli Venezia Giulia). The Italian constitution attributes autonomous regions more legislative and administrative powers than ordinary regions. The first Italian regulation of nuclear energy use and production, Decree of the President of the Republic no. 185, 1964 (DPR 185, 1964), delegated to the autonomous re-

gions the administrative organization of radioprotection, but until the regional reform of 1970 and the reform of the national health service of 1978, radioprotection and radiosurveillance remained de facto a prerogative of central expert institutions. For this reason, as is clear in the case of La Maddalena, the implementation of a more decentralized system of radiosurveillance encountered numerous practical obstacles.

2. An external emergency plan was required for all the nuclear plants operating in Italy. At the provincial level, the prefect was the authority with the final word on its design and implementation in case of accidents. In military installations, but especially in nuclear ports, military authorities were in charge of an internal emergency plan, with instructions for the maneuvers required of the military personnel in case of accidents. In Italy, where the regulation of nuclear ports was not yet formally legislated, the emergency plans were adapted case by case and in accordance with the civilian authorities. This, in fact, was a peculiar setting that became more formally regulated only after 1979. La Maddalena was the first case of a nuclear port for which a formal external emergency plan was drafted and approved.

3. In line with the reform of the national health system of 1975, according to which each Italian province should create a local presidium called "Presidio di Prevenzione Sanitaria" responsible for the administration of the services related to public health, including radioprotection.

4. Provincia di Sassari, "Verbale di deliberazione del consiglio provinciale, deliberazione n. 85. OGGETTO: Richiesta contributo ministero della sanità per rete di controllo per la tutela dell'ambiente da irradiazioni ionizzanti nel comprensorio isole di La Maddalena," September 16, 1976, Archives of the Province of Sassari. "Una convenzione per controllare la radioattività a La Maddalena," *La Nuova Sardegna*, September 17, 1976, 4.

5. The criteria for the composition of the special committee for the design radiosurveillance system of La Maddalena emerge clearly from the transcripts of the meetings. See for example, Provincia di Sassari, "Commissione Speciale per la Programmazione e Realizzazione di una Rete di Monitoraggio contro i Pericoli delle Radiazioni Ionizzanti, Verbale n. 1," October 11, 1978, Archives of the Province of Sassari.

6. Provincia di Sassari, "Capitolato Programma sul Sistema di Monitoraggio da Installare nell'Arcipelago di La Maddalena, 1978," Archives of the Province of Sassari.

7. Comune di La Maddalena, "Verbale di deliberazione del Consiglio Comunale, n. 5, March 10, 1976: Dichiarazioni programmatiche del sindaco," 9–10, Municipal Archive of La Maddalena.

8. "La Maddalena: sospeso il controllo sulla radioattività nella base USA," *L'Unità*, February 22, 1976. "La nave atomica USA lasciata senza controlli," *Paese Sera*, March 6, 1976. "L'isola col cuore atomico," *Il Messaggero*, March 6, 1976.

9. CNEN and ISS experts, including Arrigo Cigna and Gloria Campos Venuti, restated this condition during a meeting with the personnel of the Region of Sardinia held on December 18, 1975, at the National Health Institute in Rome. On that occasion, discussing the results of the preliminary radioecological campaigns conducted in La Maddalena and the data that the Region of Sardinia was able to collect on local epidemiology and meteorology, Campos Venuti lamented that Sardinian authorities were delaying the organization of the local network of radiosurveillance. Given the situation, Campos Venuti warned that if the region did not intervene immediately, the radiosurveillance of the archipelago would not be guaranteed. "Seconda riunione per il coordinamento dello svolgimento dell'indagine ambientale a La Maddalena, Rome, December 18, 1975, Istituto Superiore di Sanità," Archives of the ENEA Center for the Study of Marine Environment, Lerici.

10. Birardi's declaration was reported during an interview for the Communist newspaper *L'Unità*: "La Maddalena: sospeso il controllo sulla radioattività nella base USA," February 22, 1976.

11. "L'isola col cuore atomico," *Il Messaggero*, March 6, 1976.

12. Five of the major Italian national newspapers sent their journalists to the press conference organized by the American diplomacy to counter the alarmist climate created around the suspension of the radiosurveillance service by Italian expert institutions. Apart from the article appearing in *Corriere della Sera*, in which the journalist reconstructed the problematic history of the radiosurveillance system of La Maddalena, also reporting an ample interview with mayor Canopoli and vice-mayor Franco Tamponi, the press coverage was extremely positive for the US cause. "Cosí si vive al 'sito' Nato fra i sottomarini nucleari," *La Stampa*, May 11, 1976. "La popolazione della Maddalena teme l'inquinamento radioattivo," *Corriere della Sera*, May 11, 1976. "Pretestuose le polemiche sulla presenza USA a La Maddalena," *Il Giornale d'Italia*, May 12, 1976. "Immutata la radioattività nell'arcipelago della Maddalena," *Il Tempo*, May 12, 1976. "Atomi innocui alla Maddalena," *Il Resto del Carlino*, May 13, 1976.

13. Cable 1976 ROME08532, "La Maddalena and Sixth Fleet Visit for Italian Journalists," May 25, 1976, 18:45, https://www.wikileaks.org/plusd/cables/1976ROME08532_b.html.

14. Cable 1976 ROME08532.

15. "Nascite anormali a La Maddalena," *L'Unione Sarda*, May 28, 1976, 3.

16. "Dubbi sulla sicurezza della base atomica. Chiesto l'allontanamento," *Il Messaggero*, September 24, 1977; "La Maddalena: la popolazione non vuole fare da cavia," *Avanti!*, September 24, 1977; "Forse il sommergibile USA controllava un'unità russa," *Corriere della Sera*, September 24, 1977; "La base per I sommergibili atomici costituisce un pericolo permanente," *L'Unità*, September 24, 1977.

17. US Embassy Office in Rome, Subject: THE USS RAY INCIDENT AT LA MADDALENA, September 27, 1977 00:00, Telegram ID: 1977ROME15762_c, https://wikileaks.org/plusd/cables/1977ROME15762_c.html.

18. "La giunta chiederà al governo di tutelare la salute dei maddalenini," *La Nuova Sardegna*, September 25, 1977; "L'intera cittadinanza contro la base USA," *La Nuova Sardegna*, September 27, 1977; "Dubbi sulla sicurezza della base atomica. Chiesto l'allontanamento," *Il Messaggero*, September 24, 1977; "Via la base: troppo pericolosa," *Paese Sera*, October 8, 1977.

19. Gavino Canopoli offers his view on these events in his recently published recollections of his experience as mayor of La Maddalena, *Diario di un sindaco*.

20. Comune di La Maddalena, "Verbale di deliberazione del Consiglio Comunale, n. 22, October 10, 1977: Incidente al Sommergibile U.S.A. 'RAY,'" Municipal Archive of La Maddalena.

21. Comune di La Maddalena, "Verbali di deliberazione del Consiglio Comunale, n. 12–14, March 31, 1978: Dimissioni del Sindaco;" "n. 16–17, May 15, 1978: Presa Atto Dimissioni del Sindaco and Elezione del Sindaco"; and "n. 19, July 18, 1978: Dichiarazioni Programmatiche," Municipal Archive of La Maddalena.

22. Laboratorio Chimico Provinciale di Sassari, "Servizio di controllo sulla radioattività ambientale nell'Arcipelago di La Maddalena—Relazione Tecnica," August 28, 1975, Archives of the Province of Sassari.

23. "Convenzione tra il Ministero della Sanità e l'Amministrazione Provinciale di Sassari per lo svolgimento di studi e ricerche attinenti al controllo della radioattività nell'arcipelago di La Maddalena (prov. Sassari). Relazione no. 2, Sassari," June 20, 1980, Archives of the Province of Sassari.

24. Gian Nicola Cabizza, personal interview with the author, April 23, 2012.

25. Provincia di Sassari, "Verbale di deliberazione della giunta provinciale, n. 109 del 20 Febbraio 1979. OGGETTO: Richiesta di autorizzazione spesa registrazione ed acquisto valori bollati per n. 3 convenzioni con personale esterno per incarichi professionali connessi al controllo della radioattività ambientale nell'arcipelago di La Maddalena," Archives of the Province of Sassari. Gian Nicola Cabizza, personal interview with the author, May 2012.

26. Gian Nicola Cabizza, "Relazione sul controllo della radioattività presso l'arcipelago di La Maddalena," April 19, 1979, 1, Archives of the Province of Sassari.

27. Gian Nicola Cabizza, "Compendio dei controlli sulla radioattività ambientale alla Maddalena effettuati c/o il Laboratorio Provinciale di Fisica delle radiazioni, dalla sua entrata in funzione ad oggi. Sintesi delle precedenti comunicazioni mensili," June 20, 1980, Archives of the Province of Sassari.

28. For example, the provincial counselor Giuseppe Deligia, former mayor of La Maddalena, in 1976 asked "if it was possible to make the entire population periodically informed about the data produced by the personnel [of the future laboratory]; silence at times provokes unjustified alarms among the population." Provincia di Sassari, "Verbale di deliberazione del Consiglio Provinciale, n. 85

del 16 Settembre 1976. OGGETTO: Richiesta contributo ministero sanità per rete di controllo per la tutela dell'ambiente da irradiazioni ionizzanti nel comprensorio isole di La Maddalena, 9," Archives of the Province of Sassari.

29. This was the case in January 1980, when a city councilor of La Maddalena, Francesco Bardanzellu of Movimento Sociale Italiano (MSI), made public some indiscretions coming from anonymous sources of the Italian navy command of La Maddalena, declaring that in the previous months the signals of alarm of the monitoring stations (supposedly controlled by the navy) went on. Bardanzellu then presented an official request to the mayor to clarify the episode with the military authorities. "Fughe di radiazioni a Santo Stefano?" *La Nuova Sardegna*, January 27, 1980. "Sarebbero tre le fughe radiattive: ci sarà un'inchiesta ministeriale," *La Nuova Sardegna*, February 2, 1980. "I monitor hanno registrato solo radioattività naturale," *La Nuova Sardegna*, February 10, 1980.

30. Cabizza, interview with the author, May 2012.

31. Gian Carlo Tusceri, "Chi paga i controllori della radioattività?" *La Nuova Sardegna*, August 30, 1980.

32. Ministero della Sanità, "OGGETTO: Svolgimento studi e ricerche attinenti il controllo della radioattività ambientale nell'arcipelago delle isole di La Maddalena," Doc. N. 400.5/R-md/294, Rome, March 21, 1980, Archives of the Province of Sassari.

33. Provincia di Sassari, "OGGETTO: Svolgimento studi e ricerche attinenti il controllo della radioattività ambientale nell'arcipelago delle isole di La Maddalena," Prot. n. 11895, April 8, 1980, Letter to the Ministry of Health, General Directorate for the Services of Public Health, Archives of the Province of Sassari.

34. Provincia di Sassari, "OGGETTO: Controllo radioattività La Maddalena—Richiesta rinnovo convenzione di cui al D.M. 25.12.1977," Prot. n. 816, January 8, 1980, Letter to the Ministry of Health, Archives of the Province of Sassari.

35. Ministero della Sanità, "OGGETTO: Svolgimento di studi e ricerche attinenti il controllo della radioattività ambientale nel comprensorio delle Isole della Maddalena," Doc. N. 400.5/R-md/294, Rome, March 21, 1980, Archives of the Province of Sassari, 2.

36. For example, on December 20, 1979, Arrigo Cigna, director of the Radioprotection Laboratories of CNEN, sent a telegram to a Dr. Alamanni of the Provincial Laboratory of Hygiene and Prophylaxis, expressing his criticism of the work performed in La Maddalena: "I confirm my previous critiques about your report concerning of La Maddalena. It is contrary to correct scientific principle of radioprotection and environmental studies." Telegram 7CRC SSZ555 CAB984 33, December 20, 1979, Archives of the Province of Sassari. Also, Cabizza mentioned this telegram during our conversation in April 2012, to provide another example of his negative working experience with the

Province of Sassari due also to the unprofessionalism of some of its technical personnel.

37. Untitled telegram, Doc. Prot. n. 13749, Sassari, May 27, 1980, Archives of the Province of Sassari. "Appello al Ministro: Bloccati i controlli sulla radioattività?" *L'Unione Sarda*, May 27, 1980; "Allarme tra i partiti per il blocco del laboratorio contro le radiazioni," *L'Unione Sarda*, May 28, 1980; "Tempo di elezioni: rispunta il problema della radioattività," *La Nuova Sardegna*, June 4, 1980; "Radioattività alla Maddalena: il Ministero attende la relazione finale," *L'Unione Sarda*, June 6, 1980.

38. Provincia di Sassari, "OGGETTO: Servizio controllo radioattività—Sospensione rapporto convenzionale," Doc. Prot. n. 17775, Sassari, May 30, 1980, Letter to Dr. Gian Nicola Cabizza, Spanu Gianfranco, and Atzeni Salvatore, Archives of the Province of Sassari.

39. This was also the interpretation of one of the closest observers of the debate, *La Nuova Sardegna* correspondent Gian Carlo Tusceri: "Stanziati i fondi di gestione del laboratorio di rilevamento," *La Nuova Sardegna*, November 22, 1980.

40. "I tecnici del laboratorio di La Maddalena garantiranno i controlli anche senza lo stipendio," *L'Unione Sarda*, May 30, 1980.

41. "Il controllo della radioattività può essere affidato alla Regione," *L'Unione Sarda*, August 29, 1980; "Chi paga i controllori della radioattività," *La Nuova Sardegna*, August 30, 1980.

42. Gian Nicola Cabizza to Prof. Ennio Prozzo, Ministero della Sanità, Divisione Generale Servizio Igiene Pubblica, December 30, 1980, Archives of the Province of Sassari. Cabizza had previously sent two letters with the same content, the first dated November 24, 1980, to the Province of Sassari, the Region of Sardinia, and the mayor of La Maddalena, and the second dated December 6, 1980, to Roberto Biancastelli of ISS, and to Franco Giorcelli of CNEN, who were members of the expert consulting committee for the management of the radiosurveillance network of La Maddalena.

43. "Il laboratorio non chiude," *La Nuova Sardegna*, November 11, 1980.

44. Provincia di Sassari, "Oggetto: Richiesta informazioni Prof. Gian Nicola Cabitza [*sic*]," Doc. Prot. n. 35741, Sassari, December 2, 1980, Archives of the Province of Sassari.

45. "Il laboratorio non chiude," *La Nuova Sardegna*, November 26, 1980; "Le apparecchiature consegnate ai nuovi incaricati," *La Nuova Sardegna*, December 6, 1980.

46. Provincia di Sassari, "Servizio controllo radioattività La Maddalena—Approvazione schema di convenzione con il Prof. Angelo Parodo, docente di Fisica medica nella Università di Sassari, Verbale di deliberazione della Giunta provinciale N. 173 del giorno 13 Febbraio 1981," Doc Prot. N. 9939, Archives of the Province of Sassari.

47. For example, receipts issued by the Province of Sassari in favor of Paro-

do, Floriana Manca, and Nicolò Erre throughout 1981 and 1982 make clear that the payments were issued as reimbursements for research trips to La Maddalena. In November 1982 Manca and Erre received a stipend of 2,360,000 lire for their work in the months of September and October, including the reimbursement for six trips to La Maddalena: "Fattura N. 97—Per l'espletamento dei controlli radioattività ambientale nell'arcipelago di La Maddalena nei mesi di Settembre ed Ottobre 1982, Sassari," November 16, 1982, and similar documentation available in the Archives of the Province of Sassari. The province made similar payments (but of different amounts) during the years in which Manca and Erre worked at the laboratory under the supervision of Parodo.

48. Letter to Carlo Papucci, January 7, 1980, held in the personal archive of Carlo Papucci.

Cabizza corresponded frequently with Papucci during his time as director of the radiosurveillance laboratory. Papucci played an important role in the sociotechnical controversies about the delays in the implementation of the radiosurveillance service, both as an expert and as a union activist. In a situation of political and professional isolation, Cabizza could count on the technical and political support of Papucci. They corresponded about the technical problems given by the instruments with which the laboratory of La Maddalena was equipped. Cabizza asked Papucci practical advice on sampling procedures and instrument calibration and shared with him the results of the spectrometric analyses that the laboratory started to perform in 1979. In a letter sent on September 16, 1980, for example, the young physicist illustrated the last results of the laboratory and the problems he was having with the Province of Sassari about the renewal of his contract: Cabizza to Carlo Papucci, September 16, 1980, held in the personal archive of Carlo Papucci.

49. Nicolò Erre, personal interview with the author, October 2012.

50. Provincia di Sassari, "Convenzione per contratto d'opera senza vincolo di subordinazione per prestazioni professionali connessi al controllo della radioattività ambientale nell'arcipelago di la Maddalena. Libero professionista Dott.ssa Floriana Manca [. . .], Sassari," February 17, 1987, Archives of the Province of Sassari; "Verbale di deliberazione della Giunta Provinciale di Sassari N. 558. Oggetto: Delibera G. P. N. 2554 del 30.12.1986 Reg. Es. Integrazioni e modifiche. Servizio controllo radioattività La Maddalena. Sorveglianza e funzionamento rete rilevamento in continuo dati," Prot. N. 12591, April 7, 1987, Archives of the Province of Sassari.

51. As documented in the newly stipulated convention, the province hired a biologist, a computer technician (Erre was hired for this role), an electrical technician, a custodian, and a janitor.

52. "Adesso viviamo tutti più tranquilli, già in funzione la quinta centralina," *La Nuova Sardegna*, February 12, 1987; "Il sommergibile è radioattivo? Ora lo sappiamo con certezza," *La Nuova Sardegna*, July 29, 1987.

53. "Radioattività, controllerà la Regione," *La Nuova Sardegna*, January 21, 1988.

54. Both Erre and Manca reconstructed the history of their employment at the laboratory of La Maddalena during personal interviews. I interviewed Manca in May 2012.

55. Manca described in this way the status of the equipment in place before the installation of the electronic network in 1987. Personal interview, May 2012.

56. Cabizza, now a member of the environmentalist movement Legambiente, returned to comment on the radiosurveillance system of La Maddalena after the news about the completion of the monitoring network appeared on the newspapers: "La *Nuova* [Nuova Sardegna] recently commented with enthusiasm that 'now we can live in tranquility because the radiosurveillance system has been completed.' But I want to alert the population that things are quite different from what they appear. . . . The most dangerous and least secure nuclear plant in Italy is still the foreign plant of La Maddalena." "Quella nucleare della Maddalena è una centrale incontrollabile," *La Nuova Sardegna*, March 29, 1987.

57. Both Erre and Manca described the difficult environmental conditions that disturbed the operation of the radiosurveillance network. These difficulties continued until the beginning of the 1990s, when a major update of the entire system, including the substitution of hardware and software, made the network more efficient, stable, and reliable.

58. CO.GE.MA. s.n.c., "Rete di Monitoraggio La Maddalena—Relazione mensile, Sassari," March 3, 1989, Municipal Archives of La Maddalena; CO.GE.MA. s.n.c., "Rete di Monitoraggio La Maddalena—Relazione mensile, Sassari," April 12, 1989, Municipal Archives of La Maddalena.

59. Indeed, the local newspapers reported the episodes of false alarm and malfunction of the monitoring system on various occasions. See for example: "Se ci fossero fughe radioattive nessuno potrebbe intervenire," *La Nuova Sardegna*, March 15, 1988; "Radioattività, parla l'assessore provinciale Degortes: 'Se il controllo non funzione le colpe sono della Regione,'" *La Nuova Sardegna*, March 16, 1988; "Inchiesta della Procura sul laboratorio di Fisica," *La Nuova Sardegna*, June 14, 1988; "Sono falsi I dati sulla radioattività," *L'Unione Sarda*, June 14, 1988; "Ma il responsabile dell'igiene pubblica passa al contrattacco: 'gli apparecchi funzionano benissimo,'" *L'Unione Sarda*, June 15, 1988; "Laboratorio nel caos: guasta la centralina dell'arsenale," *La Nuova Sardegna*, June 16, 1988; "La Maddalena, c'è un rischio continuo sui controlli per la radioattività," *L'Unione Sarda*, November 2, 1988.

60. "L'arcipelago 'nucleare.' Timori e paura, nessun piano d'emergenza," *L'Unione Sarda*, November 26, 1988.

61. Erre, personal interview with the author, October 2012.

62. Erre, personal interview, October 2012.

63. In particular, the three laws established that: (1) The state had the au-

thority to individuate the sites for the construction of future nuclear plants and to enforce its decisions regardless of regional and local opposition; (2) The municipalities that hosted coal and nuclear power plants received an economic reward; (3) The Ente Nazionale Energia Elettrica (ENEL), the state-owned national electric company, could invest and make agreements with foreign partners in the nuclear sector, including technological design and construction of nuclear plants. The referendum was not exclusively on nuclear matters. Two questions concerned the so-called civil liability of the judiciary, that is, whether public prosecutors, examining magistrates, and judges should be financially liable toward the parties involved in a trial in case it was proved that they made mistakes. For a political analysis of the referendums and their consequences for the Italian political system see Hine, "Italian Referendums of 8/9 November 1987," 163–78.

64. See, for example, Cantone, Sturloni, and Brunelli, "Role Played by Stakeholders," 261–66.

65. An interesting "unconventional" reading of the 1987 referendum can be found in Ceri, "Nuclear Power Issue," 71–89.

66. "Obiettivo La Maddalena. Parte una campagna dei pacifisti contro l'atomo nelle basi militari," *L'Unione Sarda*, January 3, 1988; "Per dire no alla base. Dal 21 marzo la raccolta delle firme per il referendum," *L'Unione Sarda*, March 4, 1988; "La base USA a La Maddalena 'Violazione costituzionale,'" *L'Unione Sarda*, June 7, 1988.

67. During the Melis administration (1984–1989) the Region of Sardinia took the radiosurveillance service of La Maddalena under its responsibility, delegating the management of the local laboratory to the USL of Sassari. The transfer was announced by the regional assessor of public health in polemic with the provincial administration: "Via gli USA dalla Maddalena," *La Nuova Sardegna*, March 17, 1988.

68. "La Maddalena un 'record nucleare,'" *La Nuova Sardegna*, June 4, 1988.

69. The three questions admitted by the regional electoral commission were clearly targeting the US base of La Maddalena: (1) "Are you opposed to the presence of foreign military bases in Sardinia, established after bilateral international agreements that have not been subjected to parliamentary approval, and that offer logistic and refitting support also to nuclear ships and submarines carrying nuclear weapons?" (2) "Would you agree to a legislative proposal by the regional council . . . to prohibit the transit and the mooring of nuclear ships carrying nuclear weapons at national ports?" (3) "Would you support a legislative proposal by the regional council to institute consultative referendums for all international treaties and agreements that according to the Constitution [art. 80] require parliamentary approval?" From: "Per dire no alla base Usa," *L'Unione Sarda*, March 4, 1988; "Referendum sulla base atomica," *La Repubblica*, March 22, 1988; "Questo il testo degli interrogativi all'esame dei giudici romani," *L'Unione Sarda*, January 24, 1989.

70. This is the opinion of Andrea Pubusa, professor of public law at the University of Cagliari, Communist regional deputy, and "father" of the regional law that in 1987 introduced the institute of the consultative referendum in Sardinia. "La base Usa alla Maddalena 'violazione costituzionale,'" *La Nuova Sardegna*, June 7, 1988.

71. "Tre siluri contro la 'base Usa': I sardi alle urne per La Maddalena. Le votazioni a Dicembre o a Marzo," *La Nuova Sardegna*, October 12, 1988.

72. "E Greenpeace prepara un 'assalto' alla base Usa," *L'Unione Sarda*, January 6, 1988. The spectacular initiatives of Greenpeace in La Maddalena became frequent and later on were emulated also by Italian movements, like Legambiente. One of them is still remembered by the local residents and several US Navy sailors stationed in La Maddalena during that period. On June 25, 1988, Greenpeace activists symbolically assaulted the US Navy tender in Santo Stefano aboard a rubber boat towing a wooden yellow submarine with the phrase: "Nuclear Free Sea." See "Assalto alla 'Cable.' Incursione di Greenpeace nella base Usa di Santo Stefano," *La Nuova Sardegna*, June 25, 1988.

73. "I 'siluri' contro la base Usa bloccati dal veto romano," *La Nuova Sardegna*, October 10, 1988.

74. It is worth noting that at the national level the Partito Socialista Italiano (PSI) adhered to the antinuclear campaign promoted by the Greens, but the position of the party on La Maddalena remained contradictory. The regional PSI secretary, Antonello Cabras, justified his position with some difficulty: "We cannot say no to the American base, but we say no to nuclear power." "Nuclear no, base Usa si," *L'Unione Sarda*, November 4, 1988; "Il Psi dà ragione al governo," *La Nuova Sardegna*, October 31, 1988; "Giunta alla prova dei referendum. A confronto le due anime della maggioranza," *L'Unione Sarda*, November 2, 1988.

75. "Psd'Az contro il Psi. E Melis conferma il ricorso alla Corte," *L'Unione Sarda*, October 31, 1988.

76. "L'Isola non ama i missili. I sardi bocciano la base americana," *L'Unione Sarda*, December 12, 1988; "Tanta paura del nucleare. No di sessantotto sardi su cento alla base della Maddalena," *L'Unione Sarda*, December 12, 1988.

77. "Sentenza Corte Costituzionale N. 256, 1989," *Gazzetta Ufficiale* May 24, 1989, 21.

78. On this topic, see Codonesu, *Servitù militari modello di sviluppo e sovranità in Sardegna*.

79. See, for example, Iaccarino, "Nuclear Renaissance in Italy," 65–78, and Tonetti, "Localizzazione e consenso nel programma di rilancio del nucleare in Italia," 5–18.

80. Nuti, *La Sfida nucleare*.

81. See Frickel and Edwards, "Untangling Ignorance in Environmental Risk Assessment," 215–33.

82. For example, the Region of Sardinia formally instituted a regional radio-

surveillance program only in 1988 (with the exclusion of La Maddalena) with the help of the Physics Institute of the University of Cagliari: "Radioattività ambientale, il controllo sarà fatto dall'università," *La Nuova Sardegna*, September 9, 1988.

83. Auyero, *Patients of the State*, 2.

PART III: RISK, ACCIDENTS, AND POLITICAL MOBILIZATION

1. Campos Venuti, "Sorveglianza ambientale a La Maddalena," 16.

2. Floriana Manca, personal interview with the author, May 2012.

CHAPTER 6: MATERIAL SEMIOTICS OF RISK

1. See, for example, Kuchinskaya, *Politics of Invisibility*. In her analysis of public knowledge formation about radiation health effects in Belarus's post-Chernobyl contaminated areas, Kuchinskaya develops the concept of "articulation" to describe moments of interactions when experts and nonexperts coshape "definitions of hazards" and the explication of "the work that has to be done to mitigate it along with the conditions and resources available for this work" (9). She argues that "the theoretical significance of the dialogical approach is precisely in this acknowledgment of the co-shaping of different perspectives, as well as the situated and embodied character of interpretations" (8).

2. For example, in 1972 then director of CNEN, Ezio Clementel, signed a document, which went public, in which he excluded any sanitary concerns for the residents of the archipelago. A few days later the CGIL-Casaccia (the leftist union organization of CNEN technicians working at the laboratory for continental radioactivity) attacked Clementel's declaration by denying that any analysis on samples from La Maddalena had yet been performed.

3. Comune di La Maddalena, *Atti convegno indagine ambientale tenutosi in La Maddalena il 24.2.75*, 23, held in the personal archive of Gian Carlo Tusceri.

4. *Atti convegno*, 23.

5. *Atti convegno*, 3.

6. *Atti Convegno*, 6.

7. *Atti convegno*, 19.

8. *Atti convegno*, 21.

9. *Atti convegno*, 21.

10. *Atti convegno*, 24–27.

11. *Atti convegno*, 41.

12. *Atti convegno*, 41.

13. *Atti convegno*, 32.

14. *Atti convegno*, 43.

15. Santiago is a pseudonym. I use pseudonyms throughout this section to protect the identity of informants. As a non-US citizen doing research on confidential or classified matters regarding US nuclear technology, I could not directly access certain places and protected information. My analysis thus relies on

the generous collaboration of former US Navy personnel stationed in La Maddalena, access to documentation that the local administration made available for my research, and secondary literature. By combining these sources, I could better understand how concealment of classified information and communication strategies deployed by the US Navy in La Maddalena contributed to shaping apprehension and representations of risk at the local level.

16. Agreement between the Republic of the Philippines and the United States of America concerning Military Bases (with Annexes and Exchanges of Notes), Manila, March 14, 1947, https://digital-commons.usnwc.edu/cgi/viewcontent.cgi?article=2158&context=ils.

17. Santiago, personal interview with author, November 2012.

18. Due to the difficult conditions of their duty, the navy conceded several incentives to the personnel stationed in La Maddalena. They could extend their assignment, for example, which allowed them to have an extra month's license every year (for a total of two) and round trips paid for to every destination. All the sailors in La Madd enjoyed ninety-six hours of liberty every month, so they could leave the installation on Friday and go back to service the following Thursday.

19. Santiago, personal interview with author, November 2012.

20. They needed space for the vertical tubes nesting missiles with warheads to launch nuclear attacks undetected. While fast-attack nuclear subs—also called hunter-killers—were small and fast enough to chase other subs in reconnaissance missions, FBM subs were much bigger and used mostly as mobile, invisible, and undetectable platforms for launching missiles armed with nuclear warheads on Soviet targets. See Spinardi, *From Polaris to Trident*.

21. Department of the Navy, US Naval Support Facilities of La Maddalena, Santo Stefano North Atlantic Treaty Organization (NATO), "ENCLOSURE I: Naval Nuclear Propulsion Program—Confirmatory Survey Report in Support of the Decommissioning of U.S. Naval Facilities at the Santo Stefano North Atlantic Treaty Organization (NATO) Site," in "Site—Report on Radiological Conditions on Decommissioning, Ser. N00A/1093," January 8, 2008, 6.

22. Department of the Navy, "Site—Report on Radiological Conditions on Decommissioning, Ser. N00A/1093," January 8, 2008, ENCLOSURE I, p. 5.

23. Department of the Navy, "Site—Report on Radiological Conditions on Decommissioning, Ser. N00A/1093," January 8, 2008, ENCLOSURE I, p. 5.

24. Santiago, personal interview with the author, November 2012.

25. Jack and Laura, personal interview with the author, October 2012.

26. The navy is obviously very concerned about the kind of information that former personnel could share with the public. Breaching the barrier of confidential and classified issues may have severe repercussions even on retired sailors (from losing their pensions and benefits to conviction).

27. Information concerning the equipment and the kind of weapons on board nuclear submarines tended in La Maddalena comes from indirect sources,

according to which SUBROC, or submarine rockets that could host nuclear warheads, were present also in the archipelago. See Arkin and Fieldhouse, *Nuclear Battlefields*, 223; Buzzati-Traverso, *Morte Nucleare in Italia*, 104.

28. Goffman, *Asylums*. See also Gusterson, *Nuclear Rites*, esp. ch. 4, "Secrecy," and, more specifically on the US Navy, Zurcher, " Naval Recruit Training Center," 85–98.

29. *Nuke* is a colloquial navy term used to refer to personnel who attended the nuclear school.

30. Greg, personal interview with the author, October 2012.

31. Greg and Angela, personal interview with the author, October 2012.

32. Giulio and Roberta, personal interview with the author, September 2012.

33. Here Giulio was referring to the area known as Sasso Rosso, which the local administration used until the beginning of the 1990s as a landfill for the entire archipelago.

34. In Italy *orata* (pl. *orate*) is the commonly used name for the species *Sparus aurata*.

35. Carlo, personal interview with the author, October 2012.

36. Reno uses Harri Englund's concept of emplacement: "a phenomenological fact [that] is molded by histories of boundary making and constraint." Reno, "Beyond Risk," 516–30.

37. Irwin, Simmons, and Walker, "Faulty Environments and Risk Reasoning," 1319.

38. In their work, Kate Brown, Joseph Masco, and Olga Kuchinskaya observe that around Chernobyl and Los Alamos, and in radiocontaminated areas of Belarus, fauna and flora are thriving. They use these examples to underline the perceptive distortions that radiation introduces into the world and the sensorial and cognitive disorientation that it provokes in local communities living in contaminated areas where the healthy appearance of natural life masks the invisible hazards of radiocontamination. Brown, *Plutopia*, 2013. Masco, *Nuclear Borderlands*; Kuchinskaya, *Politics of Invisibility*.

39. Eco, "Guessing," 18.

40. Funtowicz and Ravetz, "Science for the Post-normal Age," 739–55.

41. According to CNEN and ISS experts four years were not enough for subjects exposed to low-level radiation to experience genetic effects.

42. Parmentier, "Semiotic Regimentation of Social Life," 360.

43. "Nascite anormali a La Maddalena," *L'Unione Sarda*, May 28, 1976, 3.

44. "Nascite anormali a La Maddalena," 3.

45. "Nascite anormali a La Maddalena," 3.

46. "Nascite anormali a La Maddalena," 3.

47. "Occorre un'indagine seria sulle nascite alla Maddalena," *L'Unione Sarda*, May 30, 1976, 3.

48. "Occorre un'indagine seria sulle nascite alla Maddalena," 3.

49. "Occorre un'indagine seria sulle nascite alla Maddalena," 3.

50. "Neonati deformi alla Maddalena: sono le radiazioni?" *Il Messaggero*, May 31, 1976.

51. "Neonati deformi alla Maddalena"; "La Maddalena: rilievi tecnici sulle acque," *Paese Sera*, June 1, 1976.

52. The second radioecological campaign planned by CNEN and ISS was supposed to start in early May, but for bureaucratic reasons and organizational problems Carlo Papucci and his colleagues could leave their laboratory until July.

53. "Una smentita che non smentisce," *L'Avanti*, June 2, 1976; "La Maddalena: smentite le morti da radioattività," *Corriere della Sera*, June 2, 1976; "Si studierà la radio-attività a La Maddalena," *Il Giorno*, June 4, 1976.

54. "I tre bimbi anormali: analizzate alghe, mitili della base Maddalena," *La Stampa*, June 3, 1976.

55. "Allarme alla Maddalena dopo la morte di tre bimbi," *L'Unità*, June 1, 1976.

56. "Allarme alla Maddalena dopo la morte di tre bimbi."

57. "Il ministro della Sanità rischia l'incriminazione," *L'Avanti*, June 10, 1976. Tamponi's arguments echoed and sometimes explicitly referred to statements about the status of the radiosurveillance system in La Maddalena that ISS and CNEN experts made in public debates and scientific conferences. Moreover, he used legal arguments offered by environmental activist judge Gianfranco Amendola, leader of an environmental organization named Gruppo Ambiente, who intervened quite frequently in the public debates concerning the lack of radioprotection protocols in La Maddalena also through op-eds in national newspapers. Amendola, "Basi infette," *Il Messaggero*, March 22, 1974.

58. I say especially leftist political leaders because they were a minority in La Maddalena and generally excluded from the resources that the local dominant elites monopolized thanks to their affiliation with the central government, the local church, and the Italian navy. The antibase front had to elaborate a strategy to break widespread reticence and undermine official narratives about the economic benefits of the US Navy's "innocuous" presence.

59. For the period 1972–1976 I relied on newspaper articles from *La Nuova Sardegna* and *L'Unione Sarda* available at the archive of at the municipal library of Sassari, Sardinia. For the period 1976–2008 I relied on a rich collection of articles concerning the US Navy base of La Maddalena assembled daily by the Information Office of the Italian navy command of La Maddalena. I had access to the entire collection thanks to the valuable collaboration of Francesco Nardini, director of the library at the Navy Officer Club of La Maddalena, where it was deposited after the closure of the US Navy base.

60. The base in La Maddalena was the station of the squadron.

61. During informal conversations, former US Navy servicemen mentioned

the name of Gian Carlo Tusceri as the clearest example of the antibase positions in La Maddalena. A retired US Navy officer who asked to remain anonymous told me: "That shit that G.C.T. [how Gian Carlo Tusceri was signing his articles] wrote on *La Nuova* was just political propaganda."

62. Tusceri, personal interview with the author, April 2012.

63. Dedóla, personal interview with the author, October 2012.

64. "Comitato misto paritetico per le servitú militari della Regione Sardegna."

65. The CGIL is still the largest leftist union of the country. Until 1991 CGIL was considered the union closest to the Communist Party, and other parties of the socialist constellation. The CGIL-Ricerca was the branch of the union representing the sector of public employees in national research institutions like CNEN and ISS.

66. I had access to the correspondence thanks to the collaboration of Carlo Papucci, who made his private archive available for my research.

67. A typed letter that Sanna sent to Papucci on May 5, 1977, held in the personal archive of Carlo Papucci, shows the constant flow of communication between the two: "Dear Carlo, I am attaching a copy of the most recent articles, including the one that you requested with particular interest during our last phone conversation. I also send you a copy of my note to the mayor [of La Maddalena] . . . and a copy of the letter that I sent to CNEN, hoping that it will reach [Achille] Albonetti who signed the response to my previous letter and that I also include here together with the other documents. Warm wishes, Salvatore."

68. Salvatore Sanna to Carlo Papucci, private correspondence March 11, 1977.

69. He was responsible for the energy policies sector inside CGIL-Ricerca. He often wrote op-eds in national newspapers to advocate for the continuation of the radioecological campaigns in La Maddalena or to point out critical limits of the research and the responsibilities of the government for the lack of assistance in the implementation of safety measures, like emergency plans: "Urgente un piano di emergenza a La Maddalena," *Paese Sera*, August 5, 1976, signed Carlo Papucci, Energy Committee CGIL-Ricerca.

70. "Facile ottimismo di stampa e TV sui rischi nucleari alla Maddalena," *L'Unità*, June 19, 1976.

71. "Facile ottimismo di stampa e TV sui rischi nucleari alla Maddalena."

72. "Facile ottimismo di stampa e TV sui rischi nucleari alla Maddalena."

CHAPTER 7: SCIENTIFIC CONTROVERSIES AND POLITICAL MOBILIZATION AFTER THE *HARTFORD* ACCIDENT

1. "'Cover-up' Fury after US Nuclear Sub Is Grounded," *Independent*, November 13, 2003.

2. The *Day* of New London, Connecticut, gave the news of the grounding

only three days after the accident, on October 28, http://www.theday.com/article/20031028/DAYARC/310289954/0/Search. *Stars and Stripes* reported on the accident on October 29, 2003: "Damage to USS Hartford assessed after sub is grounded." In the following weeks a series of new articles added more details about the circumstances and the consequences of the accident.

3. "'Cover-up' Fury after US Nuclear Sub Is Grounded." The *New York Times* published the news on November 19, 2003: "Nuclear Sub Heads Home after Accident," http://www.nytimes.com/2003/11/19/world/nuclear-sub-heads-home-after-accident.html.

4. "Incidente a Caprera per il sottomarino nucleare USA più potente," *Corriere della Sera*, November 12, 2003.

5. Rosanna Giudice, press conference, November 14, 2003, Municipal Archive of La Maddalena; "Un sindaco in trincea nell'isola dei militari," *L'Unione Sarda*, December 7, 2003.

6. "Giallo sul boato notturno alla Maddalena. Versione ufficiale: il bang di un aereo militare," *La Nuova Sardegna*, October 22, 2003; "Misteriosa esplosione nella notte di lunedì," *L'Unione Sarda*, October 22, 2003.

7. "Un terremoto ha scosso la Maddalena," *La Nuova Sardegna*, 25, 2003; "Troppe versioni sullo scoppio. È necessario fare chiarezza," *La Nuova Sardegna*, November 26, 2003.

8. "E dalle radio va in onda la rabbia dei maddalenini," *La Nuova Sardegna*, October 24, 2003.

9. "Un diessino rivela: 'C'è stata un'esplosione a Santo Stefano,'" *La Nuova Sardegna*, October 23, 2003.

10. "Sommergibile nucleare su una secca. Licenziati commodoro e comandante," *La Nuova Sardegna*, November 12, 2003.

11. "USS *Hartford* Gets New Skipper after Grounding," *Day*, November 15, 2003.

12. The author of the comment wants to remain anonymous. The US Navy released a detailed account of the accident months later. A summary of the report was published in the *Day*, on June 27, 2004: "Report Details How USS Hartford Failed to Steer Clear of Danger." Also, a firsthand account of the accident appeared in a book published in 2010 by a former US Navy sailor on duty on the *Hartford*. See Brownfield, *My Nuclear Family*, esp. ch. 3, "Damn the Torpedoes!!!," 37–49.

13. This is how Antonello Tovo, and other Maddalenini I interviewed in 2012, described the atmosphere of those days in the archipelago.

14. For example, the local correspondent of the regional newspaper *L'Unione Sarda* reported the alarming words of Giulio Giudice, a provincial councilor and a member of Berlusconi's party, Forza Italia. Giudice affirmed that the Americans had already asked the municipal administration of La Maddalena for the documents for the acquisition of the former Italian navy arsenal and other prop-

erties. Mentioning that "rumors inside the city hall" made clear that this was a concrete possibility, he claimed the recent change of political regime in Turkey and the imminent attack on Iraq pushed the Bush administration to consider a massive relocation of US naval contingents to La Maddalena. "Avamposto sardo contro Saddam: cinquemila soldati americani in arrivo a Santo Stefano, *L'Unione Sarda*, November 12, 2002, 29.

15. Comune di La Maddalena al Ministro della Difesa, Onorevole Martino, telegram, November 20, 2002, Doc. Prot. n. 14594, December 2, 2002, Municipal Archive of La Maddalena.

16. Risposta del Ministro della difesa al Sidaco di La Maddalena, November 29, 2002, Doc. Prot. n. 14594, December 2, 2002, Municipal Archive of La Maddalena.

17. Comitato Misto Paritetico sulle Servitù Militari della Regione Sardegna (COMIPA), "Verbale della riunione del Comitato Misto Paritetico per le servitù militari in Sardegna tenutasi alla Maddalena il giorno 08 Luglio 2003 (Convocazione di Maridipart La Spezia con MSG N. 28973/INFR del 13 Giugno 2003)," July 3, 2003, Municipal Archive of La Maddalena.

18. COMIPA, "Verbale della riunione del Comitato Misto Paritetico."

19. Tovo, personal interview with the author, November 2012.

20. Leoni, "L'Arcipelago di La Maddalena."

21. Marco Leoni provided copies of the pamphlets. I thank him for sharing his personal archive with me.

22. Pasqualino Serra, personal interview with the author, July 2009.

23. The institution of the national marine park was formally introduced through the presidential decree of May 17, 1996, and previously anticipated by the general law on national parks (Legge n.10 del 4 Gennaio 1994).

24. Serra, personal interview with the author, July 2009.

25. IZI S.p.A., "Venticinque anni di presenza della base americana nell'arcipelago della Maddalena: una valutazione economica dell'impatto," Rome, 1996, held in the personal archive of Salvatore Sanna.

26. The personnel of the local laboratory of radioecological surveillance in La Maddalena released its first official report on the levels of radioactivity only on November 30, 2003—after a full month of rumors, speculations, and insufficient explanations provided by the mayor of La Maddalena.

27. Presidio Multizonale di Prevenzione di Sassari, Laboratorio Analisi di Mongiardino, "Rapporto: campagna radioecologica straordinaria—November 2003 [Report: Extraordinary Radio-ecological Campaign—November 2003]."

28. CNEN experts identified and included these radiocontaminants in analogy with radioecological campaigns conducted around the nuclear plant of Trino Vercellese, in Piedmont. There in 1964 a pressurized water nuclear reactor was activated. Given the lack of crucial information about the discharge formula of US reactors, Italian experts worked assuming that the pressurized

water reactors propelling the US submarines had characteristics similar to Trino Vercellese's reactor.

29. Murphy, *Sick Building Syndrome*.

30. As the precise location of the submarine's impact was not released to the public, activists relied on semi-official communications provided by authorities as a proxy.

31. CRIIRAD is a nongovernmental organization that was founded in 1986 in response to widespread preoccupations outside the scientific community with the effects of the radioactive fallout after the Chernobyl accident. While the French government continued to minimize the consequences of the disaster for the population, various scientists, like the nuclear physicist Robert Bernaud, received numerous requests from citizens to analyze samples of grass, milk, and other products. In one of such occasions, Bernaud agreed to help a biology teacher and an airline pilot to analyze their samples in Bernaud's laboratory in Lyon. The teacher and the pilot, Michel Rivasi and Francois Mosnier, were among the founders of CRIIRAD. Sezin Topçu argues that "since the Chernobyl 'scandal,' the rise of mistrust in the authorities as well as the social awareness on official 'secrecy' provoked a radical shift in the manner of framing criticism of nuclear energy and risks." In this context CRIIRAD adopted a whistleblower strategy, relying on concrete experts' methodologies to demonstrate fissures in the system used by French official authorities to assess nuclear risks. Because of their independent position, made possible also through the funding of a technologically advanced radiometric laboratory, CRIIRAD established their authority as "counterexperts." This strategy allowed CRIIRAD to break the boundaries of nuclear expertise in France, by deploying qualified expertise, introducing new kind of facts, introducing new policymaking logics, and criticizing upfront the bureaucratic rules of French national nuclear agencies and the regime of nuclear secrecy. For further discussion, see Topçu, "Confronting Nuclear Risk," 225–45.

32. CRIIRAD sells its services to a vast array of clients, including individual citizens, activist groups, and other nongovernmental organizations. The modality with which Sardinian and Corsican environmentalist groups asked CRIIRAD for its services corresponds to a consolidated practice in line with CRIIRAD's post-Chernobyl legacy. In the case of La Maddalena, World Wildlife Foundation Gallura and ABCDE could afford to pay only for gamma radiospectrometry analyses on a few samples collected by their activists (2,000 euros). More complex analyses with sophisticated methods of chemical separation and material preparation would have involved much longer times and considerable expenses. CRIIRAD eventually took the initiative to ask a Belgian laboratory to perform more sophisticated analyses, whose results were publicly announced only in June 2004.

33. CRIIRAD, "Analyses d'eau de mer and algues marines dans les bouches

de Bonifacio (Corse du Sud) et le secteur de La Maddalena suite a' l'incident de navigation du sous-marin USS Hartford, 15 juin 2004," Note 04/01.

34. "Nessun allarme radioattività dalla base atomica americana, *L'Unione Sarda*, on February 6, 2004, 7.

35. "Nessun allarme radioattività."

36. Fabrizio Aumento, personal interview with the author, January 2012.

37. Aumento formally presented his research in La Maddalena in November 2004, but before then, Legambiente informally disclosed some of the preliminary results.

38. Aumento, Le Donne, and Eroe, "Transuranium Radionuclide Pollution," 81–93. Alpha autoradiographs are indeed the simplest way to individuate the emission of alpha particles even at very low intensities in a large variety of samples, which require much less expensive and time-consuming preparation. In cases when the number of samples to be compared is in the order of hundreds, this large spectrum analysis allows a quick preliminary assessment of the presence of alpha-emitting particles but requires further examinations in order to assess whether the particles belong to natural elements or anthropogenic radionuclides. The technique consists of simply exposing readily available films sensitive only to alpha rays of a particular energy to the treated sample so that the emission of alpha particles from the material examined leaves traces concentrated in portions of the film that are vertically bombarded (hot spots). The density of the tracks impressed on the film by the emission of the alpha particles gives a visible measure of the intensity of radioactivity present in the sample. Given the variety of natural and artificial radionuclides are alpha-emitting elements, identifying the exact elements that produced the tracks on the film requires a second step. Aumento found traces compatible with artificial transuranium nuclides such as plutonium and americium, but in order to test his hypothesis his team proceeded to the analysis of the hot spots through radiochemical/alpha ray spectrometry, which confirmed only the presence of uranium (while americium had concentrations below the detectable limits). What captured Aumento's attention was the fact that the supposed traces of plutonium were detectable only in the samples coming from around the US Navy base. Thus, he advanced the thesis that the source of alpha-emitting particles must have been local; that is, related to the activity of the US Navy base of La Maddalena.

39. It is important to emphasize that the initiative of the minister of the environment started only after an official request from the mayor of La Maddalena. On February 2, 2004, the City Council of La Maddalena gathered in an extraordinary session to discuss the alarming data published by CRIIRAD. Comune di La Maddalena, "Verbale di deliberazione del Consiglio Comunale," February 2, 2004, Municipal Archives of La Maddalena.

40. APAT-ICRAM-ARPAS-PMP Sassari-PMP Cagliari, *Indagini Straordinarie di Monitoraggio*.

41. CRIIRAD, "Base Militaire de San Stefano, Communiqué de Presse," June 24, 2004.

42. I learned this in November 2012 during a personal interview with Vincenzo Migaleddu, a radiologist, member of World Wildlife Foundation Gallura and one of the promoters of the CRIIRAD intervention in La Maddalena.

43. Fastame's comments on Aumento's study were published in *Il Vento*, October 1, 2004, 8, and June 4, 2005, 9. Fastame also explained his position during a personal interview in June 2012.

44. Aumento, Le Donne, and Eroe, "Transuranium Radionuclide Pollution."

45. Harsh comments were sent to the editor by P. Kershaw, Roberta Delfanti, and Carlo Papucci, and Arrigo Cigna: "Letters to the Editor," 130–33.

46. "Letters to the Editor," 131.

47. "Letters to the Editor," 132.

48. "Letters to the Editor," 133.

49. "Letters to the Editor," 133.

50. Bruschi, Cigna, Maffei, Zattera, and Zurlini, *Indagine radioecologica nell'arcipelago*.

51. Bruschi, Cigna, Maffei, Zattera, and Zurlini, *Indagine radioecologica nell'arcipelago*, 52.

52. Aumento, personal interview with the author, January 2013.

53. Knorr-Cetina, *Epistemic Cultures*.

54. Fastame, personal interview with the author, May 2012.

55. Antonello Tovo, personal interview with the author, October 2012. Other CoCiS members expressed the same opinion, in almost the exact same words, when I interviewed them individually.

56. Fabio Canu, personal interview with the author, April 2012.

57. Canu, personal interview with the author, April 2012.

58. Tovo, personal interview, October 2012.

59. See, for example, the article by Paolo Figus in *L'Unione Sarda*, on April 13, 1990: "La Guerra dei dati" [The war of data], in which the author interviews the commander of the US fleet in the Mediterranean, who for the first time responds directly to Cortellessa and other scientists and activists denouncing the presence of radioactive discharges in La Maddalena.

60. Interview by the journalist Lucio Salis, "La base non è a rischio radioattività," *L'Unione Sarda*, January 28, 2004.

61. One of the paradoxes posed for La Maddalena, and likely other communities living around nuclear submarine installations, was the invisibility of the system of radiosurveillance that supposedly monitored their safety. At first glance, this apparent paradox could be interpreted as a self-defense mechanism: removing the object of risk from daily life by simply ignoring it. However, this explanation does not take into account concrete historical processes that have made nuclear risk disappear from the map of visible threats (see, for example,

Hecht, *Being Nuclear*). At the level of general public opinion, but more dangerously among political elites, the phasing out of nuclear plants after the Chernobyl accident in Italy has made nuclear risk disappear as if problems like plants' decommissioning, spent fuel disposal and reprocessing, and transportation of highly radioactive material are outdated concerns.

62. "Una battaglia di cifre sui rischi del nucleare," *L'Unione Sarda*, February 15, 2004; "Nel granito non c'è la radioattività," *La Nuova Sardegna*, February 15, 2004.

63. Floriana Manca, who was invited to intervene during the second public assembly organized by CoCiS, described the atmosphere of that night in the city hall. Manca, personal interview with the author, May 2012.

64. Randaccio, email message to Sotgiu, February 17, 2004. I would like to thank both Professor Randaccio and Professor Sotgiu for allowing me to access their private correspondence about the post-*Hartford* debate in La Maddalena.

65. The distinction between the terms *radioactive* and *nuclear* here is crucial, in that while the first indicates a general status that encompasses both anthropogenic and natural sources of radioactivity, the second directly related the source of radioactivity to nuclear technology—therefore, "artificial" sources (both for civilian and military purposes).

66. Sotgiu, email message to Randaccio, February 17, 2004.

CONCLUSION: NATURAL HISTORIES

1. "L'U.S. Navy è andata via dalla Maddalena," *La Nuova Sardegna*, January 26, 2008.

2. "La Maddalena: pomodori contro Legambiente," *La Nuova Sardegna*, March 17, 2008.

3. Paolo Mondani, *Protezione e Benedizione*, https://www.raiplay.it/video/2011/05/Report---Protezione-e-Benedizione-4b842b28-6005-47f9-8928-a7ab2f43fe87.html.

4. Fabrizio Gatti, "La grande bugia di Bertolaso," *L'Espresso*, June 29, 2010, https://espresso.repubblica.it/attualita/cronaca/2010/06/29/news/la-grande-bugia-di-bertolaso-1.22052/.

5. The author attended the meeting on July 13, 2011, in the city hall of La Maddalena.

6. "La Maddalena: esposto per amianto dalla figlia di un dipendente dell'Arsenale morto," *La Nuova Sardegna*, July 6, 2013.

7. Gian Carlo Tusceri, "Della Renella, dei Sanculotti, e dell'epopea subalterna maddalenina," *Arcipelago di La Maddalena: Storia, Natura, Bellezza*, September 21, 2020, https://arcipelagolamaddalena.wordpress.com/2020/09/21/della-renella-dei-sanculotti-e-dellepopea-subalterna-maddalenina/.

8. "Radioattività, polemiche dopo la chiusura del laboratorio," *La Nuova Sardegna*, April 27, 2014.

9. Personal interviews with the author, September 2012.

10. Boeri, "Mineral Geopolitics," 18–23.

11. Boeri, "Mineral Geopolitics," 18–23.

12. Stefano Boeri and Barbara Cadeddu, "La Maddalena. Requalification of the ex military base," invited presentation, "Military Landscapes: A Future for Military Heritage," La Maddalena, Friday, June 23, 2017.

13. Ordonnance of the Italian Prime Minister, n. 3716, November 19, 2008, art. 12, https://www.gazzettaufficiale.it/atto/serie_generale/caricaDettaglioAtto/originario?atto.dataPubblicazioneGazzetta=2008-12-04&atto.codiceRedazionale=08A09125.

14. Bono, "Oltre la 'mala Protezione civile,'" 185–205.

Bibliography

ARCHIVES

Archive of the Province of Sassari
Archive Piani di Emergenza Esterna (formerly CNEN/ENEA Direzione Sicurezza e Protezione), ISPRA (formerly APAT), Rome
Camera dei Deputati e Senato della Repubblica, Libraries of the Italian Parliament, Rome
ENEA/CNEN
  Archive Research Center of Casaccia, Rome
  Archive Center for the Study of Marine Environment, Santa Teresa, La Spezia
  Archive of Scientific Documentaries and Films:
  http://webtv.sede.enea.it/index.php?page=listafilmcat2&idfilm=169&idcat=30
  Library Research Center of Casaccia, Rome
Library of the Italian Navy Command, Circolo Ufficiali, La Maddalena
Municipal Archive of La Maddalena
Public Library, Sassari

NEWSPAPERS AND MAGAZINES

L'Astrolabio
Avanti!

Avvenire
Corriere della Sera
Day (New London, CT)
L'Espresso
L'Europeo
Il Giornale d'Italia
Independent (UK)
KYRN (Corsica)
Mainichi Daily News (Japan)
Il Mattino
Il Messaggero
New York Times
La Nuova Sardegna (Sardinia)
Paese Sera
La Repubblica
Il Resto del Carlino
Rinascita
Rivista Italiana di Ecologia
La Stampa
Il Settimanale
Il Tempo
Tutto Quotidiano (Sardinia)
L'Unione Sarda (Sardinia)
L'Unità
Il Vento (Sardinia)
Stars and Stripes (USA)

SERIALS

Camera dei Deputati. *Resoconti stenografici.*

Comune di La Maddalena. *Verbali di deliberazione del Consiglio Comunale.*

Senato della Repubblica. *Resoconti stenografici.*

Comitato Nazionale per l'Energia Nucleare (CNEN). *Notiziario CNEN.*

Istituto Superiore di Sanità (ISS). *Annali dell'Istituto Superiore di Sanità.*

European Nuclear Energy Agency and Organization for Economic Cooperation and Development. *Nuclear Legislation Analytical Study: Regulations Governing Nuclear Installations and Radiation Protection*, 1969, 1972, and following years.

Provincia di Sassari. *Verbali di deliberazione del Consiglio Provinciale.*

PRIMARY SOURCES

Amaldi, Ugo, and Gloria Campos Venuti. "Aspetti sanitari della protezione delle popolazioni dalle radiazioni nell'intorno degli impianti nucleari." Report ISS-P-73-1. Istituto Superiore di Sanità, May 3, 1973.

Amaldi, Ugo, Gloria Campos Venuti, S. Frullani, L. Maiani, and E. Tabet. "Criteri di scelta dell'ubicazione delle installazioni nucleari." *Annali dell'Istituto Superiore di Sanità* 7, no. 4 (1971): 626–46.

Anselmi, B., et al. *Studi sui parametri geologici rilevanti ai fini della determinazione della contaminazione del territorio nazionale*. Technical report, Comitato Nazionale Energia Nucleare, CNEN RT/PROT (79) 14, Rome, 1979.

APAT-ICRAM-ARPAS-PMP Sassari-PMP Cagliari. *Indagini Straordinarie di Monitoraggio della radioattività ambientale nell'Arcipelago della Maddalena*, June 2004.

Argiero, Luigi, G. Del Corso, S. Manfredini, and G. Palmas. "Controllo sistematico della radioattività lungo le coste italiane. Organizzazione delle reti di prelievo di campioni d'acqua e fauna marina. Tecniche e dati di misure. Programma di studi e ricerche." *Minerva Nucleare* 7, no. 7 (July 1963): 261–67.

Atti del primo convegno sullo stato di avanzamento della radioecologia in Italia, organizzato dall'Istituto di Zoologia dell'Università degli Studi di Parma e dall'Associazione Italiana di Fisica Sanitaria e Protezione contro le Radiazioni, 5–6 Novembre, 1970. Parma, 1970.

Aumento, Fabrizio, K. Le Donne, and K. Eroe. "Transuranium Radionuclide Pollution in the Waters of the La Maddalena National Marine Park." *Journal of Environmental Radioactivity* 82, no. 1 (2005): 81–93.

Bernhard, M., et al. *Rapporto sull'attività scientifica e tecnica del Laboratorio per lo Studio della Contaminazione Radioattiva del Mare, Fiascherino, La Spezia (maggio1959—maggio 1962)*. Comitato Nazionale Energia Nucleare, RT/BIO (63) 8. Rome, February 1963.

Boeri, G. C., and F. Giorcelli. "Le reti nazionali per il rilevamento della radioattività ambientale in Italia." CNEN-RT/DISP (81): 28–35. 1981.

Brondi, A., and O. Ferretti. "Efficacia ed applicazione di concetti mesotipologici come elementi per un approccio globale ai problem ambientali." *Acqua-Aria*, no. 5 (1986): 451–59.

Brondi, A., O. Ferretti, B. Anselmi, and L. Rabottino. "Arcipelago della Maddalena—Studi Mineralogici e Sedimentologici." *Bollettino della Società Geologica Italiana* 95 (1976): 1293–318.

Bruschi, A., A. Cigna, P. Maffei, A. Zattera, and G. Zurlini, eds. *Indagine radioecologica nell'arcipelago della Maddalena: Rapporto conclusivo delle campagne effettuate*. ENEA RT/PAS/84/9, Rome, May 1984.

Buffoni, G., C. P. Gasparini, and A. Zattera. "Uno studio dei contaminanti in mare nell'area de 'La Maddalena.'" *ENEA RTI-INN (92)–22*, 1992.

Buzzati-Traverso, Adriano, and Michael Bernhard. "The CNRN Research Programme for the Study of Radioisotope Accumulation by Marine Organisms and its Effects as Regards Radioactive Contamination of the Ocean." In *Disposal of Radioactive Wastes*, vol. 2, Conference Proceedings, Vienna, IAEA, 1960.

Cagnetti, P., M. Bernhard, and A. Zattera. "Radiological Investigations in the Gulf of Taranto," Paper presented at the Simposio internazionale di radioecologia applicata alla protezione dell'uomo e del suo ambiente, Rome, September 7–10, 1971.

Campos Venuti, Gloria. "Sorveglianza ambientale a La Maddalena: relazione introduttiva." In AA.VV., *Atti del XX Congresso Nazionale dell'Associazione Italiana di Fisica Sanitaria a Radioprotezione*, Bologna, October 26–28, 1977.

Cigna, Arrigo. "Forty Years of Anthropogenic Radionuclides in Surface Seawater: Italian and Japanese Data." *Ocean Science Journal* 41, no. 4 (2006): 261–90.

Cigna, Arrigo. "Ripercussioni anche in Italia degli scoppi atomici di Las Vegas?" *Incontri*, no. 5 (May 1951): 84–87.

Cigna, Arrigo, G. Domenici, A. Malvicini, and L. Vido. "Radioattivita' dei prodotti di fissione nel fallout raccolto dopo l'esplosione nucleare francese nel Sahara." *Minerva Nucleare* 5, no. 4 (April 1961): 73–78.

Cigna, Arrigo, and Osvaldo Ilari. "The Role of the Radioecological Protection in the Environmental Preservation in Italy." *Proceedings of the Regional Study Group Meeting on Radiological and Environmental Protection*, Istanbul, Turkey, November 20–24, 1972, 194–218. The paper has been republished by Comitato Nazionale Energia Nucleare as a technical report: *The Role of the Radioecological Protection in the Environmental Preservation in Italy*, RT/PROT (76) 12, CNEN, 1976.

Cigna, Arrigo, and A. Malvicini. "Determinazione della data dell'esplosione di una bomba nucleare mediante analisi del decadimento radioattivo dei prodotti di fissione raccolti su filtro." *Minerva Nucleare* 4, no. 6 (June 1960): 162–69.

Comitato Nazionale per l'Energia Nucleare. "Misure di radioattivita' ambientale presso l'isola della Maddalena eseguite dal Laboratorio per la Radioattivita' Ambientale del CNEN." *Notiziario CNEN* 20, no. 5 (May 1974): 87–90.

Comitato Nazionale per l'Energia Nucleare. "Presupposti tecnici per il piano di emergenza esterna relativo alla sosta di unità navali militari a propulsione nucleare nei porti italiani." *Documento DISP, Navi nucleari (1979) 1 del C.N.E.N.* Rome, May 1979.

Comitato Nazionale per l'Energia Nucleare. *Relazione sulla campagna radioecologica attorno al sito della centrale elettronucleare di Caorso, Giugno 1977*. CNEN–RT/DISP (78) 11, 1977.

Comitato Nazionale per l'Energia Nucleare. *Relazione sulla campagna radioecologica attorno al sito della centrale elettronucleare di Latina, Aprile 1979*. CNEN–RT/DISP (81) 9, 1981.

Comitato Nazionale per le Ricerche Nucleari. *V Congresso Nucleare, 1960*, Vol. 1. Rassegna Internazionale Elettronica Nucleare, Rome, 1960.

Commission de Recherche et d'Information Indépendantes sur la Radioactivité. *Analyses d'eau de mer and algues marines dans les bouches de Bonifacio (Corse du Sud) et le secteur de La Maddalena suite à l'incident de navigation du sous-marin USS Hartford*, June 15, 2004, Note 04/01.

Comune di La Maddalena, *Atti convegno indagine ambientale tenutosi in La Maddalena il 24.2.75*.

D'Amato, S., ed. "Rapporto Annuale sulla radioattività ambientale in Italia." Vol. 2, "Reti Locali." CNEN-RT/DISP-AMB/117/79 (1976): 313–22.

De Franceschi, L., A. Gentili, P. Guidi, et al. "60°C in the Marine Mollusc *Pinna nobilis*." *Health Physics* 31, no. 4 (1976): 376–77.

Delfanti, Roberta, Carlo Papucci, and Arrigo Cigna. "Letters to the Editor." *Journal of Environmental Radioactivity* 87 (2006): 130–33.

Department of the Navy, US Naval Support Facilities of La Maddalena, Santo Stefano North Atlantic Treaty Organization (NATO). *Site—Report on Radiological Conditions on Decommissioning, Ser. N00A/1093*, January 8, 2008.

IZI, S.p.A. *Venticinque anni di presenza della base americana nell'arcipelago della Maddalena: una valutazione economica dell'impatto*. IZI S.p.A, Rome, 1996.

Partito Comunista Italiano. *La ricerca scientifica e la società in Italia. Atti del convegno nazionale del PCI—Ariccia 20/22 Marzo 1970*. PCI, Direzione–Sezione Culturale, 1970.

Partito Comunista Italiano. *L'Energia del Futuro: Atti del convegno di studio tenuto alle Frattocchie (Roma) il 9–10 luglio 1977*. Rome: Editori Riuniti e Istituto Gramsci, 1978.

Partito Comunista Italiano. *Una politica per la ricerca scientifica. Atti del Convegno di studio tenuto a Roma il 2 e 3 dicembre 1961, promosso dell'Ufficio Centrale Attività Culturali della Democrazia Cristiana*. Rome: Edizioni Cinque Lune, 1962.

Partito Comunista Italiano. *Uomo Natura Ambiente: Ecologia e rapporti sociali. Atti del convegno tenuto a Frattocchie (Roma) dal 5 al 7 novembre 1971*. Rome: Editori Riuniti—Istituto Gramsci, 1972.

Polvani, Carlo. "L'ubicazione degli impianti nucleari e la protezione della popolazione." In Comitato Nazionale per l'Energia Nucleare, *La Scelta dei siti per gli impianti nucleari: Atti del Convegno organizzato al Centro Ricerche Nucleari della Trisaia nei giorni 15–16 Settembre 1972 da CNEN, Regione Basilicata, Associazione Italiana di Fisica Sanitaria*. CNEN, Rome, 1973.

Presidio Multizonale di Prevenzione di Sassari, Laboratorio Analisi di Mongiardino, *Rapporto della Campagna Radioecologica Straordinaria—Novembre 2003*.

Pucciarelli, Lidia, ed. "Presupposti tecnici Presupposti tecnici per il piano di emergenza esterna relativo alla sosta di unità navali militari a propulsione nucleare nei porti italiani—Revisione 2000." DOC. NUC/NAVI NUCLEARI (2000) 1, May 25, 2000.

Sacripanti, Attilio, and Giuseppe Fameli, eds. *Centrale nucleare del Garigliano: Campagna radioecologica di controllo del territorio, Settembre-Ottobre 1980*, CNEN–RT/DISP, 1981.

US Navy, Naval Support Activity La Maddalena, Human Resources Management Department. *Benvenuti a La Maddalena*, Revised May 1976.

Zeppa, Paolo, ed. *Piani di emergenza esterna relativi alla sosta di unità navali militari a propulsione nucleare nei porti. Doc. EME-PIA/2004/02—Revisione 1*, APAT, Rome, February 2005.

SECONDARY SOURCES

Abate, Salvatore, and Francesco Nardini. *Il pane del governo: La Maddalena, 1946–1956*. La Maddalena: Paolo Sorba Editore, 2010.

Adams, Vincanne, Michelle Murphy, and Adele E. Clarke. "Anticipation: Technoscience, Life, Affect, Temporality." *Subjectivity* 28, no. 1 (2009): 246–65.

Agamben, Giorgio. *State of Exception*. Chicago: University of Chicago Press, 2005.

Agrawal, Arun. "Dismantling the Divide between Indigenous and Scientific Knowledge." *Development and Change* 26, no. 3 (1995): 413–39.

Alfano, Barbara, and Anna Celenza, eds. "Projecting Americanism Abroad: Italy in the Cold War." Special issue, *Italian Quarterly* 54, no. 211–14 (Winter–Fall 2017).

Allen, Barbara. *Uneasy Alchemy: Citizens and Experts in Louisiana's Chemical Corridor Disputes*. Cambridge: MIT Press, 2003.

Allen, Michael Thad, and Gabrielle Hecht, eds. *Technologies of Power: Essays in Honor of Thomas Parke Hughes and Agatha Chipley Hughes*. Cambridge: MIT Press, 2001.

Amaldi, Edoardo. *Da via Panisperna all'America: I fisici italiani e la Seconda guerra mondiale*. Rome: Editori Riuniti, 1997.

Amme, Carl H., Jr. "The Soviet Navy in the Mediterranean Sea." *Naval War College Review* 21, no. 10 (June 1969): 154–59.

Arkin, William M., and Richard W. Fieldhouse. "Focus on the Nuclear Infrastructure." *Bulletin of the Atomic Scientists* 41, no. 6 (1985): 11–15.

Arkin, William M., and Richard W. Fieldhouse. *Nuclear Battlefields: Global Links in the Arms Race*. Cambridge, MA: Ballinger, 1985.

Auyero, Javier. *Patients of the State: The Politics of Waiting in Argentina*. Durham, NC: Duke University Press, 2012.

Auyero, Javier, and Débora Alejandra Swistun. *Flammable: Environmental Suffering in an Argentine Shantytown*. Oxford: Oxford University Press, 2009.

Baldacci, Osvaldo, ed. *Ricerche sull'Arcipelago de La Maddalena. Memorie della Società Geografica Italiana*. Rome: Società Geografica Italiana, 1961.

Barca, Stefania. "Work, Bodies, Militancy: The 'Class Ecology' Debate in 1970s Italy." In *Powerless Science? Science and Politics in a Toxic World*, edited by Soraya Boudia and Nathalie Jas, 115–33. New York: Berghahn Books, 2014.

Barry, Andrew. *Material Politics: Disputes along the Pipeline*. Chichester, West Sussex: Wiley-Blackwell, 2013.

Beck, Ulrich. *Risk Society: Towards a New Modernity*. London: SAGE, 1992.

Beck, Ulrich. *World at Risk*. Cambridge: Polity, 2008.

Benjamin, Walter. *Illuminations: Essays and Reflections*. New York: Schocken Books, 1968.

Benson, Peter, and Stuart Kirsch. "Capitalism and the Politics of Resignation." *Current Anthropology* 51, no. 4 (2010): 459–86.

Berlinguer, Giovanni. *Politica della scienza*. Rome: Editori Riuniti, 1970.

Bini, Elisabetta. "Atoms for Peace (and War): US Forms of Influence on Italy's Civilian Nuclear Programs (1946–1964)." In Bini and Londero, *Nuclear Italy*, 23–40.

Bini, Elisabetta, and Igor Londero, eds. *Nuclear Italy: An International History of Italian Nuclear Policies during the Cold War*. Trieste: Edizioni Università Trieste, 2017.

Bobbio, Norberto. *Democrazia e segreto*. Turin: Einaudi, 2011.

Bobbio, Norberto. *The Future of Democracy: A Defence of the Rules of the Game*. Minneapolis: University of Minnesota Press, 1987. First published as *Il Futuro della Democrazia* (Turin: Einaudi, 1984).

Bobbio, Norberto. "Introduzione." In *Il Doppio Stato: Contributo alla teoria della dittatura*, by Ernst Fraenkel. Turin: Einaudi, 1983.

Boeri, Stefano. "Mineral Geopolitics." In *La Maddalena Effect: An Architectural Affair*, edited by Michelle Brunello and Francisca Insulza, 18–23. Milan: Rizzoli, 2009.

Bono, Irene. "Oltre la 'mala Protezione civile': l'emergenza come stile di governo." *Meridiana* 65–66 (2009): 185–205.

Boudia, Soraya. "Global Regulation: Controlling and Accepting Radioactivity Risks." *History and Technology* 23, no. 4 (2007): 389–406.

Boudia, Soraya, and Nathalie Jas. "Risk and Risk Society in Historical Perspective." *History and Technology* 23, no. 4 (2007): 317–31.

Bourdieu, Pierre. *Outline of a Theory of Practice*. Cambridge: Cambridge University Press, 1977.

Brown, Kate. *Plutopia: Nuclear Families, Atomic Cities, and the Great Soviet and American Plutonium Disasters*. Oxford: Oxford University Press, 2013.

Brown, Phil. *No Safe Place: Toxic Waste, Leukemia, and Community Action*. 2nd ed. Berkeley: University of California Press, 1997.

Brown, Phil. "Popular Epidemiology and Toxic Waste Contamination: Lay and Professional Ways of Knowing." *Journal of Health and Social Behavior* 33, no. 3 (1992): 267–81.

Brownfield, Christopher. *My Nuclear Family: A Coming-of-Age in America's Twenty-first Century Military*. New York: Alfred A. Knopf, 2010.

Bruti Liberati, Edmondo. "La magistratura: la 'radicale svolta' della metà degli anni sessanta del Novecento," *Il Politico* 251, no. 2 (2019): 77–104.

Burhop, E. H. S. "Scientists and Soldiers: America's 'Jason Group' Looks Back on Its Vietnam Involvement." *Bulletin of the Atomic Scientists* 30, no. 9 (1974): 4–8.

Burnett, Christina Duffy. "The Edges of Empire and the Limits of Sovereignty: American Guano Islands." *American Quarterly* 57, no. 3 (2005): 779–803.

Buzzati-Traverso, Adriano. *Morte Nucleare in Italia*. Bari: Laterza, 1982.

Cacace, Paolo. *L'atomica europea: i progetti della Guerra Fredda, il ruolo dell'Italia, le domande del futuro*. Rome: Fazi Editore, 2004.

Callon, Michel, Pierre Lascoumes, and Yannik Barthe. *Acting in an Uncertain World: An Essay on Technical Democracy*. Cambridge: MIT Press, 2009.

Campbell-Keller, Ann. *Science in Environmental Policy: The Politics of Objective Advice*. Cambridge: MIT Press, 2009.

Camprubí, Lino. "'No Longer an American Lake': Depth and Geopolitics in the Mediterranean." *Diplomatic History* 44, no. 3 (2020): 428–46.

Camprubí, Lino, and Sam Robinson. "A Gateway to Ocean Circulation: Surveillance and Sovereignty at Gibraltar." *Historical Studies in the Natural Sciences* 46, no. 4 (2016): 429–59.

Candela, Andrea. *Storia ambientale dell'energia nucleare. Gli anni della contestazione*. Milan: Mimesis, 2017.

Canetti, Elias. *Crowds and Power*. Translated by Carol Stewart. New York: Farrar, Straus and Giroux, 1984. First published as *Masse und Macht* (Hamburg: Claassen, 1960).

Canopoli, Gavino. *Diario di un sindaco: frammenti di vita maddalenina dal 1975 al 1982*. La Maddalena: Paolo Sorba Editore, 2018.

Cantone, Marie Claire, Giancarlo Sturloni, and Giancarlo Brunelli. "The Role Played by Stakeholders in the Public Debate that Brought Italy Out of the Club of Nuclear Energy Producers." *Health Physics* 93, no. 4 (2007): 261–66.

Capocci, Mauro, and Gilberto Corbellini. "Adriano Buzzati-Traverso and the Foundation of the International Laboratory of Genetics and Biophysics in Naples (1962–1969)." *Studies in the History and Philosophy of Biology and Biomedical Sciences* 24, no. 33 (2002): 489–513.

Cassata, Francesco. "Building the Casaccia Gamma Field: Nuclear Energy, Cold War and the Transnational Circulation of Scientific Knowledge in Italy (1955–1960)." *Italia Contemporanea*, no. 294, suppl. 1 (2020): 67–100.

Cassata, Francesco. *Le due scienze: Il "caso Lysenko" in Italia*. Turin: Bollati Boringhieri, 2008.

Cassata, Francesco. *L'Italia intelligente: Adriano Buzzati-Traverso e il laboratorio internazionale di genetica e biofisica (1962–1969)*. Rome: Donzelli, 2013.

Castangia, Isabella. *Sovranità, contiguità territoriale e isole in una controversia internazionale del XVIII secolo*. Naples: Eugenio Jovene, 1988.

Ceri, Paolo. "The Nuclear Power Issue: A New Political Cleavage within Italian Society?" *Italian Politics* 2 (1988): 71–89.

Ciafaloni, Francesco. "Le corporazioni della scienza e le lotte nelle università." *Quaderni Piacentini* 8, no. 34 (1968): 79–90.

Ciasca, Raffaele. "Corsi Colonizzatori della Sardegna nel Sec. XVIII." *Archivio Storico di Corsica* 4, no. 4 (1928): 314–90.

Ciccotti, Giovanni, Marcello Cini, Michelangelo de Maria, and Giovanni Jona-Lasinio. *L'ape e l'architetto: Paradigmi scientifici e materialismo storico*. Milan: Feltrinelli, 1976.

Ciglioni, Laura. *Culture atomiche. Gli Stati Uniti, la Francia e l'Italia di fronte alla questione nucleare (1962–68)*. Rome: Carocci, 2020.

Clô, Alberto. *Si fa presto a dire nucleare. Confessioni di un nuclearista non pentito*. Bologna: Il Mulino, 2010.

Cocco, Antonio. *Per la Patria e per il Re. Memorie di un Ammiraglio ottuagenario*. Foggia: Bastogi Editrice Italiana, 2006.

Codonesu, Fernando. *Servitù militari modello di sviluppo e sovranità in Sardegna*. Cagliari: CUEC Editrice, 2013.

Cohen, Avner. *The Worst-Kept Secret: Israel's Bargain with the Bomb*. New York: Columbia University Press, 2010.

Collins, Harry, Robert Evans, and Mike Gorman. "Trading Zones and Interactional Expertise." *Studies in the History and Philosophy of Science* 38, no. 4 (2007): 657–66.

Comaroff, John, and Jean Comaroff. *Ethnography and the Historical Imagination*. Boulder, CO: Westview, 1992.

Conti, Antonio. *Sbirizzendi pe l'Isula: appunti di etnologia e dialettologia isolana*. La Maddalena: Paolo Sorba Editore, 2014.

Cooley, Alexander. *Base Politics: Democratic Change and the U.S. Military Overseas*. Ithaca, NY: Cornell University Press, 2008.

Corburn, Jason. *Street Science: Community Knowledge and Environmental Health Justice*. Cambridge: MIT Press, 2005.

Corona, Gabriella. *Breve storia dell'ambiente in Italia*. Bologna: Il Mulino, 2015.

Coronil, Fernando. "After Empire: Reflections on Imperialism from the Americas." In *Imperial Formations*, edited by Ann Laura Stoler, Carole McGranaham, and Peter C. Perdue, 241–71. Santa Fe: School for Advanced Research Press, 2007.

Coronil, Fernando. "Beyond Occidentalism: Toward Nonimperial Geohistorical Categories." *Cultural Anthropology* 11, no. 1 (1996): 51–87.

Creager, Angela. *Life Atomic: A History of Radioisotopes in Science and Medicine*. Chicago: University of Chicago Press, 2013.

Csordas, Thomas. "Somatic Modes of Attention." *Cultural Anthropology* 8, no. 2 (1993): 135–56.

Curli, Barbara. "Energia nucleare per il Mezzogiorno. L'Italia e la Banca Mondiale (1955–1959)." *Studi Storici* 37, no. 1 (January–March 1996): 317–51.

Curli, Barbara. *Il Progetto Nuclear Italiano (1952–1964): Conversazioni con Felice Ippolito*. Soveria Mannelli: Rubbettino Editore, 2000.

De Castro Pietrangeli, Paola, Daniela Marsili, and Sara Modigliani, eds. *Storia e identità di un ente di ricerca. L'Istituto Superiore di Sanità attraverso racconti e testimonianze orali*. Rome: Istituto Superiore di Sanità, 2011.

De Felice, Franco. "Doppia lealtà e doppio stato." *Studi Storici* 30, no. 3 (1989): 493–563.

De Grazia, Victoria. *Irresistible Empire: America's Advance through Twentieth-Century Europe*. Cambridge, MA: Belknap Press of Harvard University Press, 2006.

De Martino, Renzo. *Il dialetto maddalenino. Storia, grammatica, genovesismi, il dialetto còrso*. Cagliari: Edizioni Della Torre, 1996.

Della Seta, Roberto. *La difesa dell'ambiente in Italia: storia e cultura del movimento ecologista*. Milan: FrancoAngeli, 2000.

Desiderio, Alfonso. "La fatal Vicenza." *Limes* 4 (2007): 397–406.

Desiderio, Alfonso. "Paghiamo con le basi la nostra sicurezza." *Limes* 4 (1999): 27–42.

Desiderio, Alfonso. "Viaggio nelle basi USA in Italia." *Limes* 3 (2007): 59–72.

Dessy, Ugo. *Sardegna. Un'isola per i militari*. Padova: Marsilio Editore, 1972.

Dobson, Alan P., and Charlie Whitman. "Project Lamachus: The Cold War Comes to Scotland—The Holy Loch U.S. Nuclear Base and Its Impact on Scotland, 1959–1974." In *Local Consequences of the Global Cold War*, edited by Jeffrey A. Engel, 169–92. Washington, DC: Woodrow Wilson Center Press; Stanford, CA: Stanford University Press, 2007.

Douglas, Mary. *Risk Acceptability according to the Social Sciences.* London: Russell Sage Foundation, 1985.

Douglas, Mary, and Aaron Wildavsky. *Risk and Culture: An Essay on the Selection of Technological and Environmental Dangers.* Berkeley: University of California Press, 1982.

Downey, Gary. "Risk in Culture: The American Conflict over Nuclear Power." *Cultural Anthropology* 1, no. 4 (1984): 388–412.

Duke, Simon. *United States Military Forces and Installations in Europe*. Oxford: Oxford University Press, 1989.

Duncan, Francis. *Rickover and the Nuclear Navy: The Discipline of Technology*. Annapolis, MD: Naval Institute Press, 1990.

Eco, Umberto. "Guessing: From Aristotle to Sherlock Holmes." *Versus* 30 (1981): 3–19.

Eco, Umberto. "Peirce's Analysis of Meaning." In *Proceedings of the C. S. Peirce Bicentennial International Congress Held at Amsterdam, the Hague and Utrecht, the Netherlands, 26–28 June 1976*, edited by Kenneth L. Ketner, 179–93. Lubbock: Texas Tech Press, 1981.

Edwards, Paul N. *The Closed World: Computers and the Politics of Discourse in Cold War America*. Cambridge: MIT Press, 1996.

Edwards, Paul N. *A Vast Machine: Computer Models, Climate Data, and the Politics of Global Warming*. Cambridge: MIT Press, 2010.

Elli, Mauro. *Atomi per l'Italia. La vicenda politica, industriale e tecnologica della centrale nucleare ENI di Latina, 1956–1972*. Milan: Unicopli, 2011.

Elli Mauro. "British First Nuclear Export: ENI's Atomic Power Station at Latina and Anglo-Italian Nuclear Cooperation." *Annales historiques de l'électricité* 9, no. 1 (2011): 27–42.

Engel, Jeffrey A., and Katherine Carté Engel. "On Writing the Local, within Diplomatic History—Trends, Historiography, Purpose." In *Local Consequenc-*

es of the Global Cold War, edited by Jeffrey A. Engel, 1–30. Washington, DC: Woodrow Wilson Center Press; Stanford, CA: Stanford University Press, 2007.

Epstein, Steven. *Impure Science: AIDS, Activism, and the Politics of Knowledge*. Berkeley: University of California Press, 1996.

"Esperienze e prospettive di lotta nei centri di ricerca. A cura di un collettivo di lavoro del C.N.E.N." *Quaderni Piacentini* 8, no. 39 (1969): 173–91.

Fann, K. T. *Peirce's Theory of Abduction*. The Hague: Martinus Nijhoff, 1970.

Fraenkel, Ernst. *The Dual State: A Contribution to the Theory of Dictatorship*. Oxford: Oxford University Press, 1941.

Frickel, Scott. "Absences: Methodological Note about Nothing, in Particular." *Social Epistemology* 28, no. 1 (2014): 86–95.

Frickel, Scott. "Not Here and Everywhere: The Non-production of Scientific Knowledge." In *Routledge Handbook of Science, Technology, and Society*, edited by Daniel Lee Kleinman and Kelly Moore, 263–76. London: Routledge, 2014.

Frickel, Scott. "On Missing New Orleans: Lost Knowledge and Knowledge Gaps in an Urban Hazardscape." *Environmental History* 13, no. 4 (2008): 643–50.

Frickel, Scott, and Michelle Edwards. "Untangling Ignorance in Environmental Risk Assessment." In *Powerless Science? Science and Politics in a Toxic World*, edited by Soraya Boudia and Nathalie Jas, 215–33. New York: Berghahn Books, 2014.

Frickel, Scott, Sahra Gibbon, Jeff Howard, Joanna Kempner, Gwen Ottinger, and David J. Hess. "Undone Science: Charting Social Movement and Civil Society Challenges to Research Agenda Setting." *Science, Technology and Human Values* 35, no. 4 (2010): 444–73.

Frickel, Scott, and Kelly Moore, eds. *The New Political Sociology of Science: Institutions, Networks, and Power*. Madison: University of Wisconsin Press, 2006.

Frickel, Scott, and Bess Vincent. "Hurricane Katrina, Contamination, and the Unintended Organization of Ignorance." *Technology in Society* 29, no. 2 (2007): 181–88.

Funtowicz, Silvio O., and Jerome R. Ravetz. "Science for the Post-normal Age." *Futures* 25, no. 7 (1993): 739–55.

Galison, Peter. *Image and Logic: A Material Culture of Microphysics*. Chicago: University of Chicago Press, 1997.

Galison, Peter. "Removing Knowledge: The Logic of Modern Censorship." In *Agnotology: The Making and Unmaking of Ignorance*, edited by Robert N. Proctor and Laura Schiebinger, 37–54. Stanford, CA: Stanford University Press, 2008.

Garelli, Aristide. "I Fatti della Maddalena nel Febbraio 1793." *Rivista Marittima* 31, no. 2 (May 1898): 299–333.

Garelli, Aristide. *L'isola della Maddalena: documenti e appunti storici*. Venice: Prem. Tipo-Litografia Veneziana, 1907. Reprint, Bologna: Atesa Editrice, 1987.

Garribba, Sergio, and Sergio Vaccà, eds. *Il controllo sociale dell'energia nucleare in Italia*. Milan: Franco Angeli Editore, 1978.

Gerlini, Matteo. "The Rise and Fall of Nuclear Italy." In *Pathways into and out of Nuclear Power in Western Europe: Austria, Denmark, Federal Republic of Germany, Italy and Sweden*, edited by Astrid Mignon Kirchhof, 170–237. Munich: Deutsches Museum, 2020.

Ghiretti, F., B. Salvato, S. Carlucci, and R. De Pieri. "Manganese in *Pinna nobilis*." *Experientia* 28, no. 2 (1972): 232–33.

Giddens, Anthony. *The Consequences of Modernity*. Stanford, CA: Stanford University Press, 1991.

Gieryn, Thomas. "Boundary-Work and the Demarcation of Science from Non-science: Strains and Interests in Professional Ideologies of Scientists." *American Sociological Review* 48, no. 6 (1983): 781–95.

Gillem, Mark L. *America Town: Building the Outpost of Empire*. Minneapolis: University of Minnesota Press, 2007.

Ginzburg, Carlo. "Morelli, Freud, and Sherlock Holmes: Clues and Scientific Method." *History Workshop* 9 (1980): 5–36.

Goffman, Erving. *Asylums: Essays on the Sociological Situation of Mental Patients and Other Inmates*. New York: Doubleday, 1961.

Gramsci, Antonio. *Selections from the Prison Notebooks*. Edited and translated by Quntin Hoare and Geoffrey Nowell-Smith. New York: International Publishers, 1971.

Gross, Matthias. *Ignorance and Surprise: Science, Society, and Ecological Design*. Cambridge: MIT Press, 2010.

Gross, Matthias. "The Unknown in Process: Dynamic Connections of Ignorance, Non-knowledge and Related Concepts." *Current Sociology* 55, no. 5 (2007): 742–59.

Gusterson, Hugh. *Nuclear Rites: A Weapons Laboratory at the End of the Cold War*. Berkeley: University of California Press, 1998.

Gusterson, Hugh. "Nuclear Weapons and the Other in the Western Imagination." *Cultural Anthropology* 14, no. 1 (1999): 111–43.

Hacker, Barton C. *Elements of Controversy: The Atomic Energy Commission and Radiation Safety in Nuclear Weapons Testing, 1947–1974*. Berkeley: University of California Press, 1994.

Hacking, Ian. *Historical Ontology*. Cambridge, MA: Harvard University Press, 2004.

Hagen, Joel. *An Entangled Bank: The Origins of Ecosystem Ecology*. New Brunswick, NJ: Rutgers University Press, 1992.

Hall, Stuart. "The Problem of Ideology: Marxism without Guarantees." *Journal of Communication Inquiry* 10, no. 2 (1986): 28–44.

Hamblin, Jacob D. "Environmentalism for the Atlantic Alliance: NATO's Experiment with the 'Challenges of Modern Society,'" *Environmental History* 15, no. 1 (2010): 54–75.

Hamblin, Jacob Darwin. *Poison in the Well: Radioactive Waste in the Oceans at the Dawn of the Nuclear Age*. New Brunswick, NJ: Rutgers University Press, 2008.

Hardenberg, Wilko Graf von, and Paolo Pelizzari. "The Environmental Question, Employment, and Development in Italy's Left, 1945–1990." *Left History* 13, no. 1 (2008): 77–105.

Harkavy, Robert E. *Great Power Competition for Overseas Bases: The Geopolitics of Access Diplomacy*. New York: Pergamon, 1982.

Hecht, Gabrielle. *Being Nuclear: Africans and the Global Uranium Trade*. Cambridge: MIT Press, 2012.

Hecht, Gabrielle, ed. *Entangled Geographies: Empire and Technopolitics in the Global Cold War*. Cambridge: MIT Press, 2011.

Hecht, Gabrielle. "Invisible Production and the Production of Invisibility: Cleaning, Maintenance, and Mining in the Nuclear Sector." In *Routledge Handbook of Science, Technology, and Society*, edited by Daniel Lee Kleinman and Kelly Moore, 353–68. Routledge, 2014.

Hecht, Gabrielle. *The Radiance of France: Nuclear Power and National Identity after World War II*. Cambridge: MIT Press, 1998.

Hecht, Gabrielle, and Paul Edwards. "The Technopolitics of Cold War: Toward a Transregional Perspective." In *Essays on Twentieth-Century History*, edited by Michael Adas, 271–314. Philadelphia: Temple University Press, 2010.

Heller, David, and Hans Lammerant. "U.S. Nuclear Weapons Bases in Europe." In Lutz, *Bases of Empire*, 96–130.

Hess, David. *Alternative Pathways in Science and Industry: Activism, Innovation, and the Environment in an Era of Globalization*. Cambridge: MIT Press, 2007.

Hess, David J., and Scott Frickel. "Fields of Knowledge and Theory Traditions in the Sociology of Science." In *Fields of Knowledge: Science, Politics and Publics in the Neoliberal Age*, edited by Scott Frickel and David J. Hess, 1–30. Bingley, UK: Emerald, 2014.

Hewlett, Richard G., and Francis Duncan. *Nuclear Navy, 1946–1962*. Chicago: University of Chicago Press, 1974.

Higuchi, Toshihiro. *Political Fallout: Nuclear Weapons Testing and the Making of a Global Environmental Crisis*. Stanford, CA: Stanford University Press, 2020.

Hilgartner, Stephen. *Science on Stage: Expert Advice as Public Drama*. Stanford, CA: Stanford University Press, 2000.

Hilgartner, Stephen, Richard Bell, and Rory O'Connor. *Nukespeak: The Selling of Nuclear Technology in America*. San Francisco: Sierra Club Books, 1982.

Hindmarsh, Richard, ed. *Nuclear Disaster at Fukushima Daiichi: Social, Political and Environmental Issues*. London: Routledge, 2013.

Hine, David. "The Italian Referendums of 8/9 November 1987." *Electoral Studies* 7, no. 2 (1988): 163–78.

Horowitz, Leah. "Local Environmental Knowledge." In *The Routledge Handbook of*

Political Ecology, edited by Tom Perreault, Gavin Bridge, and James McCarthy, 235–48. London: Routledge, 2015.

Hutter, Bridget M., and Michael Power. *Organizational Encounters with Risk*. Cambridge, MA: Cambridge University Press, 2005.

Iaccarino, Fabrizio. "Nuclear Renaissance in Italy." *Nuclear Law Bulletin* 2010, no. 1 (2010): 65–78.

Irwin, Alan. *Citizen Science: A Study of People, Expertise, and Sustainable Development*. London: Routledge, 1995.

Irwin, Alan, P. Simmons, and G. Walker. "Faulty Environments and Risk Reasoning: The Local Understanding of Industrial Hazards." *Environment and Planning A* 31, no. 7 (1999): 1311–26.

Jasanoff, Sheila. "Acceptable Evidence in a Pluralistic Society." In *Acceptable Evidence: Science and Values in Risk Management*, edited by Deborah G. Mayo and Rachelle D. Hollander, 29–46. Oxford: Oxford University Press, 1991.

Jasanoff, Sheila. "Civilization and Madness: The Great BSE Scare of 1996." *Public Understanding of Science* 6, no. 3 (1997): 221–32.

Jasanoff, Sheila. *Designs on Nature: Science and Democracy in Europe and the United States*. Princeton, NJ: Princeton University Press, 2005.

Jasanoff, Sheila. *The Fifth Branch: Science Advisers as Policymakers*. Cambridge, MA: Harvard University Press, 1990.

Jasanoff, Sheila. "A Mirror for Science." *Public Understanding of Science* 23, no. 1 (2014): 21–26.

Jasanoff, Sheila. "The Political Science of Risk Perception." *Reliability Engineering and System Safety* 59, no. 1 (1998): 91–99.

Jasanoff, Sheila. *Risk Management and Political Culture: A Comparative Analysis of Science*. New York: Russell Sage Foundation, 1986.

Jasanoff, Sheila, ed. *States of Knowledge: The Co-production of Science and the Social Order*. London: Routledge, 2004.

Johnson, Chalmers, ed. *Okinawa: Cold War Island*. Cardiff, CA: Japan Policy Research Institute, 1999.

Johnstone, Diana, and Ben Cramer. "The Burdens of the Glory: U.S. Bases in Europe." In *The Sun Never Sets: Confronting the Network of Foreign U.S. Military Bases*, edited by Joseph Gerson and Bruce Birchard, 199–223. Boston: South End, 1991.

Jørgensen, Dolly, Finn Arne Jørgensen, and Sara B. Pritchard, eds. *New Natures: Joining Environmental History and Science and Technology Studies*. Pittsburgh: University of Pittsburgh Press, 2013.

Jungk, Robert. *The Nuclear State*. London: John Calder, 1979.

Kagan, Robert. "The Benevolent Empire." *Foreign Policy* 111 (1998): 24–35.

Kaplan, Amy. "Where is Guantanamo?" *American Quarterly* 57, no. 3 (2005): 831–58.

Kasperson, Roger E. "The Social Amplification of Risk: Progress in Developing and Integrative Framework." in Krimsky and Golding, *Social Theories of Risk*, 153–78.

Kaur, Raminder. "Sovereignty without Hegemony, the Nuclear State, and a 'Secret Public Hearing' in India." *Theory, Culture, and Society* 30, no. 3 (2013): 3–28.

Keane, Webb. *Christian Moderns: Freedom and Fetish in the Mission Encounter*. Berkeley: University of California Press, 2007.

Keane, Webb. "On Semiotic Ideology." *Signs and Society* 6, no. 1 (2018): 64–87.

Keane, Webb. "Semiotics and the Social Analysis of Material Things." *Language and Communication* 23, no. 3 (2003): 409–25.

Keane, Webb. "Signs Are Not the Garb of Meaning: On the Social Analysis of Material Things." In *Materiality*, edited by Daniel Miller, 182–205. Durham, NC: Duke University Press, 2005.

Keane, Webb. *Signs of Recognition: Powers and Hazards of Representation in an Indonesian Society*. Berkeley: University of California Press, 1997.

Keane Webb. "Subjects and Objects." In *Handbook of Material Culture*, edited by Chris Tilley, Webb Keane, Susanne Küchler, Mike Rowlands, and Patricia Spyer, 197–202. London: SAGE, 2006.

Kelly, John Dunham. "U.S. Power, after 9/11 and before It: If Not an Empire, Then What?" *Public Culture* 15, no. 2 (2003): 347–69.

Kirsch, Stuart. *Mining Capitalism: The Relationship between Corporations and Their Critics*. Berkeley: University of California Press, 2014.

Knorr-Cetina, Karin. *Epistemic Cultures: How Sciences Make Knowledge*. Cambridge, MA: Harvard University Press, 1999.

Kramer, Mark, ed. "Italy and the Cold War." Special issue, *Journal of Cold War Studies* 4, no. 3 (2002).

Krige, John. *American Hegemony and the Postwar Reconstruction of Science in Europe*. Cambridge: MIT Press, 2008.

Krige, John, ed. *How Knowledge Moves: Writing the Transnational History of Science and Technology*. Chicago: University of Chicago Press, 2019.

Krige, John. "Hybrid Knowledge: The Transnational Co-production of the Gas Centrifuge for Uranium Enrichment in the 1960s." *British Journal for the History of Science* 45, no. 3 (2012): 337–57.

Krimsky, Sheldon, and Dominic Golding, eds. *Social Theories of Risk*. Santa Barbara, CA: Praeger, 1992.

Krupar, Shiloh R. *Hot Spotter's Report: Military Fables of Toxic Waste*. Minneapolis: University of Minnesota Press, 2013.

Kruse, Felicia E. "Indexicality and the Abductive Link." *Transactions of the Charles S. Peirce Society* 22, no. 4 (1986): 435–47.

Kuchinskaya, Olga. "Articulating the Signs of Danger: Lay Experiences of Post-Chernobyl Radiation Risks and Effects." *Public Understanding of Science* 20, no. 3 (2011): 405–21.

Kuchinskaya, Olga. *The Politics of Invisibility: Public Knowledge about Radiation Health Effects after Chernobyl*. Cambridge: MIT Press, 2014.

Kuchinskaya, Olga. "Twice Invisible: Formal Representations of Radiation Danger." *Social Studies of Science* 43, no. 1 (2013): 78–96.

Kuletz, Valerie. *The Tainted Desert: Environmental and Social Ruin in the American West*. London: Routledge, 1998.

Lash, Scott, Bronislaw Szerszynski, and Brian Wynne, eds. *Risk, Environment, and Modernity: Towards a New Ecology*. London: SAGE, 1996.

Latour, Bruno. "Give Me a Laboratory and I Will Raise the World." In *Science Observed: Perspectives on the Social Study of Science*, edited by Karin Knorr-Cetina and Michael Mulkay, 141–70. London: SAGE, 1983.

Latour, Bruno. *Pandora's Hope: Essays on the Reality of Science Studies*. Cambridge, MA: Harvard University Press, 1999.

Latour, Bruno. *The Pasteurization of France*. Cambridge, MA: Harvard University Press, 1993.

Latour, Bruno. *Reassembling the Social: An Introduction to Actor-Network-Theory*. Oxford: Oxford University Press, 2007.

Latour, Bruno. *Science in Action: How to Follow Scientists and Engineers through Society*. Cambridge, MA: Harvard University Press, 1987.

Latour, Bruno. *We Have Never Been Modern*. Cambridge, MA: Harvard University Press, 1993.

Leoni, Marco. "L'Arcipelago di La Maddalena: Nuove Prospettive in Relazione alla Realizzazione del Parco Naturale." BA thesis, University of Florence, 1998.

Liberatore, Angela. *The Management of Uncertainty: Learning from Chernobyl*. London: Routledge, 1999.

Lundestad, Geir. "'Empire by Invitation' in the American Century." *Diplomatic History* 23, no. 2 (1999): 189–217.

Lupton, Deborah. *Risk*. 2nd ed. London: Routledge, 2013.

Lutz, Catherine, ed. *The Bases of Empire: The Global Struggle against U.S. Military Posts*. New York: New York University Press, 2009.

Lutz, Catherine. "Empire Is in the Details." *American Ethnologist* 33, no. 4 (2006): 593–611.

Lutz, Catherine. *Homefront: A Military City and the American Twentieth Century*. Boston: Beacon, 2002.

Macola, Giacomo, and Derek Peterson, eds. *Recasting the Past: History Writing and Political Work in Modern Africa*. Athens: Ohio University Press, 2006.

Malvicini, Alessandro. "Le ricadute radioattive nella zona di Ispra dovute alle esplosioni nucleari durante il quinquennio, 1958–1962." *Minerva Nucleare* 7, no. 7 (July 1963): 267–76.

Manetti, Giovanni. *Theories of the Sign in Classical Antiquity*. Bloomington: Indiana University Press, 1993.

Marchese, C. "L'Ammiraglio Nelson alla Maddalena e la Marina Sarda di quei Tempi." *Rivista Marittima* 35, no. 4 (October 1902): 5–38.

Marcus, George E. "Ethnography in/of the World System: The Emergence of Multi-sited Ethnography." *Annual Review of Anthropology* 24 (1995): 95–117.

Marmonier, Henri. "La question de La Maddalena." *Revue Historique* 62, no. 1 (September–October 1896): 1–41.

Marres, Noortje. "Frontstaging Nonhumans: Publicity as a Constraint on the Political Activity of Things." In *Political Matter: Technoscience, Democracy, and Public Life*, edited by Bruce Braun and Sarah J. Whatmore, 177–210. Minneapolis: University of Minnesota Press, 2010.

Marres, Noortje. "Issues Spark a Public into Being: A Key but Often Forgotten Point of the Lippmann-Dewey Debate." In *Making Things Public: Atmospheres of Democracy*, edited by Bruno Latour and Peter Weibel, 208–17. Cambridge: MIT Press, 2005.

Marres, Noortje. *Material Participation: Technology, the Environment, and Everyday Publics*. London: Palgrave Macmillan, 2012.

Masco, Joseph. "Lie Detectors: On Secrets and Hypersecurity in Los Alamos." *Public Culture* 14, no. 3 (2002): 441–67.

Masco, Joseph. *The Nuclear Borderlands: The Manhattan Project in Post–Cold War New Mexico*. Princeton, NJ: Princeton University Press, 2006.

Masco, Joseph. *The Theater of Operations: National Security Affect from the Cold War to the War on Terror*. Durham, NC: Duke University Press, 2014.

Mateos, Gisela, and Edna Suárez-Díaz. "Technical Assistance in Movement: Nuclear Knowledge Crosses Latin American Borders." In *How Knowledge Moves: Writing the Transnational History of Science and Technology*, edited by John Krige, 345–67. Chicago: University of Chicago Press, 2019.

McCaffrey, Katherine T. *Military Power and Popular Protest: The U.S. Navy in Vieques, Puerto Rico*. New Brunswick, NJ: Rutgers University Press, 2002.

McGoey, Linsey. "The Logic of Strategic Ignorance." *British Journal of Sociology* 63, no. 3 (2012): 553–76.

McGoey, Linsey. "Strategic Unknowns: Towards a Sociology of Ignorance." *Economy and Society* 41, no. 1 (2012): 1–16.

Meniconi, Antonella. *Storia della magistratura italiana*. Bologna: Il Mulino, 2013.

Merlini, Cesare. "A Concise History of Nuclear Italy." *International Spectator* 23, no. 3 (July–September 1988): 135–52.

Messersmith, Andrene. *The American Years: Dunoon and the US Navy*. Glendaruel: Argyll, 2003.

Michael, Mike. "Comprehension, Apprehension, Prehension: Heterogeneity and the Public Understanding of Science." *Science, Technology and Human Values* 27, no. 3 (2002): 357–78.

Michael, Mike. "Ignoring Science: Discourses of Ignorance in the Public Understanding of Science." In *Misunderstanding Science? The Public Reconstruction of Science and Technology*, edited by Alan Irwin and Brian Wynne, 105–25. Cambridge: Cambridge University Press, 1996.

Michel, Ersilio. "L'Occupazione Sarda delle Isole della Maddalena." *Annali del Regio Istituto Storico Italiano per la Storia Moderna e Contemporanea* 1 (1936): 28–64.

Miller, Clark A. "Civic Epistemologies: Constituting Knowledge and Order in Political Communities." *Sociology Compass* 2, no. 6 (2008): 1896–1919.

Mitman, Gregg, Michelle Murphy, and Christopher C. Sellers, eds. 2004. "Landscapes of Exposure: Knowledge and Illness in Modern Environments," Special issue, *Osiris* 19 (2004).

Mol, Annemarie. *The Body Multiple: Ontology in Medical Practice*. Durham, NC: Duke University Press, 2003.

Moon, Katharine. *Sex among the Allies: Prostitution in U.S.-Korea Relations*. New York: Columbia University Press, 1997.

Moore, Kelly. *Disrupting Science: Social Movements, American Scientists, and the Politics of the Military, 1945–1975*. Princeton, NJ: Princeton University Press, 2008.

Moore, Kelly. "Organizing Integrity: American Science and the Creation of Public Interest Organizations, 1955–1975." *American Journal of Sociology* 101, no. 6 (1996): 1592–1627.

Morita, Atsuro, Anders Blok, and Shuhei Kimura. "Environmental Infrastructures of Emergency: The Formation of a Civic Radiation Monitoring Map during the Fukushima Disaster." In *Nuclear Disaster at Fukushima Daiichi*, edited by Richard Hindmarsh, 78–96. London: Routledge, 2013.

Moynihan, Daniel Patrick. *Secrecy: The American Experience*. New Haven, CT: Yale University Press, 1998.

Murphy, Michelle. *Sick Building Syndrome and the Problem of Uncertainty: Environmental Politics, Technoscience, and Women Workers*. Durham, NC: Duke University Press, 2006.

Nardini, Francesco. "C'era una volta l'arsenale." *Almanacco Gallurese* 2005/06 (2006): 239–43.

Nardini, Francesco. *L'imbroglio nucleare: La Maddalena, 1972–2005*. Sassari: Associazione Storica Sassarese, 2010.

Nash, Linda. *Inescapable Ecologies: A History of Environment, Disease, and Knowledge*. Berkeley: University of California Press, 2007.

Nash, Philip. *The Other Missiles of October: Eisenhower, Kennedy, and the Jupiters, 1957–1963*. Chapel Hill: University of North Carolina Press, 1997.

Nelkin, Dorothy. *Controversy: Politics of Technical Decisions*. 3rd ed. London: SAGE, 1992.

Nelkin, Dorothy. *Nuclear Power and Its Critics: The Cayuga Lake Controversy*. Ithaca, NY: Cornell University Press, 1971.

Nelkin, Dorothy, and Michael Pollack. *The Atom Besieged: Antinuclear Movements in France and Germany*. Cambridge: MIT Press, 1982.

Newell, Sasha. "The Affectiveness of Symbols: Materiality, Magicality, and the Limits of the Antisemiotic Turn." *Current Anthropology* 59, no. 1 (2018): 1–22.

Nixon, Rob. *Slow Violence and the Environmentalism of the Poor.* Cambridge, MA: Harvard University Press, 2011.

Nolan, Mary. "Anti-Americanism and Americanization in Germany." *Politics and Society* 33, no. 1 (2005): 88–122.

Nolan, Mary. *Visions of Modernity: American Business and the Modernization of Germany.* Oxford: Oxford University Press, 1994.

Nuti, Leopoldo. *La Sfida nucleare: la politica estera italiana e le armi atomiche, 1945–1991.* Bologna: Il Mulino, 2007.

Nuti, Leopoldo. "U.S. Forces in Italy, 1945–63." In *U.S. Military Forces in Europe: The Early Years, 1945–1970*, edited by Simon Duke and Wolfgang Krieger, 249–72. Boulder, CO: Westview, 1993.

Nye, Joseph. *Soft Power: The Means to Success in World Politics.* Washington, DC: PublicAffairs, 2004.

Odum, Eugene P. "Ecology and the Atomic Age." *ASB Bulletin* 4, no. 2 (June 1957): 27–29.

Oldenziel, Ruth. "Islands: The United States as a Networked Empire." In Hecht, *Entangled Geographies*, 13–41.

Oreskes, Naomi, and Erik M. Conway. "Challenging Knowledge: How Climate Science Became a Victim of the Cold War." In *Agnotology: The Making and Unmaking of Ignorance*, edited by Robert N. Proctor and Londa Schiebinger, 55–89. Stanford, CA: Stanford University Press, 2008.

Oreskes, Naomi, and Erik Conway. *Merchants of Doubt: How a Handful of Scientists Obscured the Truth on Issues from Tobacco Smoke to Global Warming.* New York: Bloomsbury, 2010.

Oreskes, Naomi, and John Krige, eds. *Science and Technology in the Global Cold War.* Cambridge: MIT Press, 2014.

Ortner, Sherry B. "Resistance and the Problem of Ethnographic Refusal." *Comparative Studies in Society and History* 37, no. 1 (1995): 173–93.

Ottinger, Gwen. "Buckets of Resistance: Standards and the Effectiveness of Citizen Science." *Science, Technology and Human Values* 35, no. 2 (2010): 244–70.

Ottinger, Gwen. "Changing Knowledge, Local Knowledge, and Knowledge Gaps: STS Insights into Procedural Justice." *Science, Technology and Human Values* 38, no. 2 (2013): 250–70.

Ottinger, Gwen. *Refining Expertise: How Responsible Engineers Subvert Environmental Justice Challenges.* New York: New York University Press, 2013.

Paoloni, Giovanni, ed. *Energia, ambiente, innovazione. Dal Cnrn all'Enea.* Rome: Laterza, 1992.

Paoloni, Giovanni. "Gli esordi del nucleare in Italia." In *Storia dell'Industria Elettrica in Italia*, vol. 4, *Dal dopoguerra alla nazionalizzazione, 1945–1962*, edited by Valerio Castronovo, 383–408. Rome: Laterza, 1994.

Paoloni, Giovanni. *Il Nucleare in Italia/Nuclear Power in Italy.* Rome: Archivio Storico ENEL, 2009.

Parmentier, Richard. "Peirce Divested for Non-intimates." *RSSI: Recherches Sémiotiques/Semiotic Inquiries* 7, no. 1 (1987): 19–39.

Parmentier, Richard. *The Pragmatic Semiotics of Cultures*. Berlin: Mouton de Gruyter, 1997.

Parmentier, Richard. "The Semiotic Regimentation of Social Life." *Semiotica* 95, no. 3/4 (1993): 357–95.

Parr, Joy. "Smells Like? Sources of Uncertainty in the History of the Great Lakes Environment." *Environmental History* 11, no. 2 (2006): 269–99.

Parr, Joy. "A Working Knowledge of the Insensible? Radiation Protection in Nuclear Generating Stations, 1962–1992." *Comparative Studies in Society and History* 48, no. 4 (2006): 820–51.

Parthasarathy, Shobita. "Breaking the Expertise Barrier: Understanding Activist Challenges to Science and Technology Policy Domains." *Science and Public Policy* 37, no. 5 (2010): 355–67.

Peirce, Charles S. *Philosophical Writings of Peirce*. Edited by Justus Buchler. New York: Dover, 2011.

Perin, Constance. *Shouldering Risks: The Culture of Control in the Nuclear Power Industry*. Princeton, NJ: Princeton University Press, 2004.

Petryna, Adriana. *Life Exposed: Biological Citizens after Chernobyl*. Princeton, NJ: Princeton University Press, 2002.

Pickering, Andrew. *The Mangle of Practice: Time, Agency, and Science*. Chicago: University of Chicago Press, 1995.

Pirina, Mario Luciano. "L'arsenale militare di La Maddalena." Thesis, Università degli Studi di Cagliari, Facoltà di Economia, 1999.

Polikarpov, Gennady G. *Radioekologiya Mosrkikh Organizmov by Atomizdat*. Doctoral dissertation, 1964. Published as *Radioecology of Aquatic Organisms* (New York: Reinhold, 1966).

Polmar, Norman, and K. J. Moore. *Cold War Submarines: The Design and Construction of U.S. and Soviet Submarines, 1945–2001*. Washington, DC: Potomac Books, 2003.

Polmar, Norman, and Edward Whitman. *Hunters and Killers: Anti-Submarine Warfare since 1943*. Washington DC: Naval Institute Press, 2016.

Prasca, Emilio. "Alcune Lettere dell'Ammiraglio Nelson ai reali di Savoia (1799–1805)." *Rivista Militare* 56, no. 1 (1923): 374–407.

Prasca, Emilio. *L'ammiraglio Giorgio des Geneys e i suoi tempi: memorie storico-marinaresche (1761–1839)*. Pinerolo: Tipografia gia Chiantore-Mascarelli, 1926.

Pritchard, Sara B. *Confluence: The Nature of Technology and the Making of the Rhône*. Cambridge, MA: Harvard University Press, 2011.

Proctor, Robert N. *Cancer Wars: How Politics Shapes What We Know and Don't Know about Cancer*. New York: Basic Books, 1995.

Proctor, Robert N., and Londa Schiebinger, eds. *Agnotology: The Making and Unmaking of Ignorance*. Stanford, CA: Stanford University Press, 2008.

Rainger, Ronald. "Science at the Crossroads: The Navy, Bikini Atoll, and American Oceanography in the1940s." *Historical Studies in the Physical and Biological Sciences* 30 (2000): 349–71.

Rainger, Ronald. "'A Wonderful Oceanographic Tool': The Atomic Bomb, Radioactivity, and American Oceanography." In *The Machine in Neptune's Garden: Historical Perspectives on Technology and the Marine Environment*, edited by Helen M. Rozwadowski and David K. van Keuren, 96–132. Sagamore Beach, MA: Science History, 2004.

Reno, Joshua. "Beyond Risk: Emplacement and the Production of Environmental Evidence." *American Ethnologist* 38, no. 3 (2011): 516–30.

Reuss, Martin, and Stephen H. Cutcliffe. *The Illusory Boundary: Environment and Technology in History*. Charlottesville: University of Virginia Press, 2010.

Riccio, Myriam. "La vittoria de la Maddalena nel 1793." *Rassegna Storica del Risorgimento* 1, no. 8 (1935): 885–900.

Ronchi, Claudio. "Quando vennero gli americani." *Almanacco Maddalenino* 3 (2004): 37–43.

Rothschild, Rachel. "Environmental Awareness in the Atomic Age: Radioecologists and Nuclear Technology." *Historical Studies in the Natural Sciences* 43, no. 4 (2013): 492–530.

Roucek, Joseph S. "The US and USSR Confrontation in the Middle East and Eastern Mediterranean." *Il Politico* 38, no. 3 (1973): 546–75.

Sandars, Christopher T. *America's Overseas Garrisons*. Oxford: Oxford University Press, 2000.

Sanna, Salvatore. "Il Centenario di Marisardegna." *Rivista Marittima* (1993): 101–8.

Sanna, Salvatore. *Il popolamento dell'Arcipelago Maddalenino prima dei Savoia (1650–1767): La colonizzazione corsa delle isole intermedie*. La Maddalena: Paolo Sorba, 2014.

Sanna, Salvatore. *La base atomica di La Maddalena-Santo Stefano: dall'Inizio alla fine*. La Maddalena: Paolo Sorba Editore, 2008.

Sanna, Salvatore. "La Maddalena del Bailo Paolo Maria Foassa (1781–1786): Da Colonia Militare a Comunità." *Almanacco Maddalenino* 4 (2006): 41–50.

Sanna, Salvatore, ed. *La Maddalena, Sardegna: storia e cronaca della base nucleare di S. Stefano, 1972–1991*. Cagliari: CUEC Editrice, 1994.

Sanna, Salvatore. "La torre di Villamarina a S. Stefano. La casamatta della pietra dura." *Almanacco Maddalenino* 5 (2007): 57–73.

Sanna, Salvatore. "Quadro storico sull'origine e l'evoluzione delle servitú militari in Sardegna." *Tessiduras de Paghe, Quaderni del Satyagraha* 9 (2006): 69–92.

Santomauro, L., and A. Cigna. "Prime misure sulla radioattivita' delle precipitazioni atmosferiche." Communication presented at the national meeting of the Italian Geophysics Association, Rome, June 17–18, 1953.

Schmid, Sonja D. "Organizational Culture and Professional Identities in the Soviet Nuclear Power Industry." *Osiris* 23 (2008): 82–111.

Schmid, Sonja D. *Producing Power: The Pre-Chernobyl History of the Soviet Nuclear Industry*. Cambridge: MIT Press, 2015.

Schultz, Vincent, and Alfred W. Klement Jr., eds. *Radioecology: Proceedings of the First National Symposium on Radioecology, Held at Colorado State University, Fort Collins, September 10–15, 1961*. New York: Reinhold; Washington DC: American Institute of Biological Sciences, 1962.

Sellers, Christopher. *Hazards of the Job: From Industrial Disease to Environmental Health Science*. Chapel Hill: University of North Carolina Press, 1997.

Shapiro, Nicholas. "Attuning to the Chemosphere: Domestic Formaldehyde, Bodily Reasoning, and the Chemical Sublime." *Cultural Anthropology* 30, no. 3 (2015): 368–93.

Shils, Edward A. *The Torment of Secrecy: The Background and Consequences of American Security Policies*. New York: Free Press, 1956.

Silvestri, Mario. *Il costo della menzogna: Italia Nucleare (1945–1968)*. Turin: Einaudi, 1968.

Simich, Laura. "Comiso: The Politics of Peace in a Sicilian Town." PhD dissertation, Columbia University, 1988.

Simich, Laura. "The Corruption of a Community's Economic and Political Life: The Cruise Missile Base in Comiso." In *The Sun Never Sets . . . : Confronting the Network of Foreign U.S. Military Bases*, edited by Joseph Gerson and Bruce Birchard, 77–94. Boston: South End, 1991.

Simmel, Georg. "The Sociology of Secrecy and of Secret Societies." *American Journal of Sociology* 11, no. 4 (1906): 441–98.

Slayton, Rebecca. *Arguments that Count: Physics, Computing, and Missile Defense, 1949–2012*. Cambridge: MIT Press, 2013.

Slovic, Paul. "Perception of Risk: Reflections on the Psychometric Paradigm." In Krimsky and Golding, *Social Theories of Risk*, 117–52.

Smith, Neil. *American Empire: Roosevelt's Geographer and the Prelude to Globalization*. Berkeley: University of California Press, 2003.

Società Italiana di Fisica. *La scienza nella società capitalistica*. Bari: De Donato, 1971.

Sole, Carlino. "Contributo alla Storia di La Maddalena (1720–1767)." *Ichnusa* 5, no. 4 (1957): 35–62.

Sole, Carlino. "Sovranità e Giurisdizione sulle Isole Intermedie (1767–1793)." *Archivio Storico Sardo* 26 (1959): 255–479.

Soneryd, Linda. "Deliberations on the Unknown, the Unsensed, and the Unsayable? Public Protests and the Development of Third-Generation Mobile Phones in Sweden." *Science, Technology and Human Values* 32, no. 3 (2007): 287–314.

Sotgiu, Giovanna. *Arcipelago di La Maddalena (1839–1843): La divisione delle terre*. La Maddalena: Lo Scoglio Editrice, 2002.

Sotgiu, Giovanna. *A Mimoria d'a petra. L'arcipelago ed il suo granito*. Sassari: J. Webber Editore, 2006.

Sotgiu, Giovanna. "Gli ultimi anni di Vincenzo Sulis: L'esilio alla Maddalena." *Almanacco Maddalenino* (2006): 51–61.

Sotgiu, Giovanna. "Il Parco Nazionale dell'arcipelago di La Maddalena: Le ragioni di una scelta." *Almanacco Maddalenino* (2007): 74–83.

Sotgiu, Giovanna. "La Maddalena luogo di prigionia ed esilio." *Almanacco Maddalenino* (2004): 77–86.

Sotgiu, Giovanna. *Moneta: Contributo allo studio della comunità maddalenina*. La Maddalena: Tipografia Rossi, 1993.

Sotgiu, Giovanna, and Francesco Del Monaco. *La Maddalena e i suoi traffici marittimi*. La Maddalena: Paolo Sorba Editore, 2014.

Sotgiu, Giovanna, and Antonio Frau. *Le Bocche di Bonifacio. Il Tempo e i luoghi di una regione di frontiera*. La Maddalena: Paolo Sorba Editore, 2018.

Sotgiu, Giovanna, and Alberto Sega. *Un'isola e il suo ammiraglio. Giorgio Andrea Desgeneys e La Maddalena*. La Maddalena: Paolo Sorba Editore, 2008.

Spaziante, Vincenzo. *Questione Nucleare e Politica Legislativa*. Rome: Officina Edizioni, 1980.

Spinardi, Graham. *From Polaris to Trident: The Development of US Fleet Ballistic Missile Technology*. Cambridge: Cambridge University Press, 1994.

Steinmetz, George. "Return to Empire: The New U.S. Imperialism in Comparative Historical Perspective." *Sociological Theory* 23, no. 4 (2005): 339–67.

Stephan, Alexander, ed. *The Americanization of Europe: Culture, Diplomacy, and Anti-Americanism after 1945*. New York: Berghahn Books, 2008.

Stoler, Ann Laura, Carole McGranaham, and Peter C. Perdue, eds. *Imperial Formations*. Santa Fe: School for Advanced Research Press, 2007.

Taussig, Michael. *Defacement: Public Secrecy and the Power of the Negative*. Stanford, CA: Stanford University Press, 1999.

Tonetti, Alessandro. "Localizzazione e consenso nel programma di rilancio del nucleare in Italia." *Giornale di Diritto Amministrativo* 1 (2011): 5–18.

Topçu, Sezin. "Confronting Nuclear Risks: Counter-expertise and Politics within the French Nuclear Energy Debate." *Nature and Culture* 3, no. 2 (2008): 225–45.

Touraine, Alain, Zsuzsa Hegedus, François Dubet, and Michael Wieviorka. *Anti-nuclear Protest: The Opposition to Nuclear Energy in France*. Cambridge: Cambridge University Press, 1983. First published as *La prophetie anti-nucleaire* (Paris: Éditions du Seuil, 1980).

Tovo, Antonello. "L'America a La Maddalena: Aspetti della presenza militare degli Stati Uniti nell'Arcipelago di La Maddalena (1972–2003)." Master's thesis, University of Pisa, 2003.

Tulloch, John, and Deborah Lupton. *Risk and Everyday Life*. London: SAGE, 2003.

Turchetti, Simone. *Greening the Alliance: The Diplomacy of NATO's Science and Environmental Initiatives*. Chicago: University of Chicago Press, 2019.

Turchetti, Simone. "A Most Active Customer: How the U.S. Administration Helped the Italian Atomic Energy Project to 'De-develop.'" *Historical Studies in the Natural Sciences* 44, no. 5 (2014): 470–502.

Turchetti, Simone. "Sword, Shield and Buoys: A History of the NATO Sub-committee on Oceanographic Research, 1959–1973." *Centaurus* 54, no. 3 (2012): 205–31.

Tusceri, Gian Carlo. *Il governatore. Storia di Mons. Salvatore Capula e della "sua" isola*. La Maddalena: Paolo Sorba Editore, 2000.

"Una corporazione della scienza: il Laboratorio Internazionale di Biofisica e Genetica di Napoli." *Quaderni Piacentini* 8, no. 38 (1969): 121–46.

Vaglini, Amerigo. *Il nucleare a Pisa: quaderno di memorie storiche sul C.A.M.E.N., Centro applicazioni militari energia nucleare, 1955–1985*. Pisa: ETS, 2009.

Varsori, Antonio. "Cold War History in Italy." *Cold War History* 8, no. 2 (2008): 157–87.

Vine, David. *Base Nation: How the U.S. Military Bases Abroad Harm America and the World*. New York: Skyhorse, 2015.

Vine, David. *Island of Shame: The Secret History of the U.S. Military Base on Diego Garcia*. Princeton, NJ: Princeton University Press, 2009.

Vološinov V. N. *Marxism and the Philosophy of Language*. Cambridge, MA: Harvard University Press, 1973.

Walker, Brett. *Toxic Archipelago: A History of Industrial Disease in Japan*. Seattle: University of Washington Press, 2010.

Walker, J. Samuel. *Permissible Dose: A History of Radiation Protection in the Twentieth Century*. Berkeley: University of California Press, 2000.

Watts, Richard J. "Pulse Type Regulated High Voltage Supply for G-M Tubes." *Review of Scientific Instruments* 21, no. 4 (April 1950): 342–43.

Weinberg, Alvin M. "Science and Trans-science." *Minerva* 10, no. 2 (1972): 209–22.

Winner, Langdon. *Autonomous Technology: Technics-out-of-Control as a Theme in Political Thought*. Cambridge: MIT Press, 1978.

Winner, Langdon. *The Whale and the Reactor: A Search for Limits in an Age of High Technology*. Chicago: University of Chicago Press, 1986.

Wise, David, and Thomas B. Ross. *The Invisible Government*. New York: Random House, 1964.

Wohlstetter, A. J., F. S. Hoffman, R. J. Lutz, and H. S. Rowen. *Selection and Use of Strategic Air Bases*. Report R-266. RAND Corporation, April 1954.

Wynne, Brian. "Elephants in the Rooms Where Publics Encounter 'Science'? A Response to Darrin Durant, 'Accounting for Expertise: Wynne and the Autonomy of the Lay Public.'" *Public Understanding of Science* 17 (2008): 21–33.

Wynne, Brian. "May the Sheep Safely Graze? A Reflexive View of the Expert–Lay Knowledge Divide." In *Risk, Environment, and Modernity: Towards a New Ecology*, edited by Scott Lash, Bronislaw Szerszynski, and Brian Wynne, 44–83. London: SAGE, 1996.

Wynne, Brian. "Misunderstood Misunderstandings: Social Identities and the Public Uptake of Science." In *Misunderstanding Science? The Public Reconstruction of*

Science and Technology, edited by Alan Irwin and Brian Wynne, 19–46. Cambridge: Cambridge University Press, 1996.

Wynne, Brian. "Public Understanding of Science." In *Handbook of Science and Technology Studies*, edited by Sheila Jasanoff, Gerald E. Markle, James C. Peterson, and Trevor Pinch, 361–89. London: SAGE, 1995.

Wynne, Brian. *Rationality and Ritual: Participation and Exclusion in Nuclear Decision-making*. 2nd ed. London: Routledge, 2010.

Zaninelli, Sergio, ed. *Ricerca, innovazione, impresa: storia del CISE, 1946–1996*. Bari: Laterza, 1996.

Zonabend, Françoise. *The Nuclear Peninsula*. Cambridge: Cambridge University Press, 1993.

Zorzoli, Giovanni Battista. *La ricerca scientifica in Italia*. Milan: Franco Angeli, 1970.

Zurcher, Louis A. "The Naval Recruit Training Center: A Study of Role Assimilation in a Total Institution." *Sociological Inquiry* 37, no. 1 (1967): 85–98.

Index

Note: Page numbers in *italics* indicate illustrative material.